T0255040

London Mathematical Society Student Texts

Managing Editor: Professor D. Benson,
Department of Mathematics, University of Aberdeen, UK

LONDON MATHEMATICAL SOCIETY STUDENT
TEXTS 74

Representation Theorems
in Hardy Spaces

JAVAD MASHREGHI
Université Laval

CAMBRIDGE
UNIVERSITY PRESS

CAMBRIDGE
UNIVERSITY PRESS

University Printing House, Cambridge CB2 8BS, United Kingdom

One Liberty Plaza, 20th Floor, New York, NY 10006, USA

477 Williamstown Road, Port Melbourne, VIC 3207, Australia

314-321, 3rd Floor, Plot 3, Splendor Forum, Jasola District Centre, New Delhi - 110025, India

103 Penang Road, #05-06/07, Visioncrest Commercial, Singapore 238467

Cambridge University Press is part of the University of Cambridge.

It furthers the University's mission by disseminating knowledge in the pursuit of education, learning and research at the highest international levels of excellence.

www.cambridge.org
Information on this title: www.cambridge.org/9780521732017

© J.Mashreghi 2009

First published 2009

A catalogue record for this publication is available from the British Library

ISBN   978-0-521-51768-3   Hardback
ISBN   978-0-521-73201-7   Paperback

To My Parents: Masoumeh Farzaneh
and
Ahmad Mashreghi

Contents

Preface

In 1915 Godfrey Harold Hardy, in a famous paper published in the *Proceedings of the London Mathematical Society*, answered in the affirmative a question of Landau [7]. In this paper, not only did Hardy generalize Hadamard's three-circle theorem, but he also put in place the first brick of a new branch of mathematics which bears his name: the theory of Hardy spaces. For three decades afterwards Hardy, alone or with others, wrote many more research articles on this subject.

The theory of Hardy spaces has close connections to many branches of mathematics, including Fourier analysis, harmonic analysis, singular integrals, probability theory and operator theory, and has found essential applications in robust control engineering. I have had the opportunity to give several courses on Hardy spaces and some related topics. A part of these lectures concerned the various representations of harmonic or analytic functions in the open unit disc or in the upper half plane. This topic naturally leads to the representation theorems in Hardy spaces.

There are excellent books [5, 10, 13] and numerous research articles on Hardy spaces. Our main concern here is only to treat the representation theorems. Other subjects are not discussed and the reader should consult the classical textbooks. A rather complete description of representation theorems of $H^p(\mathbb{D})$, the family of Hardy spaces of the open unit disc, is usually given in all books. To study the corresponding theorems for $H^p(\mathbb{C}_+)$, the family of Hardy spaces of the upper half plane, a good amount of Fourier analysis is required. As a consequence, representation theorems for the upper half plane are not discussed thoroughly in textbooks mainly devoted to Hardy spaces. Moreover, quite often it is mentioned that they can be derived by a conformal mapping from the corresponding theorems on the open unit disc. This is a useful technique in certain cases and we will also apply it at least on one occasion. However, in the present text, our main goal is to give a complete description of the representation theorems with *direct proofs* for both classes of Hardy spaces. Hence, certain topics from Fourier analysis have also been discussed. But this is not a book about Fourier analysis, and we have been content with the minimum required to obtain the representation theorems. For further studies on Fourier analysis many interesting references are available, e.g. [1, 8, 9, 15, 21].

I express my appreciation to the many colleagues and students who made valuable comments and improved the quality of this book. I deeply thank Colin Graham and Mostafa Nasri, who read the entire manuscript and offered several suggestions and Masood Jahanmir, who drew all the figures. I am also grateful to Roger Astley of the Cambridge University Press for his great management and kind help during the publishing procedure.

I have benefited from various lectures by Arsalan Chademan, Galia Dafni, Paul Gauthier, Kohur GowriSankaran, Victor Havin, Ivo Klemes and Paul Koosis on harmonic analysis, potential theory and the theory of Hardy spaces. As a matter of fact, the first draft of this manuscript dates back to 1991, when I attended Dr Chademan's lectures on the theory of H^p spaces. I take this opportunity to thank them with all my heart.

Thanks to the generous support of Kristian Seip, I visited the Norwegian University of Science and Technology (NTNU) in the fall semester of 2007–2008. During this period, I was able to concentrate fully on the manuscript and prepare it for final submission to the Cambridge University Press. I am sincerely grateful to Kristian, Yurii Lyubarski and Eugenia Malinnikova for their warm hospitality in Trondheim.

I owe profound thanks to my friends at McGill University, Université Laval and Université Claude Bernard Lyon 1 for their constant support and encouragement. In particular, Niky Kamran and Kohur GowriSankaran have played a major role in establishing my mathematical life, Thomas Ransford helped me enormously in the early stages of my career, and Emmanuel Fricain sends me his precious emails on a daily basis. The trace of their efforts is visible in every single page of this book.

Québec
August 2008

Chapter 1

Fourier series

1.1   The Laplacian

An open connected subset of the complex plane $\mathbb{C}$ is called a *domain*. In particular, $\mathbb{C}$ itself is a domain. But, for our discussion, we are interested in two special domains: the open *unit disc*

$$\mathbb{D} = \{\, z \in \mathbb{C} \,:\, |z| < 1 \,\}$$

whose boundary is the *unit circle*

$$\mathbb{T} = \{\, \zeta \in \mathbb{C} \,:\, |\zeta| = 1 \,\}$$

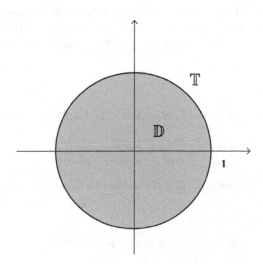

Fig. 1.1. The open unit disc $\mathbb{D}$ and its boundary $\mathbb{T}$.

and the *upper half plane*

$$\mathbb{C}_+ = \{\, z \in \mathbb{C} \,:\, \Im z > 0 \,\}$$

whose boundary is the *real line* $\mathbb{R}$ (see Figures 1.1 and 1.2). They are essential domains in studying the theory of Hardy spaces.

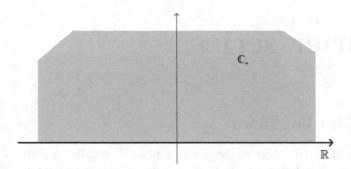

Fig. 1.2. The upper half plane $\mathbb{C}_+$ and its boundary $\mathbb{R}$.

The notations

$$
\begin{aligned}
D(a,r) &= \{\, z \in \mathbb{C} \,:\, |z-a| < r \,\}, \\
\overline{D(a,r)} &= \{\, z \in \mathbb{C} \,:\, |z-a| \leq r \,\}, \\
\partial D(a,r) &= \{\, z \in \mathbb{C} \,:\, |z-a| = r \,\}
\end{aligned}
$$

for the open or closed discs and their boundaries will be used frequently too. We will also use D_r for a disc whose center is the origin, with radius r.

The *Laplacian* of a twice continuously differentiable function $U : \Omega \longrightarrow \mathbb{C}$ is defined by

$$\nabla^2 U = \frac{\partial^2 U}{\partial x^2} + \frac{\partial^2 U}{\partial y^2}.$$

If $0 \notin \Omega$ and we use polar coordinates, then the Laplacian becomes

$$\nabla^2 U = \frac{\partial^2 U}{\partial r^2} + \frac{1}{r} \frac{\partial U}{\partial r} + \frac{1}{r^2} \frac{\partial^2 U}{\partial \theta^2}.$$

We say that U is *harmonic* on Ω if it satisfies the *Laplace equation*

$$\nabla^2 U = 0 \tag{1.1}$$

at every point of Ω. By direct verification, we see that

$$U(re^{i\theta}) = r^n \cos(n\theta), \qquad (n \geq 0),$$

and
$$U(re^{i\theta}) = r^n \, \sin(n\theta), \qquad (n \geq 1),$$

are real harmonic functions on $\mathbb{C}$. Since (1.1) is a linear equation, a complex-valued function is harmonic if and only if its real and imaginary parts are real harmonic functions. The complex version of the preceding family of real harmonic functions is

$$U(re^{i\theta}) = r^{|n|} \, e^{in\theta}, \qquad (n \in \mathbb{Z}).$$

A special role is played by the constant function $U \equiv 1$ since it is the only member of the family whose integral means

$$\frac{1}{2\pi} \int_{-\pi}^{\pi} U(re^{i\theta}) \, d\theta$$

are not zero. This fact is a direct consequence of the elementary identity

$$\frac{1}{2\pi} \int_{-\pi}^{\pi} e^{in\theta} \, d\theta = \begin{cases} 1 & \text{if} \quad n = 0, \\ 0 & \text{if} \quad n \neq 0, \end{cases} \qquad (1.2)$$

which will be used frequently throughout the text.

Let F be analytic on a domain Ω and let U and V represent respectively its real and imaginary parts. Then U and V are infinitely continuously differentiable and satisfy the *Cauchy–Riemann equations*

$$\frac{\partial U}{\partial x} = \frac{\partial V}{\partial y} \qquad \text{and} \qquad \frac{\partial U}{\partial y} = -\frac{\partial V}{\partial x}. \qquad (1.3)$$

Using these equations, it is straightforward to see that U and V are real harmonic functions. Hence an analytic function is a complex harmonic function. In the following we will define certain classes of harmonic functions, and in each class there is a subclass containing only analytic elements. Therefore, any representation formula for members of the larger class will automatically be valid for the corresponding subclass of analytic functions.

Exercises

Exercise 1.1.1 Let F be analytic on Ω and let $U = \Re F$ and $V = \Im F$. Show that U and V are real harmonic functions on Ω.
Hint: Use (1.3).

Exercise 1.1.2 Let $F : \Omega \longmapsto \mathbb{C}$ be analytic on Ω, and suppose that $F(z) \neq 0$ for all $z \in \Omega$. Show that $\log |F|$ is harmonic on Ω.

Exercise 1.1.3   Let U be harmonic on the annular domain

$$A(R_1, R_2) = \{\, z \in \mathbb{C} \,:\, 0 \le R_1 < |z| < R_2 \le \infty \,\}.$$

Show that $U(\frac{1}{r}\, e^{i\theta})$ is harmonic on the annular domain $A(\frac{1}{R_2}, \frac{1}{R_1})$.

Exercise 1.1.4   Let $F : \Omega_1 \longmapsto \Omega_2$ be analytic on Ω_1, and let $U : \Omega_2 \longmapsto \mathbb{C}$ be harmonic on Ω_2. Show that $U \circ F : \Omega_1 \longmapsto \mathbb{C}$ is harmonic on Ω_1.

Exercise 1.1.5   Define the differential operators

$$\partial = \frac{1}{2}\left(\frac{\partial}{\partial x} - i\,\frac{\partial}{\partial y} \right),$$

$$\bar{\partial} = \frac{1}{2}\left(\frac{\partial}{\partial x} + i\,\frac{\partial}{\partial y} \right).$$

Show that

$$\nabla^2 = 4\partial\bar{\partial}.$$

Exercise 1.1.6   Let $F = U + iV$ be analytic. Show that the Cauchy–Riemann equations (1.3) are equivalent to the equation

$$\bar{\partial} F = 0.$$

Remark: We also have $F' = \partial F$.

Exercise 1.1.7   Let U_1 and U_2 be real harmonic functions on a domain Ω. Under what conditions is $U_1\, U_2$ also harmonic on Ω?
Remark: We emphasize that U_1 and U_2 are *real* harmonic functions. The answer to this question changes dramatically if we consider complex harmonic functions. For example, if F_1 and F_2 are analytic functions, then, under no extra condition, $F_1\, F_2$ is analytic. Note that an analytic function is certainly harmonic.

Exercise 1.1.8   Let U be a real harmonic function on a domain Ω. Suppose that U^2 is also harmonic on Ω. Show that U is constant.
Hint: Use Exercise 1.1.7.

Exercise 1.1.9   Let F be analytic on a domain Ω, and let Φ be a twice continuously differentiable function on the range of F. Show that

$$\nabla^2(\Phi \circ F) = \big((\nabla^2 \Phi) \circ F \big) \, |F'|^2.$$

Exercise 1.1.10 Let F be analytic on a domain Ω, and let $\alpha \in \mathbb{R}$. Suppose that F has no zeros on Ω. Show that

$$\nabla^2(|F|^\alpha) = \alpha^2 |F|^{\alpha-2} |F'|^2.$$

Hint: Apply Exercise 1.1.9 with $\Phi(z) = |z|^\alpha$.

Exercise 1.1.11 Let F be analytic on a domain Ω. Under what conditions is $|F|^2$ harmonic on Ω?
Hint: Apply Exercise 1.1.10.

Exercise 1.1.12 Let F be a complex function on a domain Ω such that F and F^2 are both harmonic on Ω. Show that either F or $\bar{F}$ is analytic on Ω.
Hint: Use Exercise 1.1.7.

1.2   Some function spaces and sequence spaces

Let f be a measurable function on $\mathbb{T}$, and let

$$\|f\|_p = \left(\frac{1}{2\pi} \int_{-\pi}^{\pi} |f(e^{it})|^p \, dt \right)^{\frac{1}{p}}, \qquad (0 < p < \infty),$$

and

$$\|f\|_\infty = \inf_{M>0} \left\{ M \, : \, |\{ e^{it} \, : \, |f(e^{it})| > M \}| = 0 \right\},$$

where $|E|$ denotes the Lebesgue measure of the set E. Then *Lebesgue spaces* $L^p(\mathbb{T})$, $0 < p \leq \infty$, are defined by

$$L^p(\mathbb{T}) = \{ f \, : \, \|f\|_p < \infty \}.$$

If $1 \leq p \leq \infty$, then $L^p(\mathbb{T})$ is a Banach space. In particular, $L^2(\mathbb{T})$, equipped with the inner product

$$\langle f, g \rangle = \frac{1}{2\pi} \int_{-\pi}^{\pi} f(e^{it}) \, \overline{g(e^{it})} \, dt,$$

is a Hilbert space. It is easy to see that

$$L^\infty(\mathbb{T}) \subset L^p(\mathbb{T}) \subset L^1(\mathbb{T})$$

for each $p \in (1, \infty)$. In the following, we will mostly study $L^p(\mathbb{T})$, $1 \leq p \leq \infty$, and their subclasses and thus $L^1(\mathbb{T})$ is the largest function space that enters our discussion. This simple fact has important consequences. For example, we will define the Fourier coefficients of functions in $L^1(\mathbb{T})$, and thus the Fourier coefficients of elements of $L^p(\mathbb{T})$, for all $1 \leq p \leq \infty$, are automatically defined too. We will appreciate this fact when we study the Fourier transform on the real line. Spaces $L^p(\mathbb{R})$ do not form a chain, as is the case on the unit circle, and thus after defining the Fourier transform on $L^1(\mathbb{R})$, we need to take further steps in order to define the Fourier transform for some other $L^p(\mathbb{R})$ spaces.

A continuous function on $\mathbb{T}$, a compact set, is necessarily bounded. The space of all continuous functions on the unit circle $\mathcal{C}(\mathbb{T})$ can be considered as a subspace of $L^\infty(\mathbb{T})$. As a matter of fact, in this case the maximum is attained and we have

$$\|f\|_\infty = \max_{e^{it} \in \mathbb{T}} |f(e^{it})|.$$

On some occasions we also need the smaller subspace $\mathcal{C}^n(\mathbb{T})$ consisting of all n times continuously differentiable functions, or even their intersection $\mathcal{C}^\infty(\mathbb{T})$ consisting of functions having derivatives of all orders. The space $\mathcal{C}^n(\mathbb{T})$ is equipped with the norm

$$\|f\|_{\mathcal{C}^n(\mathbb{T})} = \sum_{k=1}^{n} \frac{\|f^{(k)}\|_\infty}{k!}.$$

Lipschitz classes form another subfamily of $\mathcal{C}(\mathbb{T})$. Fix $\alpha \in (0,1]$. Then $\mathrm{Lip}_\alpha(\mathbb{T})$ consists of all $f \in \mathcal{C}(\mathbb{T})$ such that

$$\sup_{\substack{t,\tau \in \mathbb{R} \\ \tau \neq 0}} \frac{|f(e^{i(t+\tau)}) - f(e^{it})|}{|\tau|^\alpha} < \infty.$$

This space is equipped with the norm

$$\|f\|_{\mathrm{Lip}_\alpha(\mathbb{T})} = \|f\|_\infty + \sup_{\substack{t,\tau \in \mathbb{R} \\ \tau \neq 0}} \frac{|f(e^{i(t+\tau)}) - f(e^{it})|}{|\tau|^\alpha}.$$

The space of all complex *Borel measures* on $\mathbb{T}$ is denoted by $\mathcal{M}(\mathbb{T})$. This space equipped with the norm

$$\|\mu\| = |\mu|(\mathbb{T}),$$

where $|\mu|$ denotes the total variation of μ, is a Banach space. Remember that the total variation $|\mu|$ is the smallest positive Borel measure satisfying

$$|\mu(E)| \leq |\mu|(E)$$

for all Borel sets $E \subset \mathbb{T}$. To each function $f \in L^1(\mathbb{T})$ there corresponds a Borel measure

$$d\mu(e^{it}) = \frac{1}{2\pi} f(e^{it}) \, dt.$$

Clearly we have $\|\mu\| = \|f\|_1$, and thus the map

$$\begin{array}{ccc} L^1(\mathbb{T}) & \longrightarrow & \mathcal{M}(\mathbb{T}) \\ f & \longmapsto & f(e^{it}) \, dt/2\pi \end{array}$$

is an embedding of $L^1(\mathbb{T})$ into $\mathcal{M}(\mathbb{T})$.

In our discussion, we also need some *sequence spaces*. For a sequence of complex numbers $\mathcal{Z} = (z_n)_{n \in \mathbb{Z}}$, let

$$\|\mathcal{Z}\|_p = \left(\sum_{n=-\infty}^{\infty} |z_n|^p \right)^{\frac{1}{p}}, \qquad (0 < p < \infty),$$

and

$$\|\mathcal{Z}\|_\infty = \sup_{n \in \mathbb{Z}} |z_n|.$$

Then, for $0 < p \leq \infty$, we define

$$\ell^p(\mathbb{Z}) = \{ \mathcal{Z} : \|\mathcal{Z}\|_p < \infty \}$$

and

$$c_0(\mathbb{Z}) = \{ \mathcal{Z} \in \ell^\infty(\mathbb{Z}) : \lim_{|n| \to \infty} |z_n| = 0 \}.$$

If $1 \leq p \leq \infty$, then $\ell^p(\mathbb{Z})$ is a Banach space and $c_0(\mathbb{Z})$ is a closed subspace of $\ell^\infty(\mathbb{Z})$. The space $\ell^2(\mathbb{Z})$, equipped with the inner product

$$\langle \mathcal{Z}, \mathcal{W} \rangle = \sum_{n=-\infty}^{\infty} z_n \overline{w}_n,$$

is a Hilbert space. The subspaces

$$\ell^p(\mathbb{Z}^+) = \{ \mathcal{Z} \in \ell^p(\mathbb{Z}) : z_{-n} = 0, \, n \geq 1 \}$$

and

$$c_0(\mathbb{Z}^+) = \{ \mathcal{Z} \in c_0(\mathbb{Z}) : z_{-n} = 0, \, n \geq 1 \}$$

will also appear when we study the Fourier transform of certain subclasses of $L^p(\mathbb{T})$.

Exercises

Exercise 1.2.1 Let f be a measurable function on $\mathbb{T}$, and let

$$f_\tau(e^{it}) = f(e^{i(t-\tau)}).$$

Show that

$$\lim_{\tau \to 0} \|f_\tau - f\|_X = 0$$

if $X = L^p(\mathbb{T})$, $1 \leq p < \infty$, or $X = \mathcal{C}^n(\mathbb{T})$. Provide examples to show that this property does not hold if $X = L^\infty(\mathbb{T})$ or $X = \text{Lip}_\alpha(\mathbb{T})$, $0 < \alpha \leq 1$. However, show that

$$\|f_\tau\|_X = \|f\|_X$$

in all function spaces mentioned above.

Exercise 1.2.2 Show that $\ell^p(\mathbb{Z}^+)$, $0 < p \leq \infty$, is closed in $\ell^p(\mathbb{Z})$.

Exercise 1.2.3 Show that $c_0(\mathbb{Z})$ is closed in $\ell^\infty(\mathbb{Z})$.

Exercise 1.2.4 Show that $c_0(\mathbb{Z}^+)$ is closed in $c_0(\mathbb{Z})$.

Exercise 1.2.5 Let $c_c(\mathbb{Z})$ denote the family of sequences of compact support, i.e. for each $\mathcal{Z} = (z_n)_{n\in\mathbb{Z}} \in c_c(\mathbb{Z})$ there is $N = N(\mathcal{Z})$ such that $z_n = 0$ for all $|n| \geq N$. Show that $c_c(\mathbb{Z})$ is dense in $c_0(\mathbb{Z})$.

1.3   Fourier coefficients

Let $f \in L^1(\mathbb{T})$. Then the nth *Fourier coefficient* of f is defined by

$$\hat{f}(n) = \frac{1}{2\pi} \int_{-\pi}^{\pi} f(e^{it})\, e^{-int}\, dt, \qquad (n \in \mathbb{Z}).$$

The two-sided sequence $\hat{f} = (\hat{f}(n))_{n\in\mathbb{Z}}$ is called the *Fourier transform* of f. We clearly have

$$|\hat{f}(n)| \leq \frac{1}{2\pi} \int_{-\pi}^{\pi} |f(e^{it})|\, dt, \qquad (n \in \mathbb{Z}),$$

which can be rewritten as $\hat{f} \in \ell^\infty(\mathbb{Z})$ with

$$\|\hat{f}\|_\infty \leq \|f\|_1.$$

Therefore, the Fourier transform

$$\begin{aligned} L^1(\mathbb{T}) &\longrightarrow \ell^\infty(\mathbb{Z}) \\ f &\longmapsto \hat{f} \end{aligned}$$

is a linear map whose norm is at most one. The constant function shows that the norm is indeed equal to one. We will show that this map is one-to-one and its range is included in $c_0(\mathbb{Z})$. But first, we need to develop some techniques.

The *Fourier series* of f is formally written as

$$\sum_{n=-\infty}^{\infty} \hat{f}(n)\, e^{int}.$$

The central question in Fourier analysis is to determine *when, how* and *toward what* this series converges. We will partially address these questions in the following. Any formal series of the form

$$\sum_{n=-\infty}^{\infty} a_n\, e^{int}$$

is called a *trigonometric series*. Hence a Fourier series is a special type of trigonometric series. However, there are trigonometric series which are not Fourier series. In other words, the coefficients a_n are not the Fourier coefficients of any integrable function. Using Euler's identity

$$e^{int} = \cos(nt) + i\sin(nt),$$

a trigonometric series can be rewritten as

$$\alpha_0 + \sum_{n=1}^{\infty} \alpha_n \cos(nt) + \beta_n \sin(nt).$$

It is easy to find the relation between α_n, β_n and a_n.

An important example which plays a central role in the theory of harmonic functions is the *Poisson kernel*

$$P_r(e^{it}) = \frac{1 - r^2}{1 + r^2 - 2r\cos t}, \qquad (0 \le r < 1). \qquad (1.4)$$

(See Figure 1.3.)

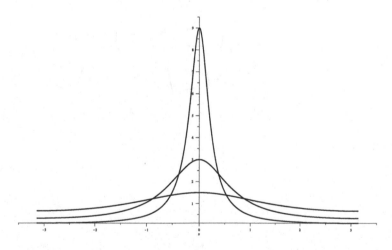

Fig. 1.3. The Poisson kernel $P_r(e^{it})$ for $r = 0.2, 0.5, 0.8$.

Clearly, for each fixed $0 \le r < 1$,

$$P_r \in C^{\infty}(\mathbb{T}) \subset L^1(\mathbb{T}).$$

Direct computation of $\widehat{P}_r$ is somehow difficult. But, the following observation makes its calculation easier. We have

$$
\begin{aligned}
\frac{1 - r^2}{1 + r^2 - 2r\cos t} &= \frac{1 - r^2}{1 + r^2 - r(e^{it} + e^{-it})} \\
&= \frac{1 - r^2}{(1 - re^{it})(1 - re^{-it})} \\
&= \frac{1}{1 - re^{it}} + \frac{1}{1 - re^{-it}} - 1
\end{aligned}
$$

and thus, using the geometric series

$$
1 + w + w^2 + \cdots = \frac{1}{1 - w}, \qquad (|w| < 1),
$$

we obtain

$$
P_r(e^{it}) = \sum_{n=-\infty}^{\infty} r^{|n|}\, e^{int}. \tag{1.5}
$$

Moreover, for each fixed $r < 1$, the partial sums are *uniformly* convergent to P_r. The uniform convergence is the key to this shortcut method. Therefore, for each $n \in \mathbb{Z}$,

$$
\begin{aligned}
\widehat{P}_r(n) &= \frac{1}{2\pi} \int_{-\pi}^{\pi} P_r(e^{it})\, e^{-int}\, dt \\
&= \frac{1}{2\pi} \int_{-\pi}^{\pi} \left(\sum_{m=-\infty}^{\infty} r^{|m|}\, e^{imt} \right) e^{-int}\, dt \\
&= \sum_{m=-\infty}^{\infty} r^{|m|} \left(\frac{1}{2\pi} \int_{-\pi}^{\pi} e^{i(m-n)t}\, dt \right) \\
&= r^{|n|}.
\end{aligned} \tag{1.6}
$$

(See Figure 1.4.)

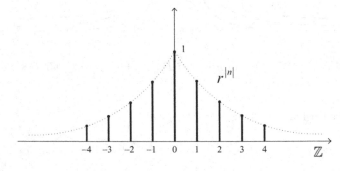

Fig. 1.4. The spectrum of P_r.

We emphasize that the uniform convergence of the series enables us to change the order of $\sum$ and $\int$ in the third equality above. This phenomenon will appear frequently in our discussion. The identity (1.5) also shows that P_r is equal to its Fourier series at all points of $\mathbb{T}$.

It is rather easy to extend the definition of Fourier transform for Borel measures on $\mathbb{T}$. Let $\mu \in \mathcal{M}(\mathbb{T})$. The nth Fourier coefficient of μ is defined by

$$\hat{\mu}(n) = \int_{\mathbb{T}} e^{-int} \, d\mu(e^{it}), \qquad (n \in \mathbb{Z}),$$

and the Fourier transform of μ is the two-sided sequence $\hat{\mu} = (\hat{\mu}(n))_{n \in \mathbb{Z}}$. Considering the embedding

$$\begin{array}{ccc} L^1(\mathbb{T}) & \longrightarrow & \mathcal{M}(\mathbb{T}) \\ f & \longmapsto & f(e^{it}) \, dt/2\pi, \end{array}$$

if we think of $L^1(\mathbb{T})$ as a subspace of $\mathcal{M}(\mathbb{T})$, it is easy to see that the two definitions of Fourier coefficients are consistent. In other words, if

$$d\mu(e^{it}) = \frac{1}{2\pi} \, f(e^{it}) \, dt,$$

where $f \in L^1(\mathbb{T})$, then we have

$$\hat{\mu}(n) = \hat{f}(n), \qquad (n \in \mathbb{Z}).$$

Lemma 1.1 *Let $\mu \in \mathcal{M}(\mathbb{T})$. Then $\hat{\mu} \in \ell^\infty(\mathbb{Z})$ and*

$$\| \hat{\mu} \|_\infty \leq \| \mu \|.$$

In particular, for each $f \in L^1(\mathbb{T})$,

$$\| \hat{f} \|_\infty \leq \| f \|_1.$$

Proof. For each $n \in \mathbb{Z}$, we have

$$\begin{aligned} |\hat{\mu}(n)| &= \left| \int_{\mathbb{T}} e^{-int} \, d\mu(e^{it}) \right| \\ &\leq \int_{\mathbb{T}} |e^{-int}| \, d|\mu|(e^{it}) \\ &= \int_{\mathbb{T}} d|\mu|(e^{it}) = |\mu|(\mathbb{T}) = \|\mu\|. \end{aligned}$$

The second inequality is a special case of the first one with $d\mu(e^{it}) = f(e^{it}) \, dt/2\pi$. In this case, $\hat{\mu}(n) = \hat{f}(n)$ and $\|\mu\| = \|f\|_1$. It was also proved directly at the beginning of this section. $\qquad\square$

Based on the preceding lemma, the Fourier transform

$$\begin{array}{ccc} \mathcal{M}(\mathbb{T}) & \longrightarrow & \ell^{\infty}(\mathbb{Z}) \\ \mu & \longmapsto & \hat{\mu} \end{array}$$

is a linear map whose norm is at most one. The Dirac measure δ_1 shows that the norm is actually equal to one. We will show that this map is also one-to-one. However, since

$$\hat{\delta}_1(n) = 1, \qquad (n \in \mathbb{Z}),$$

its range is not included in $c_0(\mathbb{Z})$.

Exercises

Exercise 1.3.1   Let $f \in L^p(\mathbb{T})$, $1 \leq p \leq \infty$. Show that $\| \hat{f} \|_\infty \leq \| f \|_p$.

Exercise 1.3.2   Let $z = re^{i\theta} \in \mathbb{D}$. Show that

$$P_r(e^{i(\theta - t)}) = \Re\left(\frac{e^{it} + z}{e^{it} - z} \right).$$

Exercise 1.3.3   Let $f \in L^1(\mathbb{T})$. Define

$$F(z) = \frac{1}{2\pi} \int_{-\pi}^{\pi} \frac{e^{it} + z}{e^{it} - z} f(e^{it}) \, dt, \qquad (z \in \mathbb{D}).$$

Show that

$$F(z) = \hat{f}(0) + 2 \sum_{n=1}^{\infty} \hat{f}(n) \, z^n, \qquad (z \in \mathbb{D}).$$

Hint: Note that

$$\frac{e^{it} + z}{e^{it} - z} = 1 + 2 \sum_{n=1}^{\infty} z^n \, e^{-int}.$$

Exercise 1.3.4   Let $f \in L^1(\mathbb{T})$ and define

$$g(e^{it}) = f(e^{i2t}).$$

Show that

$$\hat{g}(n) = \begin{cases} \hat{f}(\frac{n}{2}) & \text{if} \quad 2 | n, \\ \\ 0 & \text{if} \quad 2 \nmid n. \end{cases}$$

Consider a similar question if we define

$$g(e^{it}) = f(e^{ikt}),$$

where k is a fixed positive integer.

Exercise 1.3.5 Let $f \in \mathrm{Lip}_\alpha(\mathbb{T})$, $0 < \alpha \leq 1$. Show that

$$\hat{f}(n) = O(1/n^\alpha),$$

as $|n| \to \infty$.

Hint: If $n \neq 0$, we have

$$\hat{f}(n) = \frac{1}{4\pi} \int_{-\pi}^{\pi} \left(f(e^{it}) - f(e^{i(t+\pi/n)}) \right) e^{-int} \, dt.$$

1.4 Convolution on $\mathbb{T}$

Let $f, g \in L^1(\mathbb{T})$. Then we cannot conclude that $fg \in L^1(\mathbb{T})$. Indeed, it is easy to manufacture an example such that

$$\int_{-\pi}^{\pi} |f(e^{it}) g(e^{it})| \, dt = \infty.$$

Nevertheless, by Fubini's theorem,

$$\int_{-\pi}^{\pi} \left(\int_{-\pi}^{\pi} |f(e^{i\tau}) g(e^{i(t-\tau)})| \, d\tau \right) dt$$

$$= \int_{-\pi}^{\pi} |f(e^{i\tau})| \left(\int_{-\pi}^{\pi} |g(e^{i(t-\tau)})| \, dt \right) d\tau$$

$$= \left(\int_{-\pi}^{\pi} |f(e^{i\tau})| \, d\tau \right) \left(\int_{-\pi}^{\pi} |g(e^{is})| \, ds \right) < \infty.$$

Therefore, we necessarily have

$$\int_{-\pi}^{\pi} |f(e^{i\tau}) g(e^{i(t-\tau)})| \, d\tau < \infty$$

for almost all $e^{it} \in \mathbb{T}$. This observation enables us to define

$$(f * g)(e^{it}) = \int_{-\pi}^{\pi} f(e^{i\tau}) g(e^{i(t-\tau)}) \, d\tau$$

for almost all $e^{it} \in \mathbb{T}$, and besides the previous calculation shows that

$$f * g \in L^1(\mathbb{T})$$

with

$$\|f * g\|_1 \leq \|f\|_1 \, \|g\|_1. \tag{1.7}$$

The function $f * g$ is called the *convolution* of f and g. It is straightforward to see that the convolution is

(i) commutative: $f * g = g * f$,

(ii) associative: $f * (g * h) = (f * g) * h$,

(iii) distributive: $f * (g + h) = f * g + f * h$,

(iv) homogenous: $f * (\alpha g) = (\alpha f) * g = \alpha(f * g)$,

for all $f, g, h \in L^1(\mathbb{T})$ and $\alpha \in \mathbb{C}$. In technical terms, $L^1(\mathbb{T})$, equipped with the convolution as its product, is a *Banach algebra*. As a matter of fact, this concrete example inspired most of the abstract theory of Banach algebras.

Let $f, g \in L^1(\mathbb{T})$ and let $\varphi \in C(\mathbb{T})$. Then, by Fubini's theorem,

$$\int_{\mathbb{T}} \varphi(e^{is}) \, (f * g)(e^{is}) \, \frac{ds}{2\pi} = \int_{\mathbb{T}} \int_{\mathbb{T}} \varphi(e^{is}) \left(f(e^{i(s-\tau)}) g(e^{i\tau}) \, \frac{d\tau}{2\pi} \right) \frac{ds}{2\pi}$$

$$= \int_{\mathbb{T}} \int_{\mathbb{T}} \varphi(e^{i(t+\tau)}) \left(f(e^{it}) \, \frac{dt}{2\pi} \right) \left(g(e^{i\tau}) \, \frac{d\tau}{2\pi} \right).$$

This fact enables us to define the convolution of two Borel measures on $\mathbb{T}$ such that if we consider $L^1(\mathbb{T})$ as a subset of $\mathcal{M}(\mathbb{T})$, the two definitions are consistent. Let $\mu, \nu \in \mathcal{M}(\mathbb{T})$, and define $\Lambda : C(\mathbb{T}) \longrightarrow \mathbb{C}$ by

$$\Lambda(\varphi) = \int_{\mathbb{T}} \int_{\mathbb{T}} \varphi(e^{i(t+\tau)}) \, d\mu(e^{it}) \, d\nu(e^{i\tau}), \qquad (\varphi \in C(\mathbb{T})).$$

The functional Λ is clearly linear and satisfies

$$|\Lambda(\varphi)| \leq \|\mu\| \, \|\nu\| \, \|\varphi\|_\infty, \qquad (\varphi \in C(\mathbb{T})),$$

which implies

$$\|\Lambda\| \leq \|\mu\| \, \|\nu\|.$$

Therefore, by the Riesz representation theorem for bounded linear functionals on $C(\mathbb{T})$, there exists a unique Borel measure, which we denote by $\mu * \nu$ and call the convolution of μ and ν, such that

$$\Lambda(\varphi) = \int_{\mathbb{T}} \varphi(e^{it}) \, d(\mu * \nu)(e^{it}), \qquad (\varphi \in C(\mathbb{T})),$$

and moreover,

$$\|\Lambda\| = \|\mu * \nu\|.$$

Hence, $\mu * \nu$ is defined such that

$$\int_{\mathbb{T}} \varphi(e^{it}) \, d(\mu * \nu)(e^{it}) = \int_{\mathbb{T}} \int_{\mathbb{T}} \varphi(e^{i(t+\tau)}) \, d\mu(e^{it}) \, d\nu(e^{i\tau}), \qquad (1.8)$$

for all $\varphi \in C(\mathbb{T})$, and it satisfies

$$\|\mu * \nu\| \leq \|\mu\| \, \|\nu\|. \qquad (1.9)$$

The following result is easy to prove. Nevertheless, it is the most fundamental connection between convolution and the Fourier transform. Roughly speaking, it says that the Fourier transform changes convolution to multiplication.

Theorem 1.2 *Let* $\mu, \nu \in \mathcal{M}(\mathbb{T})$. *Then*

$$\widehat{\mu * \nu}(n) = \hat{\mu}(n)\,\hat{\nu}(n), \qquad (n \in \mathbb{Z}).$$

In particular, if $f, g \in L^1(\mathbb{T})$, *then*

$$\widehat{f * g}(n) = \hat{f}(n)\,\hat{g}(n), \qquad (n \in \mathbb{Z}).$$

Proof. Fix $n \in \mathbb{Z}$ and put

$$\varphi(e^{it}) = e^{-int}$$

in (1.8). Hence,

$$
\begin{aligned}
\widehat{\mu * \nu}(n) &= \int_{\mathbb{T}} e^{-int}\, d(\mu * \nu)(e^{it}) \\
&= \int_{\mathbb{T}} \int_{\mathbb{T}} e^{-in(t+\tau)}\, d\mu(e^{it})\, d\nu(e^{i\tau}) \\
&= \int_{\mathbb{T}} e^{-int}\, d\mu(e^{it}) \times \int_{\mathbb{T}} e^{-in\tau}\, d\nu(e^{i\tau}) = \hat{\mu}(n)\,\hat{\nu}(n).
\end{aligned}
$$

$\square$

We saw that if we consider $L^1(\mathbb{T})$ as a subset of $\mathcal{M}(\mathbb{T})$, the two definitions of convolution are consistent. Therefore, $\mathcal{M}(\mathbb{T})$ contains $L^1(\mathbb{T})$ as a subalgebra. But, we can say more in this case. We show that $L^1(\mathbb{T})$ is actually an ideal in $\mathcal{M}(\mathbb{T})$.

Theorem 1.3 *Let* $\mu \in \mathcal{M}(\mathbb{T})$ *and let* $f \in L^1(\mathbb{T})$. *Let*

$$d\nu(e^{it}) = f(e^{it})\, dt.$$

Then $\mu * \nu$ *is also absolutely continuous with respect to Lebesgue measure and we have*

$$d(\mu * \nu)(e^{it}) = \left(\int_{\mathbb{T}} f(e^{i(t-\tau)})\, d\mu(e^{i\tau}) \right) dt.$$

Proof. According to (1.8), for each $\varphi \in \mathcal{C}(\mathbb{T})$ we have

$$
\begin{aligned}
\int_{\mathbb{T}} \varphi(e^{it})\, d(\mu * \nu)(e^{it}) &= \int_{\mathbb{T}} \int_{\mathbb{T}} \varphi(e^{i(t+\tau)})\, d\mu(e^{it})\, d\nu(e^{i\tau}) \\
&= \int_{\mathbb{T}} \int_{\mathbb{T}} \varphi(e^{i(t+\tau)})\, d\mu(e^{it})\, f(e^{i\tau})\, d\tau \\
&= \int_{\mathbb{T}} \varphi(e^{is}) \left(\int_{\mathbb{T}} f(e^{i(s-t)})\, d\mu(e^{it}) \right) ds.
\end{aligned}
$$

Therefore, by the uniqueness part of the Riesz representation theorem,

$$d(\mu * \nu)(e^{is}) = \left(\int_{\mathbb{T}} f(e^{i(s-t)})\, d\mu(e^{it}) \right) ds.$$

$\square$

Let $\mu \in \mathcal{M}(\mathbb{T})$ and $f \in L^1(\mathbb{T})$. Considering f as a measure, by the preceding theorem, $\mu * f$ is absolutely continuous with respect to Lebesgue measure and we may write

$$(\mu * f)(e^{it}) = \int_{\mathbb{T}} f(e^{i(t-\tau)}) \, d\mu(e^{i\tau}). \tag{1.10}$$

Theorem 1.3 ensures that $(\mu * f)(e^{it})$ is well-defined for almost all $e^{it} \in \mathbb{T}$, $\mu * f \in L^1(\mathbb{T})$ and, by (1.9),

$$\|\mu * f\|_1 \le \|f\|_1 \, \|\mu\|. \tag{1.11}$$

We will need a very special case of (1.10) where $f \in \mathcal{C}(\mathbb{T})$. In this case, $(\mu * f)(e^{it})$ is defined for *all* $e^{it} \in \mathbb{T}$.

Exercises

Exercise 1.4.1   Let $\chi_n(e^{it}) = e^{int}$, $n \in \mathbb{Z}$. Show that

$$f * \chi_n = \hat{f}(n) \, \chi_n$$

for any $f \in L^1(\mathbb{T})$.

Exercise 1.4.2   Show that $\mathcal{M}(\mathbb{T})$, equipped with convolution as its product, is a commutative Banach algebra. What is its unit?

Exercise 1.4.3   Are you able to show that the Banach algebra $L^1(\mathbb{T})$ does not have a unit?
Hint: Use Theorem 1.2. Come back to this exercise after studying the Riemann–Lebesgue lemma in Section 2.5.

1.5   Young's inequality

Since $L^p(\mathbb{T}) \subset L^1(\mathbb{T})$, for $1 \le p \le \infty$, and since the convolution was defined on $L^1(\mathbb{T})$, then a priori $f * g$ is well-defined whenever $f \in L^r(\mathbb{T})$ and $g \in L^s(\mathbb{T})$ with $1 \le r, s \le \infty$. The following result gives more information about $f * g$, when we restrict f and g to some smaller subclasses of $L^1(\mathbb{T})$.

Theorem 1.4 (Young's inequality) *Let $f \in L^r(\mathbb{T})$, and let $g \in L^s(\mathbb{T})$, where $1 \le r, s \le \infty$ and*

$$\frac{1}{r} + \frac{1}{s} \ge 1.$$

Let

$$\frac{1}{p} = \frac{1}{r} + \frac{1}{s} - 1.$$

Then $f * g \in L^p(\mathbb{T})$ and

$$\|f * g\|_p \le \|f\|_r \, \|g\|_s.$$

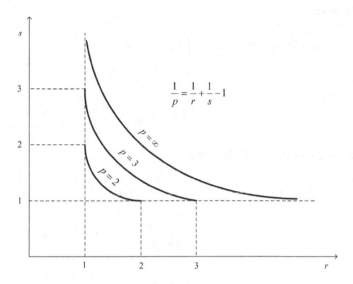

Fig. 1.5. The level curves of p.

Proof. (Figure 1.5 shows the level curves of p.) If $p = \infty$, or equivalently $1/r + 1/s = 1$, then $f * g$ is well-defined for *all* $e^{i\theta} \in \mathbb{T}$ and Young's inequality reduces to Hölder's inequality. Now, suppose that $1/r + 1/s > 1$. We need a generalized form of Hölder's inequality. Let $1 < p_1, \ldots, p_n < \infty$ such that

$$\frac{1}{p_1} + \cdots + \frac{1}{p_n} = 1,$$

and let $f_1, \ldots, f_n$ be measurable functions on a measure space $(X, \mathfrak{M}, \mu)$. Then

$$\int_X |f_1 \cdots f_n| \, d\mu \le \left(\int_X |f_1|^{p_1} \, d\mu \right)^{\frac{1}{p_1}} \cdots \left(\int_X |f_n|^{p_n} \, d\mu \right)^{\frac{1}{p_n}}.$$

This inequality can be proved by induction and the ordinary Hölder's inequality.

Let r' and s' be respectively the conjugate exponents of r and s, i.e.

$$\frac{1}{r} + \frac{1}{r'} = 1 \qquad \text{and} \qquad \frac{1}{s} + \frac{1}{s'} = 1.$$

Then, according to the definition of p, we have

$$\frac{1}{r'} + \frac{1}{s'} + \frac{1}{p} = 1.$$

Fix $e^{i\theta} \in \mathbb{T}$. To apply the generalized Hölder's inequality, we write the integrand $|f(e^{it}) \, g(e^{i(\theta-t)})|$ as the product of three functions respectively in $L^{r'}(\mathbb{T})$, $L^{s'}(\mathbb{T})$

and $L^p(\mathbb{T})$. Write

$$|f(e^{i\tau})\, g(e^{i(t-\tau)})| \;=\; \left(|g(e^{i(t-\tau)})|^{1-\frac{s}{p}} \right)$$

$$\times \left(|f(e^{i\tau})|^{1-\frac{r}{p}} \right)$$

$$\times \left(|f(e^{i\tau})|^{\frac{r}{p}}\, |g(e^{i(t-\tau)})|^{\frac{s}{p}} \right).$$

Hence, by the generalized Hölder's inequality,

$$\frac{1}{2\pi}\int_{-\pi}^{\pi} |f(e^{i\tau})\, g(e^{i(t-\tau)})|\, d\tau \;\leq\; \left(\frac{1}{2\pi}\int_{-\pi}^{\pi} |g(e^{i(t-\tau)})|^{r'(1-\frac{s}{p})}\, d\tau \right)^{\frac{1}{r'}}$$

$$\times \left(\frac{1}{2\pi}\int_{-\pi}^{\pi} |f(e^{i\tau})|^{s'(1-\frac{r}{p})}\, d\tau \right)^{\frac{1}{s'}}$$

$$\times \left(\frac{1}{2\pi}\int_{-\pi}^{\pi} |f(e^{i\tau})|^{r}\, |g(e^{i(t-\tau)})|^{s}\, d\tau \right)^{\frac{1}{p}}.$$

But $r'(1 - s/p) = s$ and $s'(1 - r/p) = r$. Thus

$$|(f * g)(e^{it})| \leq \|g\|_s^{s/r'}\, \|f\|_r^{r/s'} \left(\frac{1}{2\pi}\int_{-\pi}^{\pi} |f(e^{i\tau})|^{r}\, |g(e^{i(t-\tau)})|^{s}\, d\tau \right)^{\frac{1}{p}}$$

for almost all $e^{it} \in \mathbb{T}$. Finally, by Fubini's theorem,

$$\|f * g\|_p \;=\; \left(\frac{1}{2\pi}\int_{-\pi}^{\pi} |(f * g)(e^{it})|^p\, dt \right)^{\frac{1}{p}}$$

$$\leq\; \|g\|_s^{s/r'}\, \|f\|_r^{r/s'} \left(\frac{1}{(2\pi)^2}\int_{-\pi}^{\pi}\int_{-\pi}^{\pi} |f(e^{i\tau})|^{r}\, |g(e^{i(t-\tau)})|^{s}\, dt\, d\tau \right)^{\frac{1}{p}}$$

$$=\; \|g\|_s^{s/r'}\, \|f\|_r^{r/s'} \times \|g\|_s^{s/p}\, \|f\|_r^{r/p} = \|f\|_r\, \|g\|_s.$$

□

Another proof of Young's inequality is based on the Riesz–Thorin interpolation theorem and will be discussed in Chapter 8. The following two special cases of Young's inequality are what we need later on.

Corollary 1.5 *Let $f \in L^p(\mathbb{T})$, $1 \leq p \leq \infty$, and let $g \in L^1(\mathbb{T})$. Then $f * g \in L^p(\mathbb{T})$, and*

$$\|f * g\|_p \leq \|f\|_p\, \|g\|_1.$$

Corollary 1.6 *Let $f \in L^p(\mathbb{T})$, and let $g \in L^q(\mathbb{T})$, where q is the conjugate exponent of p. Then $(f * g)(e^{it})$ is well-defined for all $e^{it} \in \mathbb{T}$, $f * g \in C(\mathbb{T})$, and*

$$\|f * g\|_\infty \leq \|f\|_p\, \|g\|_q.$$

Proof. As we mentioned in the proof of Theorem 1.4, Hölder's inequality ensures that $(f * g)(e^{it})$ is well-defined for all $e^{it} \in \mathbb{T}$. The only new fact to prove is that $f * g$ is a continuous function on $\mathbb{T}$.

At least one of p or q is not infinity. Without loss of generality, assume that $p \neq \infty$. This assumption ensures that $\mathcal{C}(\mathbb{T})$ is dense in $L^p(\mathbb{T})$ (see Section A.4). Thus, given $\varepsilon > 0$, there is $\varphi \in \mathcal{C}(\mathbb{T})$ such that

$$\| f - \varphi \|_p < \varepsilon.$$

Hence

$$
\begin{aligned}
|(f * g)(e^{it}) - (f * g)(e^{is})| \quad &\leq \quad |((f - \varphi) * g)(e^{it})| + |((f - \varphi) * g)(e^{is})| \\
&+ \quad |(\varphi * g)(e^{it}) - (\varphi * g)(e^{is})| \\
&\leq \quad 2 \| f - \varphi \|_p \| g \|_q + \omega_\varphi(|t - s|) \| g \|_q,
\end{aligned}
$$

where

$$\omega_\varphi(\delta) = \sup_{|t-s| \leq \delta} |\varphi(e^{it}) - \varphi(e^{is})|$$

is the *modulus of continuity* of φ. Since φ is uniformly continuous on $\mathbb{T}$,

$$\omega_\varphi(\delta) \longrightarrow 0$$

as $\delta \to 0$. Therefore, if $|t - s|$ is small enough, we have

$$|(f * g)(e^{it}) - (f * g)(e^{is})| \leq 3\varepsilon \| g \|_q.$$

$\square$

Exercises

Exercise 1.5.1 What can we say about $f * g$ if $f \in L^r(\mathbb{T})$ and $g \in L^s(\mathbb{T})$ with $1 \leq r, s \leq \infty$ and

$$\frac{1}{r} + \frac{1}{s} \leq 1?$$

(See Figure 1.6.)

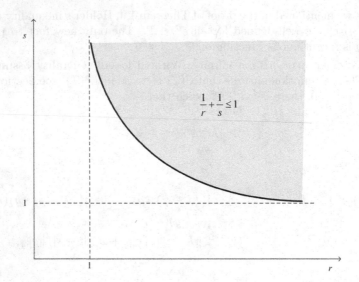

Fig. 1.6. The region $\frac{1}{r} + \frac{1}{s} \leq 1$.

Exercise 1.5.2 Show that Young's inequality is sharp in the following sense. Given r, s with $1 \leq r, s \leq \infty$ and

$$\frac{1}{r} + \frac{1}{s} \geq 1,$$

there are $f \in L^r(\mathbb{T})$ and $g \in L^s(\mathbb{T})$ such that $f * g \in L^p(\mathbb{T})$, where

$$\frac{1}{p} = \frac{1}{r} + \frac{1}{s} - 1,$$

but $f * g \notin L^t(\mathbb{T})$ for any $t > p$.

Hint: Start with the case $r = s = 1$ and a function $\varphi \in L^1(\mathbb{T})$ such that $\varphi \notin L^t(\mathbb{T})$ for any $t > 1$.

Chapter 2

Abel–Poisson means

2.1 Abel–Poisson means of Fourier series

Let $\{F_r\}_{0 \leq r < 1}$ be a family of functions on the unit circle $\mathbb{T}$. Define

$$F(re^{it}) = F_r(e^{it}), \qquad (re^{it} \in \mathbb{D}).$$

Hence, instead of looking at the family as a collection of individual functions F_r which are defined on $\mathbb{T}$, we deal with *one* single function defined on the open unit disc $\mathbb{D}$. On the other hand, if $F(re^{it})$ is given first, for each fixed $r \in [0, 1)$, we can define F_r by considering the values of F on the circle $\{|z| = r\}$. This dual interpretation will be encountered many times in what follows. An important example of this phenomenon is the Poisson kernel which was defined as a family of functions on the unit circle by (1.4). This kernel can also be considered as one function

$$P(re^{it}) = \frac{1 - r^2}{1 + r^2 - 2r \cos t}$$

on $\mathbb{D}$.

Let $\mu \in \mathcal{M}(\mathbb{T})$. Then, by (1.10), we have

$$(P_r * \mu)(e^{i\theta}) = \int_{\mathbb{T}} \frac{1 - r^2}{1 + r^2 - 2r \cos(\theta - t)} \, d\mu(e^{it}) \qquad (2.1)$$

which is called the *Poisson integral* of μ. Moreover, by (1.6) and Theorem 1.2, the Fourier coefficients of $P_r * \mu$ are given by

$$\widehat{P_r * \mu}(n) = r^{|n|} \, \hat{\mu}(n), \qquad (n \in \mathbb{Z}).$$

Thus the formal Fourier series of $P_r * \mu$ is

$$\sum_{n=-\infty}^{\infty} \hat{\mu}(n) \, r^{|n|} \, e^{in\theta}.$$

These sums are called the *Abel–Poisson means* of the Fourier series

$$\sum_{n=-\infty}^{\infty} \hat{\mu}(n) \, e^{in\theta}.$$

The Fourier series of μ is not necessarily pointwise convergent. However, we show that its Abel–Poisson means behave much better. The following theorem reveals the relation between the Abel–Poisson means of μ and its Poisson integral.

Theorem 2.1 *Let $\mu \in \mathcal{M}(\mathbb{T})$, and let*

$$U(re^{i\theta}) = (P_r * \mu)(e^{i\theta}) = \int_{\mathbb{T}} \frac{1 - r^2}{1 + r^2 - 2r\cos(\theta - t)} \, d\mu(e^{it}), \qquad (re^{i\theta} \in \mathbb{D}).$$

Then

$$U(re^{i\theta}) = \sum_{n=-\infty}^{\infty} \hat{\mu}(n) \, r^{|n|} \, e^{in\theta}, \qquad (re^{i\theta} \in \mathbb{D}).$$

The series is absolutely and uniformly convergent on compact subsets of $\mathbb{D}$, and U is harmonic on $\mathbb{D}$.

Proof. Since

$$| \, \hat{\mu}(n) \, r^{|n|} \, e^{in\theta} \, | \leq \|\mu\| \, r^{|n|},$$

the series $\sum \hat{\mu}(n) \, r^{|n|} \, e^{in\theta}$ is absolutely and uniformly convergent on compact subsets of $\mathbb{D}$. Fix $0 \leq r < 1$ and θ. Then, by (1.5),

$$\begin{aligned}
U(re^{i\theta}) &= \int_{\mathbb{T}} \frac{1 - r^2}{1 + r^2 - 2r\cos(\theta - t)} \, d\mu(e^{it}) \\
&= \int_{\mathbb{T}} \left(\sum_{n=-\infty}^{\infty} r^{|n|} \, e^{in(\theta - t)} \right) d\mu(e^{it}).
\end{aligned}$$

Since the series is uniformly convergent (as a function of e^{it}), and since $|\mu|$ is a finite positive Borel measure on $\mathbb{T}$, we can change the order of summation and integration. Hence,

$$U(re^{i\theta}) = \sum_{n=-\infty}^{\infty} \left(\int_{\mathbb{T}} e^{-int} \, d\mu(e^{it}) \right) r^{|n|} \, e^{in\theta} = \sum_{n=-\infty}^{\infty} \hat{\mu}(n) \, r^{|n|} \, e^{in\theta}.$$

There are several ways to verify that U is harmonic on $\mathbb{D}$. We give a direct proof. Fix $k \geq 0$. Then the absolute and uniform convergence of $\sum_{n=-\infty}^{\infty} n^k \hat{\mu}(n) \, r^{|n|} \, e^{in\theta}$ on compact subsets of $\mathbb{D}$ enables us to change the order of summation and any linear differential operator. In particular, let us apply the Laplace operator. Hence, remembering that each term $r^{|n|} \, e^{in\theta}$ is a harmonic function, we obtain

$$\nabla^2 U = \nabla^2 \left(\sum_{n=-\infty}^{\infty} \hat{\mu}(n) \, r^{|n|} \, e^{in\theta} \right) = \sum_{n=-\infty}^{\infty} \hat{\mu}(n) \, \nabla^2 (r^{|n|} \, e^{in\theta}) = 0.$$

$\square$

As a special case, if the measure μ in Theorem 2.1 is absolutely continuous with respect to the Lebesgue measure, i.e. $d\mu(e^{it}) = u(e^{it})\, dt/2\pi$ with $u \in L^1(\mathbb{T})$, then

$$
\begin{aligned}
U(re^{i\theta}) &= \frac{1}{2\pi} \int_{-\pi}^{\pi} \frac{1 - r^2}{1 + r^2 - 2r\cos(\theta - t)}\, u(e^{it})\, dt \\
&= \sum_{n=-\infty}^{\infty} \hat{u}(n)\, r^{|n|}\, e^{in\theta},
\end{aligned}
\tag{2.2}
$$

where the series is absolutely and uniformly convergent on compact subsets of $\mathbb{D}$, and U represents a harmonic function there.

Exercises

Exercise 2.1.1 Let $(a_n)_{n\geq 0}$ be a sequence of complex numbers. Suppose that the series

$$
S = \sum_{n=0}^{\infty} a_n
$$

is convergent. For each $0 \leq r < 1$, define

$$
S(r) = \sum_{n=0}^{\infty} a_n\, r^n.
$$

Show that $S(r)$ is absolutely convergent and moreover

$$
\lim_{r \to 1} S(r) = S.
$$

Hint: Let

$$
S_m = \sum_{n=0}^{m} a_n, \qquad (m \geq 0).
$$

Then

$$
S(r) = S + (1 - r) \sum_{n=0}^{\infty} (S_n - S)\, r^n.
$$

Exercise 2.1.2 Let $(a_n)_{n\geq 0}$ be a bounded sequence of complex numbers and let

$$
S(r) = \sum_{n=0}^{\infty} a_n\, r^n, \qquad (0 \leq r < 1).
$$

Find $(a_n)_{n\geq 0}$ satisfying the following properties:

(i) the series $\sum_{n=0}^{\infty} a_n$ is divergent;

(ii) for each $0 \leq r < 1$, $S(r)$ is absolutely convergent;

(iii) $\lim_{r \to 1} S(r)$ exists.

Exercise 2.1.3   Let $(a_n)_{n \geq 0}$ be a sequence of complex numbers. Suppose that the series

$$S = \sum_{n=0}^{\infty} a_n$$

is convergent. Let

$$S_n = \sum_{k=0}^{n} a_k$$

and define

$$C_n = \frac{S_0 + S_1 + \cdots + S_n}{n+1} = \sum_{k=0}^{n} \left(1 - \frac{k}{n+1} \right) a_k.$$

Show that

$$\lim_{n \to \infty} C_n = S.$$

Remark: The numbers C_n, $n \geq 0$, are called the *Cesàro means* of S_n.

Exercise 2.1.4   Let $(a_n)_{n \geq 0}$ be a sequence of complex numbers and define

$$C_n = \sum_{k=0}^{n} \left(1 - \frac{k}{n+1} \right) a_k.$$

Find $(a_n)_{n \geq 0}$ such that

$$\lim_{n \to \infty} C_n$$

exists, but the sequence

$$\sum_{n=0}^{\infty} a_n$$

is divergent.

Exercise 2.1.5   Let $(a_n)_{n \geq 0}$ be a sequence of complex numbers and define

$$C_n = \sum_{k=0}^{n} \left(1 - \frac{k}{n+1} \right) a_k.$$

Suppose that the series $\sum_{n=0}^{\infty} C_n$ is convergent and

$$\sum_{n=0}^{\infty} n |a_n|^2 < \infty.$$

Show that the series $\sum_{n=0}^{\infty} a_n$ is also convergent.

Remark: Compare with Exercises 2.1.3 and 2.1.4.

2.2 Approximate identities on $\mathbb{T}$

We saw that $L^1(\mathbb{T})$, equipped with convolution, is a commutative Banach algebra. This algebra does not have a unit element since such an element must satisfy

$$\hat{f}(n) = 1, \qquad (n \in \mathbb{Z}),$$

and we will see that the nth Fourier coefficient of any integrable function tends to zero as $|n| \to \infty$. To overcome this difficulty, we consider a family of integrable functions $\{\Phi_\iota\}$ satisfying

$$\lim_\iota \hat{\Phi}_\iota(n) = 1 \tag{2.3}$$

for each fixed $n \in \mathbb{Z}$. The condition (2.3) alone is not enough to obtain a family that somehow plays the role of a unit element. For example, the Dirichlet kernel satisfies this property but it is not a proper replacement for the unit element (see Exercise 2.2.2). We choose three other properties to define our family and then we show that (2.3) is fulfilled.

Let $\Phi_\iota \in L^1(\mathbb{T})$, where the index ι ranges over a directed set. In the examples given below, it ranges either over the set of integers $\{1, 2, 3, \dots\}$ or over the interval $[0, 1)$. Therefore, in the following, $\lim_\iota$ means either $\lim_{n \to \infty}$ or $\lim_{r \to 1-}$. Similarly, $\iota \succ \iota_0$ means $n > n_0$ or $r > r_0$. The family $\{\Phi_\iota\}$ is called an *approximate identity* on $\mathbb{T}$ if it satisfies the following properties:

(a) for all ι,

$$\frac{1}{2\pi} \int_{-\pi}^{\pi} \Phi_\iota(e^{it}) \, dt = 1;$$

(b)

$$C_\Phi = \sup_\iota \left(\frac{1}{2\pi} \int_{-\pi}^{\pi} |\Phi_\iota(e^{it})| \, dt \right) < \infty;$$

(c) for each fixed δ, $0 < \delta < \pi$,

$$\lim_\iota \int_{\delta \leq |t| \leq \pi} |\Phi_\iota(e^{it})| \, dt = 0.$$

The condition (a) forces $C_\Phi \geq 1$. If $\Phi_\iota(e^{it}) \geq 0$, for all ι and for all $e^{it} \in \mathbb{T}$, then $\{\Phi_\iota\}$ is called a *positive approximate identity*. In this case, (b) follows from (a) with

$$C_\Phi = 1.$$

We give three examples of a positive approximate identity below. Further examples are provided in the exercises. Our main example of a positive approximate identity is the Poisson kernel

$$P_r(e^{it}) = \frac{1 - r^2}{1 + r^2 - 2r\cos\theta} = \sum_{n=-\infty}^{\infty} r^{|n|} e^{int}, \qquad (0 \leq r < 1).$$

It is easy to verify that the *Fejér kernel*

$$\mathbf{K}_n(e^{it}) = \frac{1}{n+1} \left(\frac{\sin(\frac{(n+1)t}{2})}{\sin(\frac{t}{2})} \right)^2 = \sum_{k=-n}^{n} \left(1 - \frac{|k|}{n+1} \right) e^{ikt}, \qquad (n \geq 0),$$

is also a positive approximate identity. (See Figure 2.1.)

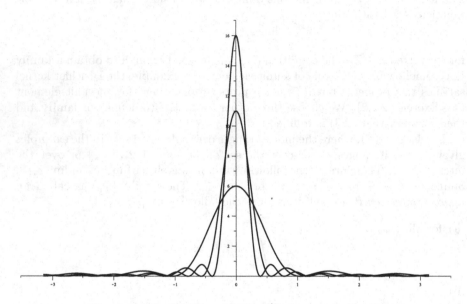

Fig. 2.1. The Fejér kernel $\mathbf{K}_n(e^{it})$ for $n = 5, 10, 15$.

A less familiar example is the family

$$F_r(e^{it}) = \frac{(1+r)^2 (1-r) t \sin t}{(1+r^2 - 2r \cos t)^2}, \qquad (0 \leq r < 1, \ -\pi \leq t \leq \pi).$$

We will apply F_r in studying the radial limits of harmonic functions. Note that the three examples given above satisfy the following stronger property:

(c′) For each fixed δ, $0 < \delta < \pi$,

$$\lim_{\iota} \left(\sup_{\delta \leq |t| \leq \pi} |\Phi_\iota(e^{it})| \right) = 0.$$

We now show that an approximate identity fulfils (2.3). Fix $n \in \mathbb{Z}$. Given $\varepsilon > 0$, there is δ such that

$$|e^{int} - 1| < \varepsilon$$

for all $t \in [-\delta, \delta]$. Hence, by (a), we have

$$\hat{\Phi}_\iota(n) - 1 = \frac{1}{2\pi} \int_{-\pi}^{\pi} \Phi_\iota(e^{it}) (e^{int} - 1) \, dt$$

and thus, by (b),

$$|\hat{\Phi}_\iota(n) - 1| \leq \frac{1}{2\pi} \left(\int_{-\delta}^{\delta} + \int_{\delta < |t| < \pi} \right) |\Phi_\iota(e^{it})| \, |e^{int} - 1| \, dt$$

$$\leq \varepsilon\, C_\Phi + \frac{1}{\pi} \int_{\delta < |t| < \pi} |\Phi_\iota(e^{it})| \, dt.$$

The property (c) ensures that the last integral tends to zero as ι grows. More precisely, there is ι_ε such that

$$|\hat{\Phi}_\iota(n) - 1| \leq (C_\Phi + 1)\, \varepsilon$$

for all $\iota > \iota_\varepsilon$.

Given $f \in L^1(\mathbb{T})$ and an approximate identity $\{\Phi_\iota\}$ on $\mathbb{T}$, we form the new family $\{\Phi_\iota * f\}$. In the rest of this chapter, assuming f belongs to one of the Lebesgue spaces $L^p(\mathbb{T})$ or to $\mathcal{C}(\mathbb{T})$, we explore the way in which $\Phi_\iota * f$ approaches f as ι grows. Similarly, for an arbitrary Borel measure $\mu \in \mathcal{M}(\mathbb{T})$, we find the relation between the measures $(\Phi_\iota * \mu)(e^{it})\, dt/2\pi$ and μ. According to Theorem 1.2, the Fourier series of $\Phi_\iota * \mu$ and $\Phi_\iota * f$ are respectively

$$\sum_{n=-\infty}^{\infty} \hat{\Phi}_\iota(n)\, \hat{\mu}(n)\, e^{in\theta} \qquad \text{and} \qquad \sum_{n=-\infty}^{\infty} \hat{\Phi}_\iota(n)\, \hat{f}(n)\, e^{in\theta}.$$

Hence, they are respectively the weighted Fourier series of μ and f. That is why we can also say that we study the weighted Fourier series in the following.

Exercises

Exercise 2.2.1 Let

$$F_n(e^{it}) = \begin{cases} 2\mathbf{K}_n(e^{it}) & \text{if } \quad 0 \leq t \leq \pi, \\ 0 & \text{if } \quad -\pi < t < 0. \end{cases}$$

Show that $(F_n)_{n \geq 0}$ is a positive approximate identity on $\mathbb{T}$. Find $\hat{F}_n$.

Exercise 2.2.2 The Dirichlet kernel is defined by

$$\mathbf{D}_n(e^{it}) = \sum_{k=-n}^{n} e^{ikt}, \qquad (n \geq 0).$$

(See Figure 2.2.) Show that

$$\mathbf{D}_n(e^{it}) = 1 + 2\sum_{k=1}^{n} \cos(kt) = \frac{\sin(\frac{(2n+1)t}{2})}{\sin(\frac{t}{2})}$$

and that $(\mathbf{D}_n)_{n\geq 0}$ is not an approximate identity on $\mathbb{T}$. Find $\widehat{\mathbf{D}}_n$. (See Figure 2.3, which shows the spectrum of $\mathbf{D}_4$.)

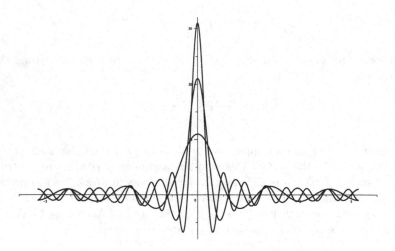

Fig. 2.2. The Dirichlet kernel $\mathbf{D}_n(e^{it})$ for $n = 5, 10, 15$.

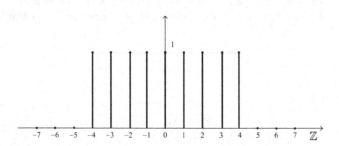

Fig. 2.3. The spectrum of $\mathbf{D}_4$.

Exercise 2.2.3   Lebesgue constants are defined by

$$L_n = \frac{1}{2\pi} \int_{-\pi}^{\pi} |\mathbf{D}_n(e^{it})| \, dt, \qquad (n \geq 0).$$

Show that $L_n \to \infty$ as $n \to \infty$.

Remark: By a more precise estimation one can show that

$$\frac{4}{\pi^2} \log n < L_n < 3 + \frac{4}{\pi^2} \log n.$$

Exercise 2.2.4 Show that

$$\mathbf{K}_n = \frac{\mathbf{D}_0 + \mathbf{D}_1 + \cdots + \mathbf{D}_n}{n+1}, \qquad (n \geq 0).$$

Exercise 2.2.5 Let

$$\varphi_n(e^{it}) = -2n \, \mathbf{K}_{n-1}(e^{it}) \sin(nt), \qquad (n \geq 1).$$

Show that $\|\varphi_n\|_1 < 2n$.

Exercise 2.2.6 Let p be a trigonometric polynomial of degree at most n. Show that

$$\|p'\|_\infty \leq 2n \, \|p\|_\infty.$$

Hint: Note that $p' = p * \varphi_n$, where φ_n is given in Exercise 2.2.5. Now, apply Corollary 1.6.

Remark: This inequality is not sharp. Bernstein showed that $\|p'\|_\infty \leq n \, \|p\|_\infty$, where p is any trigonometric polynomial of degree at most n.

Exercise 2.2.7 Show that the Fourier coefficients of the Fejér kernel $\mathbf{K}_n$ are given by

$$\widehat{\mathbf{K}}_n(m) = \begin{cases} 1 - \dfrac{|m|}{n+1} & \text{if} \quad |m| \leq n, \\ \\ 0 & \text{if} \quad |m| \geq n+1. \end{cases}$$

(Figure 2.4 shows the spectrum of $\mathbf{K}_4$.)

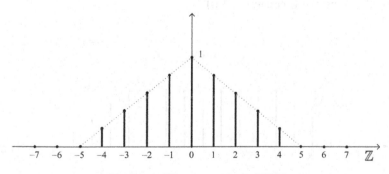

Fig. 2.4. The spectrum of $\mathbf{K}_4$.

Exercise 2.2.8   [de la Vallée Poussin's kernel] Let
$$\mathbf{V}_n(e^{it}) = 2\mathbf{K}_{2n+1}(e^{it}) - \mathbf{K}_n(e^{it}).$$
(See Figure 2.5.) Show that $(\mathbf{V}_n)_{n\geq 0}$ is an approximate identity on $\mathbb{T}$, and that

$$\widehat{\mathbf{V}}_n(m) = \begin{cases} 1 & \text{if} & |m| \leq n+1, \\[2mm] 2 - \dfrac{|m|}{n+1} & \text{if} & n+2 \leq |m| \leq 2n+1, \\[2mm] 0 & \text{if} & |m| \geq 2n+2. \end{cases}$$

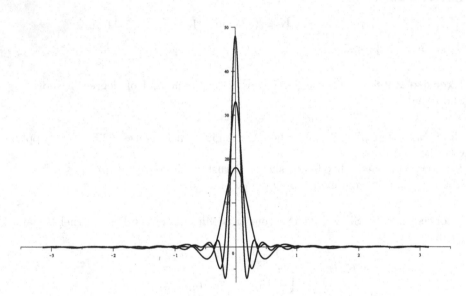

Fig. 2.5. The de la Vallée Poussin kernel $\mathbf{V}_n(e^{it})$ for $n = 5, 10, 15$.

(Figure 2.6 shows the spectrum of V_3.)

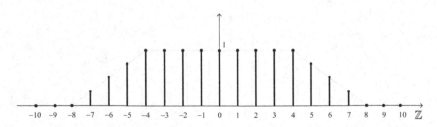

Fig. 2.6. The spectrum of $\mathbf{V}_3$.

Exercise 2.2.9 Show that

$$\|\mathbf{K}_n\|_2^2 = \frac{1}{2\pi} \int_{-\pi}^{\pi} |\mathbf{K}_n(e^{it})|^2 \, dt = \frac{2n^2 + 4n + 3}{3(n+1)}.$$

Exercise 2.2.10 [Jackson's kernel] Show that the family

$$\mathbf{J}_n(e^{it}) = \frac{\mathbf{K}_n^2(e^{it})}{\|\mathbf{K}_n\|_2^2} = \frac{3}{(2n^2 + 4n + 3)(n+1)} \left(\frac{\sin(\frac{(n+1)t}{2})}{\sin(\frac{t}{2})} \right)^4$$

is a positive approximate identity on $\mathbb{T}$. (See Figure 2.7.)

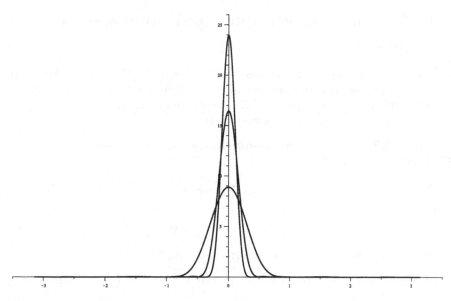

Fig. 2.7. The Jackson kernel $\mathbf{J}_n(e^{it})$ for $n = 5, 10, 15$.

Exercise 2.2.11 Let

$$F_n(e^{it}) = \begin{cases} n & \text{if} & |t| < \dfrac{\pi}{n}, \\[2mm] 0 & \text{if} & \dfrac{\pi}{n} \le |t| \le \pi. \end{cases}$$

Show that $(F_n)_{n\ge 1}$ is a positive approximate identity on $\mathbb{T}$. Find $\widehat{F}_n$.

Exercise 2.2.12   Let

$$F_r(e^{it}) = \frac{(1+r)^2\,(1-r)\,t\sin t}{(1+r^2 - 2r\cos t)^2}, \qquad (0 \le r < 1, \ -\pi \le t \le \pi).$$

Show that $(F_r)_{0 \le r < 1}$ is a positive approximate identity on $\mathbb{T}$.

Exercise 2.2.13   Let

$$F_r(e^{it}) = \frac{2(1-r^2)\,\sin^2 t}{(1+r^2 - 2r\cos t)^2}.$$

Show that $(F_r)_{0 \le r < 1}$ is a positive approximate identity on $\mathbb{T}$.

2.3   Uniform convergence and pointwise convergence

In this section we study the relation between $\Phi_\iota * f$ and f itself. We start with the simple but very important case of a continuous function. The crucial property which is exploited below is that a continuous function on $\mathbb{T}$, a compact set, is automatically uniformly continuous there.

Theorem 2.2 *Let* $\{\Phi_\iota\}$ *be an approximate identity on* $\mathbb{T}$, *and let* $f \in \mathcal{C}(\mathbb{T})$. *Then, for each* ι, $\Phi_\iota * f \in \mathcal{C}(\mathbb{T})$ *with*

$$\|\Phi_\iota * f\|_\infty \le C_\Phi \,\|f\|_\infty,$$

and moreover

$$\lim_\iota \|\Phi_\iota * f - f\|_\infty = 0.$$

In other words, $\Phi_\iota * f$ *converges uniformly to* f *on* $\mathbb{T}$.

Proof. By Corollary 1.6, for each ι, we certainly have $\Phi_\iota * f \in \mathcal{C}(\mathbb{T})$ and

$$\|\Phi_\iota * f\|_\infty \le \|\Phi_\iota\|_1 \,\|f\|_\infty \le C_\Phi \,\|f\|_\infty.$$

Since f is *uniformly continuous* on $\mathbb{T}$, given $\varepsilon > 0$, there exists $\delta = \delta(\varepsilon) > 0$ such that

$$|\,f(e^{it_2}) - f(e^{it_1})\,| < \varepsilon$$

whenever $|t_2 - t_1| < \delta$. Therefore, for all t,

$$|\,(\Phi_\iota * f)(e^{it}) - f(e^{it})\,| = \left| \frac{1}{2\pi} \int_{-\pi}^{\pi} \Phi_\iota(e^{i\tau}) \left(f(e^{i(t-\tau)}) - f(e^{it}) \right) d\tau \right|$$

$$\le \frac{1}{2\pi} \left(\int_{-\pi}^{-\delta} + \int_{-\delta}^{\delta} + \int_{\delta}^{\pi} \right) |\Phi_\iota(e^{i\tau})| \,|f(e^{i(t-\tau)}) - f(e^{it})| \,d\tau$$

$$\le \frac{\|f\|_\infty}{\pi} \int_{\delta \le |\tau| \le \pi} |\Phi_\iota(e^{i\tau})| \,d\tau + \varepsilon\,C_\Phi.$$

Pick $\iota(\varepsilon)$ so large that

$$\frac{1}{\pi} \int_{\delta \leq |\tau| \leq \pi} |\Phi_\iota(e^{i\tau})| \, d\tau < \varepsilon$$

for $\iota > \iota(\varepsilon)$. Thus, for $\iota > \iota(\varepsilon)$ and for all t,

$$|(\Phi_\iota * f)(e^{it}) - f(e^{it})| < (\|f\|_\infty + C_\Phi)\,\varepsilon.$$

$\square$

As a special case, let $\{\Phi_\iota\}$ be the Poisson kernel. By using this kernel we extend a continuous function on $\mathbb{T}$ into the unit disc $\mathbb{D}$. The outcome is a function continuous on the closed unit disc $\overline{\mathbb{D}}$ and harmonic on the open unit disc $\mathbb{D}$.

Corollary 2.3 *Let $u \in C(\mathbb{T})$, and let*

$$U(re^{i\theta}) = \begin{cases} \dfrac{1}{2\pi} \displaystyle\int_{-\pi}^{\pi} \dfrac{1 - r^2}{1 + r^2 - 2r\cos(\theta - t)}\, u(e^{it})\, dt & \text{if} \quad 0 \leq r < 1, \\[4mm] u(e^{i\theta}) & \text{if} \quad r = 1. \end{cases}$$

Then

(a) U is continuous on $\overline{\mathbb{D}}$,

(b) U is harmonic on $\mathbb{D}$,

(c) for each $0 \leq r < 1$, $\|U_r\|_\infty \leq \|u\|_\infty$.

Proof. By Theorem 2.1, U is harmonic on $\mathbb{D}$. Hence, U is at least continuous on $\mathbb{D}$. On the other hand, Theorem 2.2 ensures that, as $r \to 1$, U_r converges uniformly to u, and $\|U_r\|_\infty \leq \|u\|_\infty$. Therefore, U is also continuous at all points of $\mathbb{T}$. $\square$

By Theorem 2.1, we also have

$$U(re^{i\theta}) = \sum_{n=-\infty}^{\infty} \hat{u}(n)\, r^{|n|}\, e^{in\theta}, \quad (0 \leq r < 1). \tag{2.4}$$

Therefore, Corollary 2.3 says that the Abel–Poisson means of the Fourier series of a continuous function u converge uniformly to u.

Let $\mathcal{P}(\mathbb{T})$ denote the space of all trigonometric polynomials:

$$\mathcal{P}(\mathbb{T}) = \{p : p(e^{it}) = \sum_{n=-N}^{N} a_n\, e^{int}, \ a_n \in \mathbb{C}\}.$$

Clearly

$$\mathcal{P}(\mathbb{T}) \subset C(\mathbb{T}).$$

A celebrated theorem of Weierstrass says that $\mathcal{P}(\mathbb{T})$ is dense in $C(\mathbb{T})$. Using the Fejér kernel we are able to give a constructive proof of this result.

Corollary 2.4 (Weierstrass–Fejér) *Let $f \in C(\mathbb{T})$, and let*

$$p_n(e^{it}) = \sum_{k=-n}^{n} \left(1 - \frac{|k|}{n+1}\right) \hat{f}(k)\, e^{ikt}, \qquad (e^{it} \in \mathbb{T}).$$

Then, for each $n \geq 0$,

$$\|p_n\|_\infty \leq \|f\|_\infty$$

and

$$\lim_{n \to \infty} \|p_n - f\|_\infty = 0.$$

Proof. It is enough to observe that

$$p_n = f * \mathbf{K}_n,$$

where $\mathbf{K}_n$ is the Fejér kernel. Then apply Theorem 2.2. $\qquad\square$

In the rest of this section we study a local version of Theorem 2.2 by assuming that f is continuous at a fixed point on $\mathbb{T}$. Thus we obtain some results about the pointwise convergence of Fourier series.

Theorem 2.5 *Let $\{\Phi_\iota\}$ be an approximate identity on $\mathbb{T}$. Suppose that, for each ι,*

$$\Phi_\iota \in L^\infty(\mathbb{T}),$$

and that $\{\Phi_\iota\}$ satisfies the stronger property (c'):

$$\lim_{\iota} \left(\sup_{0 < \delta \leq |t| \leq \pi} |\Phi_\iota(e^{it})| \right) = 0.$$

Let $f \in L^1(\mathbb{T})$, and suppose that f is continuous at $e^{it_0} \in \mathbb{T}$. Then, given $\varepsilon > 0$, there exists $\iota(\varepsilon, t_0)$ and $\delta = \delta(\varepsilon, t_0) > 0$ such that

$$|(\Phi_\iota * f)(e^{it}) - f(e^{it_0})| < \varepsilon,$$

if $\iota > \iota(\varepsilon, t_0)$ and $|t - t_0| < \delta$. In particular,

$$\lim_{\iota} (\Phi_\iota * f)(e^{it_0}) = f(e^{it_0}).$$

Proof. Since $\Phi_\iota \in L^\infty(\mathbb{T})$ and $f \in L^1(\mathbb{T})$, by Corollary 1.6, $(\Phi_\iota * f)(e^{it})$ is well-defined at all $e^{it} \in \mathbb{T}$.

By assumption, given $\varepsilon > 0$, there exists $\delta = \delta(\varepsilon, t_0) > 0$ such that

$$|f(e^{i\eta}) - f(e^{it_0})| < \varepsilon$$

whenever $|\eta - t_0| < 2\delta$. Therefore, for $|t - t_0| < \delta$,

$$
\left| (\Phi_\iota * f)(e^{it}) - f(e^{it_0}) \right| = \left| \frac{1}{2\pi} \int_{-\pi}^{\pi} \Phi_\iota(e^{i\tau}) \left(f(e^{i(t-\tau)}) - f(e^{it_0}) \right) d\tau \right|
$$

$$
\leq \frac{1}{2\pi} \left(\int_{-\pi}^{-\delta} + \int_{-\delta}^{\delta} + \int_{\delta}^{\pi} \right) |\Phi_\iota(e^{i\tau})| \, |f(e^{i(t-\tau)}) - f(e^{it_0})| \, d\tau
$$

$$
\leq \frac{1}{2\pi} \int_{\delta \leq |\tau| \leq \pi} |\Phi_\iota(e^{i\tau})| \left(|f(e^{i(t-\tau)})| + |f(e^{it_0})| \right) d\tau + \varepsilon \, C_\Phi
$$

$$
\leq \left(\|f\|_1 + |f(e^{it_0})| \right) \left(\sup_{\delta \leq |t| \leq \pi} |\Phi_\iota(e^{it})| \right) + \varepsilon \, C_\Phi.
$$

Pick $\iota(\varepsilon, t_0)$ so large that

$$
\sup_{\delta \leq |t| \leq \pi} |\Phi_\iota(e^{it})| < \varepsilon
$$

whenever $\iota > \iota(\varepsilon, t_0)$. Thus, for $\iota > \iota(\varepsilon, t_0)$ and for $|t - t_0| < \delta$, we have

$$
\left| (\Phi_\iota * f)(e^{it}) - f(e^{it_0}) \right| < \left(\|f\|_1 + |f(e^{it_0})| + C_\Phi \right) \varepsilon.
$$

□

The following result is a local version of Corollary 2.3. Note that the Poisson kernels fulfils all the requirements of Theorem 2.5.

Corollary 2.6 *Let $u \in L^1(\mathbb{T})$, and let*

$$
U(re^{i\theta}) = \frac{1}{2\pi} \int_{-\pi}^{\pi} \frac{1 - r^2}{1 + r^2 - 2r\cos(\theta - t)} \, u(e^{it}) \, dt, \qquad (re^{i\theta} \in \mathbb{D}).
$$

Suppose that u is continuous at $e^{it_0} \in \mathbb{T}$. Then U is harmonic on $\mathbb{D}$ and besides

$$
\lim_{\substack{z \to e^{it_0} \\ z \in \mathbb{D}}} U(z) = u(e^{it_0}).
$$

In particular,

$$
\lim_{r \to 1} U(re^{it_0}) = u(e^{it_0}).
$$

In Theorem 2.5 and Corollary 2.6, among other things, we assumed that our function f is continuous at a fixed point e^{it_0}. Hence, we implicitly take it as granted that $f(e^{it_0})$ is a *finite* complex number and $f(e^{it})$ converges to this value as $t \to t_0$. The finiteness of $f(e^{it_0})$ is crucial. If f is a complex-valued function and

$$
\lim_{t \to t_0} |f(e^{it})| = |f(e^{it_0})| = +\infty,
$$

we may still say that f is continuous at e^{it_0}. However, the preceding results are not valid for this class of functions. Nevertheless, if our function is real-valued and

$$
\lim_{t \to t_0} f(e^{it}) = f(e^{it_0}) = +\infty,
$$

we are able to find a proper generalization. As a matter of fact, in a very essential step in the theory of Hardy spaces we apply this result.

Theorem 2.7 *Let $\{\Phi_\iota\}$ be a positive approximate identity on $\mathbb{T}$. Suppose that, for each ι,*

$$\Phi_\iota \in L^\infty(\mathbb{T}),$$

and that $\{\Phi_\iota\}$ satisfies the stronger property (c'):

$$\lim_\iota \left(\sup_{0 < \delta \le |t| \le \pi} \Phi_\iota(e^{it}) \right) = 0.$$

Let f be a real function in $L^1(\mathbb{T})$ such that

$$\lim_{t \to t_0} f(e^{it}) = +\infty.$$

Then, given $M > 0$, there exists $\iota(M, t_0)$ and $\delta = \delta(M, t_0) > 0$ such that

$$(\Phi_\iota * f)(e^{it}) > M$$

if $\iota > \iota(M, t_0)$ and $|t - t_0| < \delta$. In particular,

$$\lim_\iota (\Phi_\iota * f)(e^{it_0}) = +\infty.$$

Proof. As we mentioned before, since $\Phi_\iota \in L^\infty(\mathbb{T})$ and $f \in L^1(\mathbb{T})$, by Corollary 1.6, $(\Phi_\iota * f)(e^{it})$ is well-defined for all $e^{it} \in \mathbb{T}$.

By assumption, given $M > 0$, there exists $\delta = \delta(M, t_0) > 0$ such that

$$f(e^{i\eta}) > 2M$$

whenever $|\eta - t_0| < 2\delta$. Therefore, for $|t - t_0| < \delta$,

$$
\begin{aligned}
(\Phi_\iota * f)(e^{it}) &= \frac{1}{2\pi} \int_{-\pi}^{\pi} \Phi_\iota(e^{i\tau})\, f(e^{i(t-\tau)})\, d\tau \\
&= \frac{1}{2\pi} \left(\int_{-\pi}^{-\delta} + \int_{-\delta}^{\delta} + \int_{\delta}^{\pi} \right) \Phi_\iota(e^{i\tau})\, f(e^{i(t-\tau)})\, d\tau \\
&\ge \frac{M}{\pi} \int_{-\delta}^{\delta} \Phi_\iota(e^{i\tau})\, d\tau - \frac{1}{2\pi} \int_{\delta \le |\tau| \le \pi} \Phi_\iota(e^{i\tau})\, |f(e^{i(t-\tau)})|\, d\tau \\
&= 2M - \frac{1}{2\pi} \int_{\delta \le |\tau| \le \pi} \Phi_\iota(e^{i\tau})\, (2M + |f(e^{i(t-\tau)})|)\, d\tau \\
&\ge 2M - (2M + \|f\|_1) \left(\sup_{\delta \le |\tau| \le \pi} \Phi_\iota(e^{i\tau}) \right).
\end{aligned}
$$

Pick $\iota(M, t_0)$ so large that

$$(2M + \|f\|_1) \sup_{\delta \le |\tau| \le \pi} \Phi_\iota(e^{i\tau}) < M$$

whenever $\iota > \iota(M, t_0)$. Thus, for $\iota > \iota(M, t_0)$ and for $|t - t_0| < \delta$,

$$(\Phi_\iota * f)(e^{it}) > M.$$

$\square$

Corollary 2.8 *Let u be a real function in $L^1(\mathbb{T})$, and let*

$$U(re^{i\theta}) = \frac{1}{2\pi} \int_{-\pi}^{\pi} \frac{1-r^2}{1+r^2-2r\cos(\theta-t)} \, u(e^{it}) \, dt, \qquad (re^{i\theta} \in \mathbb{D}).$$

Suppose that

$$\lim_{t \to t_0} u(e^{it}) = +\infty.$$

Then U is harmonic in $\mathbb{D}$ and besides

$$\lim_{\substack{z \to e^{it_0} \\ z \in \mathbb{D}}} U(z) = +\infty.$$

In particular,

$$\lim_{r \to 1} U(re^{it_0}) = +\infty.$$

Let A be a subset of $\mathbb{C}$ and consider a function

$$f : A \longmapsto [0, \infty].$$

We say that f is continuous at $z_0 \in A$ whenever

$$\lim_{\substack{z \to z_0 \\ z \in A}} f(z) = f(z_0).$$

If $f(z_0) < \infty$, then there is nothing new in this definition. The other case should be familiar too. If $f(z_0) = \infty$, by continuity at z_0 we simply mean that $\lim_{z \to z_0} f(z) = \infty$. The following result is an immediate consequence of Corollaries 2.6 and 2.8. It plays a *vital role* in Fatou's construction and thereafter in F. and M. Riesz's theorem, a cornerstone of function theory (see Section 5.5).

Corollary 2.9 *Let $u : \mathbb{T} \longmapsto [0, \infty]$ be continuous on $\mathbb{T}$, and suppose that $u \in L^1(\mathbb{T})$. Let*

$$U(re^{i\theta}) = \begin{cases} \dfrac{1}{2\pi} \displaystyle\int_{-\pi}^{\pi} \dfrac{1-r^2}{1+r^2-2r\cos(\theta-t)} \, u(e^{it}) \, dt & if \quad 0 \le r < 1, \\[2ex] u(e^{i\theta}) & if \quad r = 1. \end{cases}$$

Then $U : \overline{\mathbb{D}} \longmapsto [0, \infty]$ is continuous on $\overline{\mathbb{D}}$ and harmonic on $\mathbb{D}$.

Exercises

Exercise 2.3.1 Use Corollary 2.3 and (2.4) to give another proof of Weierstrass's theorem.

Exercise 2.3.2 Let $f \in \text{Lip}_\alpha(\mathbb{T})$, $0 < \alpha < 1$. Show that

$$\|\mathbf{K}_n * f - f\|_\infty \leq C_\alpha \frac{\|f\|_{\text{Lip}_\alpha(\mathbb{T})}}{n^\alpha},$$

where C_α is an absolute constant just depending on α.

Exercise 2.3.3 Let $f \in \text{Lip}_1(\mathbb{T})$. Show that

$$\|\mathbf{K}_n * f - f\|_\infty \leq C \|f\|_{\text{Lip}_1(\mathbb{T})} \frac{\log n}{n},$$

where C is an absolute constant.

Exercise 2.3.4 Let $f \in \text{Lip}_1(\mathbb{T})$. Show that

$$\|\mathbf{J}_n * f - f\|_\infty \leq C \frac{\|f\|_{\text{Lip}_1(\mathbb{T})}}{n},$$

where C is an absolute constant.

Exercise 2.3.5 Let $f \in L^1(\mathbb{T})$, and let

$$p_n(e^{it}) = \sum_{k=-n}^{n} \left(1 - \frac{|k|}{n+1}\right) \hat{f}(k)\, e^{ikt}, \qquad (e^{it} \in \mathbb{T}).$$

Suppose that f is continuous at $e^{it_0} \in \mathbb{T}$. Show that

$$\lim_{n \to \infty} p_n(e^{it_0}) = f_n(e^{it_0}).$$

Hint: Apply Theorem 2.5.

Exercise 2.3.6 Let f be a real function in $L^1(\mathbb{T})$ such that

$$\lim_{t \to t_0} f(e^{it}) = +\infty.$$

Let

$$p_n(e^{it}) = \sum_{k=-n}^{n} \left(1 - \frac{|k|}{n+1}\right) \hat{f}(k)\, e^{ikt}, \qquad (e^{it} \in \mathbb{T}).$$

Show that

$$\lim_{n \to \infty} p_n(e^{it_0}) = +\infty.$$

Hint: Apply Theorem 2.7.

2.4 Weak* convergence of measures

According to Theorem 1.3, $f * \mu$, the convolution of an $L^1(\mathbb{T})$ function f and a Borel measure μ, is a well-defined $L^1(\mathbb{T})$ function. In this section, we explore the relation between $f * \mu$ and μ where f ranges over the elements of an approximate identity.

Theorem 2.10 *Let $\{\Phi_\iota\}$ be an approximate identity on $\mathbb{T}$, and let $\mu \in \mathcal{M}(\mathbb{T})$. Then, for all ι, $\Phi_\iota * \mu \in L^1(\mathbb{T})$ with*

$$\|\Phi_\iota * \mu\|_1 \leq C_\Phi \|\mu\|$$

and

$$\|\mu\| \leq \sup_\iota \|\Phi_\iota * \mu\|_1.$$

Moreover, the measures $d\mu_\iota(e^{it}) = (\Phi_\iota * \mu)(e^{it}) \, dt/2\pi$ converge to $d\mu(e^{it})$ in the weak* topology, i.e.

$$\lim_\iota \frac{1}{2\pi} \int_{-\pi}^{\pi} \varphi(e^{it}) \, (\Phi_\iota * \mu)(e^{it}) \, dt = \int_{\mathbb{T}} \varphi(e^{it}) \, d\mu(e^{it})$$

for all $\varphi \in \mathcal{C}(\mathbb{T})$.

Proof. By Theorem 1.3 and (1.11), $\Phi_\iota * \mu \in L^1(\mathbb{T})$ with

$$\|\Phi_\iota * \mu\|_1 \leq \|\Phi_\iota\|_1 \|\mu\| \leq C_\Phi \|\mu\|$$

for all ι. Let $\varphi \in \mathcal{C}(\mathbb{T})$, and define $\psi(e^{it}) = \varphi(e^{-it})$. Then, $\psi \in \mathcal{C}(\mathbb{T})$ and by Fubini's theorem,

$$
\begin{aligned}
\frac{1}{2\pi} \int_{-\pi}^{\pi} \varphi(e^{it}) \, (\Phi_\iota * \mu)(e^{it}) \, dt &= \frac{1}{2\pi} \int_{-\pi}^{\pi} \varphi(e^{it}) \left(\int_{\mathbb{T}} \Phi_\iota(e^{i(t-\tau)}) \, d\mu(e^{i\tau}) \right) dt \\
&= \int_{\mathbb{T}} \left(\frac{1}{2\pi} \int_{-\pi}^{\pi} \Phi_\iota(e^{i(t-\tau)}) \, \varphi(e^{it}) \, dt \right) d\mu(e^{i\tau}) \\
&= \int_{\mathbb{T}} \left(\frac{1}{2\pi} \int_{-\pi}^{\pi} \Phi_\iota(e^{i(-s-\tau)}) \, \varphi(e^{-is}) \, ds \right) d\mu(e^{i\tau}) \\
&= \int_{\mathbb{T}} \left(\frac{1}{2\pi} \int_{-\pi}^{\pi} \Phi_\iota(e^{i(-\tau-s)}) \, \psi(e^{is}) \, ds \right) d\mu(e^{i\tau}) \\
&= \int_{\mathbb{T}} (\Phi_\iota * \psi)(e^{-i\tau}) \, d\mu(e^{i\tau}).
\end{aligned}
$$

Theorem 2.2 assures that $(\Phi_\iota * \psi)(e^{-i\tau})$ converges uniformly to $\psi(e^{-i\tau})$ on $\mathbb{T}$. Since $|\mu|$ is a finite positive Borel measure, we thus have

$$
\begin{aligned}
\lim_\iota \frac{1}{2\pi} \int_{-\pi}^{\pi} \varphi(e^{it}) \, (\Phi_\iota * \mu)(e^{it}) \, dt &= \lim_\iota \int_{\mathbb{T}} (\Phi_\iota * \psi)(e^{-i\tau}) \, d\mu(e^{i\tau}) \\
&= \int_{\mathbb{T}} \psi(e^{-i\tau}) \, d\mu(e^{i\tau}) \\
&= \int_{\mathbb{T}} \varphi(e^{i\tau}) \, d\mu(e^{i\tau}).
\end{aligned}
$$

Since

$$\left| \frac{1}{2\pi} \int_{-\pi}^{\pi} \varphi(e^{it}) \, (\Phi_\iota * \mu)(e^{it}) \, dt \right| \leq \left(\sup_\iota \| \Phi_\iota * \mu \|_1 \right) \| \varphi \|_\infty,$$

the last identity implies

$$\left| \int_{\mathbb{T}} \varphi(e^{i\tau}) \, d\mu(e^{i\tau}) \right| \leq \left(\sup_\iota \| \Phi_\iota * \mu \|_1 \right) \| \varphi \|_\infty$$

for all $\varphi \in \mathcal{C}(\mathbb{T})$. Hence, by the Riesz representation theorem,

$$\| \mu \| \leq \sup_\iota \| \Phi_\iota * \mu \|_1.$$

$\square$

In the last theorem, if $\{ \Phi_\iota \}$ is a positive approximate identity on $\mathbb{T}$ then $C_\Phi = 1$ and thus we necessarily have

$$\| \mu \| = \sup_\iota \| \Phi_\iota * \mu \|_1.$$

As a matter of fact, slightly modifying the proof of the theorem, we see that

$$\| \mu \| = \lim_\iota \| \Phi_\iota * \mu \|_1.$$

In particular, if we choose the Poisson kernel then, by Theorem 2.1, we are able to extend μ to a harmonic function U on $\mathbb{D}$ such that the measures $U(re^{it}) \, dt/2\pi$ are uniformly bounded and, as $r \to 1$, converge to μ in the weak* topology.

Corollary 2.11 *Let $\mu \in \mathcal{M}(\mathbb{T})$, and let*

$$U(re^{i\theta}) = \int_{\mathbb{T}} \frac{1 - r^2}{1 + r^2 - 2r\cos(\theta - t)} \, d\mu(e^{it}), \qquad (re^{i\theta} \in \mathbb{D}). \qquad (2.5)$$

Then U is harmonic on $\mathbb{D}$, and

$$\| \mu \| = \sup_{0 \leq r < 1} \| U_r \|_1 = \lim_{r \to 1} \| U_r \|_1.$$

Moreover, the measures $d\mu_r(e^{it}) = U(re^{it}) dt/2\pi$ converge to $d\mu(e^{it})$, as $r \to 1^-$, in the weak topology, i.e.*

$$\lim_{r \to 1^-} \int_{\mathbb{T}} \varphi(e^{it}) \, d\mu_r(e^{it}) = \int_{\mathbb{T}} \varphi(e^{it}) \, d\mu(e^{it})$$

for all $\varphi \in \mathcal{C}(\mathbb{T})$.

The preceding result has interesting and profound consequences. For example, the identity $\| \mu \| = \sup_{0 \leq r < 1} \| U_r \|_1$ immediately implies some uniqueness theorems. We give two slightly different versions below.

Corollary 2.12 (Uniqueness theorem) *Let $\mu \in \mathcal{M}(\mathbb{T})$. Suppose that*

$$\int_{\mathbb{T}} \frac{1 - r^2}{1 + r^2 - 2r\cos(\theta - t)} \, d\mu(e^{it}) = 0$$

for all $re^{i\theta} \in \mathbb{D}$. Then $\mu = 0$.

Corollary 2.13 (Uniqueness theorem) *Let $\mu \in \mathcal{M}(\mathbb{T})$ and suppose that*

$$\hat{\mu}(n) = 0$$

for all $n \in \mathbb{Z}$. Then $\mu = 0$.

Proof. Define U by (2.5) and note that, by Theorem 2.1,

$$U(re^{i\theta}) = \sum_{n=-\infty}^{\infty} \hat{\mu}(n) \, r^{|n|} \, e^{in\theta} = 0$$

for all $re^{i\theta} \in \mathbb{T}$. $\qquad\qquad\square$

The last *uniqueness theorem* says that the map

$$\begin{aligned} \mathcal{M}(\mathbb{T}) &\longrightarrow \ell^{\infty}(\mathbb{Z}) \\ \mu &\longmapsto \hat{\mu} \end{aligned}$$

is one-to-one. In particular, if $f \in L^1(\mathbb{T})$ and $\hat{f}(n) = 0$, for all $n \in \mathbb{Z}$, then $f = 0$.

Exercises

Exercise 2.4.1 Let $\{\Phi_n\}$ be an approximate identity on $\mathbb{T}$, and let $\mu \in \mathcal{M}(\mathbb{T})$. Show that

$$\|\mu\| \leq \liminf_{n \to \infty} \|\Phi_n * \mu\|_1.$$

Hint: By Theorem 2.10, for each $N \geq 1$,

$$\|\mu\| \leq \sup_{n \geq N} \|\Phi_n * \mu\|_1.$$

Exercise 2.4.2 Let $\{\Phi_n\}$ be a positive approximate identity on $\mathbb{T}$, and let $\mu \in \mathcal{M}(\mathbb{T})$. Show that

$$\lim_{n \to \infty} \|\Phi_n * \mu\|_1 = \|\mu\|.$$

Hint: Use Exercise 2.4.1 and the fact that $\|\Phi_n * \mu\|_1 \leq \|\mu\|$ for all $n \geq 1$.

Exercise 2.4.3 Let $0 \leq r, \rho < 1$. Show that $P_r * P_\rho = P_{r\rho}$.
Hint: Use (1.6), Theorem 1.2 and the uniqueness theorem.

Exercise 2.4.4   Show that the map

$$\mathcal{M}(\mathbb{T}) \longrightarrow \ell^{\infty}(\mathbb{Z})$$
$$\mu \longmapsto \hat{\mu}$$

is not surjective.

Exercise 2.4.5   Let $\mu \in \mathcal{M}(\mathbb{T})$, and let

$$p_n(e^{it}) = \sum_{k=-n}^{n} \left(1 - \frac{|k|}{n+1}\right) \hat{\mu}(k)\, e^{ikt}, \qquad (e^{it} \in \mathbb{T}).$$

Show that the measures $d\mu_n(e^{it}) = p_n(e^{it})\, dt/2\pi$ converge to $d\mu(e^{it})$, as $n \to \infty$, in the weak* topology, i.e.

$$\lim_{n \to \infty} \int_{\mathbb{T}} \varphi(e^{it})\, d\mu_n(e^{it}) = \int_{\mathbb{T}} \varphi(e^{it})\, d\mu(e^{it})$$

for all $\varphi \in \mathcal{C}(\mathbb{T})$.
Hint: Apply Theorem 2.10.

Exercise 2.4.6   Let $\{a_n\}_{n \in \mathbb{Z}}$ be a sequence of complex numbers. Suppose that the measures

$$dp_n(e^{it}) = \frac{1}{2\pi} \left\{ \sum_{k=-n}^{n} \left(1 - \frac{|k|}{n+1}\right) a_k\, e^{ikt} \right\} dt$$

are convergent in the weak* topology, say to the measure $\mu \in \mathcal{M}(\mathbb{T})$. Show that a_n are in fact the Fourier coefficients of μ. (This result can be regarded as the converse of Exercise 2.4.5.)

Exercise 2.4.7   Let $\mu \in \mathcal{M}(\mathbb{T})$ and let $r_0 \in (0,1)$. Suppose that

$$\int_{\mathbb{T}} \frac{1 - r_0^2}{1 + r_0^2 - 2r_0 \cos(\theta - t)}\, d\mu(e^{it}) = 0$$

for all $e^{i\theta} \in \mathbb{T}$. Show that $\mu = 0$.

Exercise 2.4.8   Let $\mu \in \mathcal{M}(\mathbb{T})$ and let $r_0 \in (0,1)$. Suppose that $\hat{\mu}(0) = 0$ and that

$$\int_{\mathbb{T}} \frac{2r_0 \sin(\theta - t)}{1 + r_0^2 - 2r_0 \cos(\theta - t)}\, d\mu(e^{it}) = 0$$

for all $e^{i\theta} \in \mathbb{T}$. Show that $\mu = 0$.
Hint: For a fixed r, find the Fourier series of

$$\frac{2r \sin t}{1 + r^2 - 2r \cos t}.$$

The relation

$$\frac{2r \sin t}{1 + r^2 - 2r \cos t} = \frac{i}{1 - re^{it}} - \frac{i}{1 - re^{-it}}$$

might be useful.

2.5 Convergence in norm

The space of continuous functions $\mathcal{C}(\mathbb{T})$ is dense in $L^p(\mathbb{T})$, $1 \leq p < \infty$. This assertion is not true when $p = \infty$, since the uniform limit of a sequence of continuous functions has to be continuous and a typical element of $L^\infty(\mathbb{T})$ is not necessarily continuous. We exploit this fact to study the behavior of $\Phi_\iota * f$, where $f \in L^p(\mathbb{T})$ and $\{\Phi_\iota\}$ is an approximate identity.

Theorem 2.14 *Let $\{\Phi_\iota\}$ be an approximate identity on $\mathbb{T}$, and let $f \in L^p(\mathbb{T})$, $1 \leq p < \infty$. Then, for all ι, $\Phi_\iota * f \in L^p(\mathbb{T})$ with*

$$\|\Phi_\iota * f\|_p \leq C_\Phi \|f\|_p$$

and besides,

$$\lim_\iota \|\Phi_\iota * f - f\|_p = 0.$$

Proof. By Corollary 1.5, $\Phi_\iota * f \in L^p(\mathbb{T})$ and

$$\|\Phi_\iota * f\|_p \leq \|\Phi_\iota\|_1 \|f\|_p \leq C_\Phi \|f\|_p$$

for all ι.

Fix $\varepsilon > 0$. Given $f \in L^p(\mathbb{T})$, pick $\varphi \in \mathcal{C}(\mathbb{T})$ such that $\|f - \varphi\|_p < \varepsilon$. Hence

$$
\begin{aligned}
\|\Phi_\iota * f - f\|_p &= \|\Phi_\iota * (f - \varphi) - (f - \varphi) + (\Phi_\iota * \varphi - \varphi)\|_p \\
&\leq \|\Phi_\iota * (f - \varphi)\|_p + \|f - \varphi\|_p + \|\Phi_\iota * \varphi - \varphi\|_p \\
&\leq (1 + C_\Phi) \|f - \varphi\|_p + \|\Phi_\iota * \varphi - \varphi\|_p \\
&\leq (1 + C_\Phi) \varepsilon + \|\Phi_\iota * \varphi - \varphi\|_\infty.
\end{aligned}
$$

However, by Theorem 2.2, there is an $\iota(\varepsilon)$ such that $\|\Phi_\iota * \varphi - \varphi\|_\infty < \varepsilon$, for $\iota > \iota(\varepsilon)$. Therefore

$$\|\Phi_\iota * f - f\|_p < (2 + C_\Phi) \varepsilon$$

whenever $\iota > \iota(\varepsilon)$. $\qquad\square$

Since $\lim_\iota \|\Phi_\iota * f - f\|_p = 0$, we clearly have $\lim_\iota \|\Phi_\iota * f\|_p = \|f\|_p$. Hence, if $\{\Phi_\iota\}$ is a positive approximate identity on $\mathbb{T}$,

$$\lim_\iota \|\Phi_\iota * f\|_p = \sup_\iota \|\Phi_\iota * f\|_p = \|f\|_p.$$

Therefore, if we use the Poisson kernel to extend a function $u \in L^p(\mathbb{T})$ to the open unit disc we obtain a harmonic function U whose mean values $\|U_r\|_p$ are uniformly bounded, with U_r converging to u in the $L^p(\mathbb{T})$ norm.

Corollary 2.15 *Let $u \in L^p(\mathbb{T})$, $1 \leq p < \infty$, and let*

$$U(re^{i\theta}) = \frac{1}{2\pi} \int_{-\pi}^{\pi} \frac{1 - r^2}{1 + r^2 - 2r\cos(\theta - t)} \, u(e^{it}) \, dt, \qquad (re^{i\theta} \in \mathbb{D}).$$

Then U is harmonic on $\mathbb{D}$,

$$\sup_{0 \le r < 1} \|U_r\|_p = \lim_{r \to 1} \|U_r\|_p = \|u\|_p$$

and

$$\lim_{r \to 1} \|U_r - u\|_p = 0.$$

We can also exploit the Fejér kernel $\mathbf{K}_n$ in Theorem 2.14. The main advantage is that $f * \mathbf{K}_n$ is a trigonometric polynomial.

Corollary 2.16 *Let* $f \in L^p(\mathbb{T})$, $1 \le p < \infty$, *and let*

$$p_n(e^{it}) = \sum_{k=-n}^{n} \left(1 - \frac{|k|}{n+1} \right) \hat{f}(k)\, e^{ikt}, \qquad (e^{it} \in \mathbb{T}).$$

Then, for each $n \ge 1$,

$$\|p_n\|_p \le \|f\|_p$$

and

$$\lim_{n \to \infty} \|p_n - f\|_p = 0.$$

Proof. It is enough to observe that

$$p_n = f * \mathbf{K}_n,$$

where $\mathbf{K}_n$ is the Fejér kernel. Now, apply Theorem 2.14. $\square$

The constructive method of Fejér shows that the trigonometric polynomials are dense in $L^p(\mathbb{T})$, $1 \le p < \infty$. On the other hand, in Lemma 1.1, we saw that the Fourier coefficients of an integrable function are uniformly bounded. We are now in a position to improve this result by showing that the Fourier coefficients actually tend to zero.

Corollary 2.17 (Riemann–Lebesgue lemma) *Let* $f \in L^1(\mathbb{T})$. *Then*

$$\lim_{|n| \to \infty} \hat{f}(n) = 0.$$

Proof. Given $\varepsilon > 0$, by Corollary 2.16, there is n_0 such that

$$\|p_{n_0} - f\|_1 < \varepsilon,$$

where

$$p_{n_0}(e^{it}) = \sum_{k=-n_0}^{n_0} \left(1 - \frac{|k|}{n_0+1} \right) \hat{f}(k)\, e^{ikt}, \qquad (e^{it} \in \mathbb{T}).$$

Hence, by Lemma 1.1,

$$|\hat{p}_{n_0}(n) - \hat{f}(n)| \le \|p_{n_0} - f\|_1 < \varepsilon$$

for all $n \in \mathbb{Z}$. However, $\hat{p}_{n_0}(n) = 0$ if $|n| > n_0$. Therefore,

$$|\hat{f}(n)| < \varepsilon$$

for all $|n| > n_0$. $\square$

We saw that the Fourier transform

$$\begin{array}{ccc} \mathcal{M}(\mathbb{T}) & \longrightarrow & \ell^\infty(\mathbb{Z}) \\ \mu & \longmapsto & \hat{\mu} \end{array}$$

is injective but not surjective. As long as $\ell^p(\mathbb{Z})$ spaces or their well-known subspaces are concerned, $\ell^\infty(\mathbb{Z})$ is the best possible choice in this mapping. However, if instead of $\mathcal{M}(\mathbb{T})$ we consider the smaller subclass $L^1(\mathbb{T})$, by the Riemann–Lebesgue lemma (Corollary 2.17), we can slightly improve the preceding mapping and exhibit the Fourier transform on $L^1(\mathbb{T})$ as

$$\begin{array}{ccc} L^1(\mathbb{T}) & \longrightarrow & c_0(\mathbb{Z}) \\ f & \longmapsto & \hat{f}. \end{array}$$

We will see that this map is not surjective (see Exercise 2.5.6). As a matter of fact, there is no satisfactory description for the image of $L^1(T)$ under the Fourier transformation.

Exercises

Exercise 2.5.1   Let $\{a_n\}_{n\in\mathbb{Z}}$ be a sequence of complex numbers, and let

$$p_n(e^{it}) = \sum_{k=-n}^{n} \left(1 - \frac{|k|}{n+1}\right) a_k\, e^{ikt}, \qquad (e^{it} \in \mathbb{T}).$$

Let $1 \leq p < \infty$. Suppose that the sequence $(p_n)_{n\geq 0}$ is convergent in the $L^p(\mathbb{T})$ norm to a function $f \in L^p(\mathbb{T})$. Show that $\{a_n\}_{n\in\mathbb{Z}}$ are in fact the Fourier coefficients of f.
Remark: This result can be regarded as the converse of Corollary 2.16.

Exercise 2.5.2   Show that $L^p(\mathbb{T})$, $1 \leq p < \infty$, is separable. Is $L^\infty(\mathbb{T})$ separable?
Hint: Use Corollary 2.16.

Exercise 2.5.3   [Fejér's lemma] Let $f \in L^1(\mathbb{T})$ and let $g \in L^\infty(\mathbb{T})$. Show that

$$\lim_{n\to\infty} \frac{1}{2\pi} \int_{-\pi}^{\pi} f(e^{it})\, g(e^{int})\, dt = \hat{f}(0)\, \hat{g}(0).$$

Hint: First suppose that f is a trigonometric polynomial and apply Exercise 1.3.4. Then use Corollary 2.16 to prove the general case.

Exercise 2.5.4   Let $g \in L^1(\mathbb{T})$ and let $1 \leq p \leq \infty$. Consider the operator

$$\begin{array}{ccc} \Lambda : L^p(\mathbb{T}) & \longrightarrow & L^p(\mathbb{T}) \\ f & \longmapsto & g * f. \end{array}$$

By Corollary 1.5, Λ is well-defined. Show that

$$\|\Lambda\| = \|g\|_1.$$

Hint: Apply Corollaries 1.5 and 2.16.

Exercise 2.5.5   Let $f \in L^1(\mathbb{T})$. Suppose that

$$\hat{f}(n) = -\hat{f}(-n) \geq 0$$

for all $n \geq 0$. Show that

$$\sum_{n=1}^{\infty} \frac{\hat{f}(n)}{n} < \infty.$$

Hint: Let

$$F(e^{it}) = \int_0^t f(e^{i\tau})\, d\tau.$$

Then $F \in \mathcal{C}(\mathbb{T})$ and $\hat{F}(n) = \hat{f}(n)/in$, $n \neq 0$. Hence, by Corollary 2.4,

$$\lim_{n \to \infty} (\mathbf{K}_n * F)(1) = F(1) = 0.$$

Exercise 2.5.6   Show that the mapping

$$\begin{array}{ccc} L^1(\mathbb{T}) & \longrightarrow & c_0(\mathbb{Z}) \\ f & \longmapsto & \hat{f} \end{array}$$

is not surjective.
Hint: Let

$$a_n = -a_{-n} = \frac{1}{\log n}, \qquad (n \geq 2),$$

and $a_{\pm 1} = a_0 = 0$. Use Exercise 2.5.5 to show that $(a_n)_{n \in \mathbb{Z}}$ is not in the range.

Exercise 2.5.7   Let $(a_n)_{n \in \mathbb{Z}}$ be such that

(i) $a_n \geq 0$, for all $n \in \mathbb{Z}$,

(ii) $a_{-n} = a_n$, for all $n \geq 1$,

(iii) $\lim_{n \to \infty} a_n = 0$,

(iv) $a_n \leq (a_{n-1} + a_{n+1})/2$, for all $n \geq 1$.

Let

$$f(e^{it}) = \sum_{n=1}^{\infty} n(a_{n-1} + a_{n+1} - 2a_n)\, \mathbf{K}_{n-1}(e^{it}).$$

Show that $f \in L^1(\mathbb{T})$ and that

$$\hat{f}(n) = a_n$$

for all $n \in \mathbb{Z}$.

Remark: This result convinces us that $c_0(\mathbb{Z})$ is somehow optimal as the codomain of

$$
\begin{array}{ccc}
L^1(\mathbb{T}) & \longrightarrow & c_0(\mathbb{Z}) \\
f & \longmapsto & \hat{f}.
\end{array}
$$

2.6 Weak* convergence of bounded functions

Since $\mathcal{C}(\mathbb{T})$ is not uniformly dense in $L^\infty(\mathbb{T})$, the results of the preceding section are not entirely valid if $p = \infty$. However, a slightly weaker version holds in this case too.

Theorem 2.18 *Let $\{\Phi_\iota\}$ be an approximate identity on $\mathbb{T}$, and let $f \in L^\infty(\mathbb{T})$. Then, for all ι, $\Phi_\iota * f \in \mathcal{C}(\mathbb{T})$ with*

$$\|\Phi_\iota * f\|_\infty \leq C_\Phi \|f\|_\infty$$

and

$$\|f\|_\infty \leq \sup_\iota \|\Phi_\iota * f\|_\infty.$$

Moreover, $\Phi_\iota * f$ converges to f in the weak* topology, i.e.

$$\lim_\iota \int_{-\pi}^{\pi} \varphi(e^{it}) (\Phi_\iota * f)(e^{it}) \, dt = \int_{-\pi}^{\pi} \varphi(e^{it}) f(e^{it}) \, dt$$

for all $\varphi \in L^1(\mathbb{T})$.

Proof. By Corollary 1.6, $\Phi_\iota * f \in \mathcal{C}(\mathbb{T})$ and

$$\|\Phi_\iota * f\|_\infty \leq \|\Phi_\iota\|_1 \|f\|_\infty \leq C_\Phi \|f\|_\infty$$

for all ι.

Let $\varphi \in L^1(\mathbb{T})$, and let $\psi(e^{it}) = \varphi(e^{-it})$. Then, by Fubini's theorem,

$$
\begin{aligned}
\int_{-\pi}^{\pi} \varphi(e^{it}) (\Phi_\iota * f)(e^{it}) \, dt &= \int_{-\pi}^{\pi} \varphi(e^{it}) \left(\int_{-\pi}^{\pi} \Phi_\iota(e^{i(t-\tau)}) f(e^{i\tau}) \, d\tau \right) dt \\
&= \int_{-\pi}^{\pi} \left(\int_{-\pi}^{\pi} \Phi_\iota(e^{i(t-\tau)}) \varphi(e^{it}) \, dt \right) f(e^{i\tau}) \, d\tau \\
&= \int_{-\pi}^{\pi} \left(\int_{-\pi}^{\pi} \Phi_\iota(e^{i(-\tau-s)}) \psi(e^{is}) \, ds \right) f(e^{i\tau}) \, d\tau \\
&= \int_{-\pi}^{\pi} (\Phi_\iota * \psi)(e^{-i\tau}) f(e^{i\tau}) \, d\tau.
\end{aligned}
$$

Theorem 2.14 ensures that $(\Phi_\iota * \psi)(e^{-i\tau})$ converges to $\psi(e^{-i\tau})$ in $L^1(\mathbb{T})$. Since f is a bounded function, we thus have

$$\lim_\iota \int_{-\pi}^{\pi} \varphi(e^{it})\,(\Phi_\iota * f)(e^{it})\,dt \;=\; \lim_\iota \int_{-\pi}^{\pi} (\Phi_\iota * \psi)(e^{-i\tau})\,f(e^{i\tau})\,d\tau$$

$$= \int_{-\pi}^{\pi} \psi(e^{-i\tau})\,f(e^{i\tau})\,d\tau$$

$$= \int_{-\pi}^{\pi} \varphi(e^{i\tau})\,f(e^{i\tau})\,d\tau.$$

Since

$$\left| \frac{1}{2\pi} \int_{-\pi}^{\pi} \varphi(e^{it})\,(\Phi_\iota * f)(e^{it})\,dt \right| \leq \left(\sup_\iota \|\Phi_\iota * f\|_\infty \right) \|\varphi\|_1,$$

the last identity implies

$$\left| \int_{-\pi}^{\pi} \varphi(e^{i\tau})\,f(e^{i\tau})\,d\tau \right| \leq \left(\sup_\iota \|\Phi_\iota * f\|_\infty \right) \|\varphi\|_1$$

for all $\varphi \in L^1(\mathbb{T})$. Hence, by the Riesz representation theorem,

$$\|f\|_\infty \leq \sup_\iota \|\Phi_\iota * f\|_\infty.$$

$\square$

If $\{\Phi_\iota\}$ is a positive approximate identity on $\mathbb{T}$ then $C_\Phi = 1$ and thus we have

$$\|f\|_\infty = \sup_\iota \|\Phi_\iota * f\|_\infty.$$

As a matter of fact, by slightly modifying the proof of the theorem, we obtain

$$\|f\|_\infty = \lim_\iota \|\Phi_\iota * f\|_\infty.$$

Corollary 2.19 *Let $u \in L^\infty(\mathbb{T})$, and let*

$$U(re^{i\theta}) = \frac{1}{2\pi} \int_{-\pi}^{\pi} \frac{1 - r^2}{1 + r^2 - 2r\cos(\theta - t)}\, u(e^{it})\,dt, \qquad (re^{i\theta} \in \mathbb{D}).$$

Then U is bounded and harmonic on $\mathbb{D}$ and

$$\sup_{0 \leq r < 1} \|U_r\|_\infty = \lim_{r \to 1} \|U_r\|_\infty = \|u\|_\infty.$$

Moreover, as $r \to 1$, U_r converges to u in the weak* topology, i.e.

$$\lim_{r \to 1^-} \int_{-\pi}^{\pi} \varphi(e^{it})\,U(re^{it})\,dt = \int_{-\pi}^{\pi} \varphi(e^{it})\,u(e^{it})\,dt$$

for all $\varphi \in L^1(\mathbb{T})$.

Exercises

Exercise 2.6.1   Let $\{\Phi_n\}$ be an approximate identity on $\mathbb{T}$, and let $f \in L^\infty(\mathbb{T})$. Show that

$$\|f\|_\infty \leq \liminf_{n \to \infty} \|\Phi_n * f\|_\infty.$$

Exercise 2.6.2   Let $\{\Phi_n\}$ be a positive approximate identity on $\mathbb{T}$, and let $f \in L^\infty(\mathbb{T})$. Show that

$$\lim_{n \to \infty} \|\Phi_n * f\|_\infty = \|f\|_\infty.$$

Exercise 2.6.3   Let $f \in L^\infty(\mathbb{T})$ and let

$$p_n(e^{it}) = \sum_{k=-n}^{n} \left(1 - \frac{|k|}{n+1}\right) \hat{f}(k) \, e^{ikt}, \qquad (e^{it} \in \mathbb{T}).$$

Show that the sequence $(p_n)_{n \geq 0}$ converges to f in the weak* topology, i.e.

$$\lim_{n \to \infty} \int_{-\pi}^{\pi} \varphi(e^{it}) \, p_n(e^{it}) \, dt = \int_{-\pi}^{\pi} \varphi(e^{it}) \, f(e^{it}) \, dt$$

for all $\varphi \in L^1(\mathbb{T})$.
Hint: Apply Theorem 2.18.

Exercise 2.6.4   Let $\{a_n\}_{n \in \mathbb{Z}}$ be a sequence of complex numbers, and let

$$p_n(e^{it}) = \sum_{k=-n}^{n} \left(1 - \frac{|k|}{n+1}\right) a_k \, e^{ikt}, \qquad (e^{it} \in \mathbb{T}).$$

Suppose that the sequence $(p_n)_{n \geq 0}$ is convergent in the weak* topology to a function $f \in L^\infty(\mathbb{T})$. Show that $\{a_n\}_{n \in \mathbb{Z}}$ are in fact the Fourier coefficients of f.
Remark: This result can be regarded as the converse of Exercise 2.6.3.

2.7   Parseval's identity

In the preceding sections we mainly studied the Abel–Poisson means of the Fourier series of a measure or a function and, among other things, we saw how these means converge in an appropriate topology to the given measure or function. However, we have not yet considered the convergence of the Fourier series itself. The pointwise or uniform convergence of the Fourier series is a more subtle problem. The whole story is unveiled by a celebrated theorem of Carleson-Hunt, which says that the Fourier series of a function in $L^p(\mathbb{T})$,

$p > 1$, converges almost everywhere to the function, and a difficult construction
of Kolmogorov giving a function in $L^1(\mathbb{T})$ whose Fourier series diverges almost
everywhere. We do not need these results in the following. However, we explore
further the Fourier series of $L^2(\mathbb{T})$ functions.

First of all, for the subclass $L^2(\mathbb{T}) \subset L^1(\mathbb{T})$, the uniqueness theorem (Corollary 2.13) can be stated differently. A family $\{\varphi_\iota\}$ in $L^2(\mathbb{T})$ is *complete* provided
that

$$\int_{-\pi}^{\pi} f(e^{it}) \,\overline{\varphi_\iota(e^{it})}\, dt = 0, \qquad \text{for all } \iota$$

holds only if $f = 0$. Therefore, the uniqueness theorem says that the sequence
$\{e^{int}\}_{n \in \mathbb{Z}}$ is complete in $L^2(\mathbb{T})$. Secondly, $L^2(\mathbb{T})$ equipped with the inner product

$$\langle f, g \rangle = \frac{1}{2\pi} \int_{-\pi}^{\pi} f(e^{it}) \,\overline{g(e^{it})}\, dt$$

is a Hilbert space. Two functions $f, g \in L^2(\mathbb{T})$ are said to be *orthogonal* if

$$\langle f, g \rangle = 0.$$

A subset $S \subset L^2(\mathbb{T})$ is called an *orthonormal set* if every element of S has norm
one and every two distinct elements of S are orthogonal. Using this terminology,
the relation (1.2) along with the uniqueness theorem tells us that the sequence
$\{e^{int}\}_{n \in \mathbb{Z}}$ is a complete orthonormal set.

Lemma 2.20 (Bessel's inequality) *Let* $f \in L^2(\mathbb{T})$, *and let* $\{\varphi_\iota\}$ *be an orthonormal family in* $L^2(\mathbb{T})$. *Then*

$$\sum_\iota |\langle f, \varphi_\iota \rangle|^2 \leq \|f\|_2^2.$$

Proof. Let

$$\varphi = \sum \langle f, \varphi_\iota \rangle \,\varphi_\iota,$$

where the sum is over a finite subset of indices $\{\iota\}$. Then

$$\begin{aligned}
\|f - \varphi\|_2^2 &= \langle f - \varphi, f - \varphi \rangle \\
&= \langle f, f \rangle - \langle f, \varphi \rangle - \langle \varphi, f \rangle + \langle \varphi, \varphi \rangle.
\end{aligned}$$

But

$$\begin{aligned}
\langle f, \varphi \rangle &= \frac{1}{2\pi} \int_{-\pi}^{\pi} f(e^{it}) \,\overline{\varphi(e^{it})}\, dt \\
&= \frac{1}{2\pi} \int_{-\pi}^{\pi} f(e^{it}) \left(\sum \overline{\langle f, \varphi_\iota \rangle} \,\overline{\varphi_\iota(e^{it})} \right) dt \\
&= \sum \overline{\langle f, \varphi_\iota \rangle} \left(\frac{1}{2\pi} \int_{-\pi}^{\pi} f(e^{it}) \,\overline{\varphi_\iota(e^{it})}\, dt \right) \\
&= \sum \overline{\langle f, \varphi_\iota \rangle} \,\langle f, \varphi_\iota \rangle = \sum |\langle f, \varphi_\iota \rangle|^2
\end{aligned}$$

and similarly,

$$
\begin{aligned}
\langle \varphi, \varphi \rangle &= \frac{1}{2\pi} \int_{-\pi}^{\pi} \varphi(e^{it})\, \overline{\varphi(e^{it})}\, dt \\
&= \frac{1}{2\pi} \int_{-\pi}^{\pi} \left(\sum \langle f, \varphi_\iota \rangle \varphi_\iota(e^{it}) \right) \left(\sum \overline{\langle f, \varphi_{\iota'} \rangle}\; \overline{\varphi_{\iota'}(e^{it})} \right) dt \\
&= \sum \sum \langle f, \varphi_\iota \rangle \overline{\langle f, \varphi_{\iota'} \rangle} \left(\frac{1}{2\pi} \int_{-\pi}^{\pi} \varphi_\iota(e^{it})\, \overline{\varphi_{\iota'}(e^{it})}\, dt \right) \\
&= \sum \langle f, \varphi_\iota \rangle \overline{\langle f, \varphi_\iota \rangle} = \sum |\langle f, \varphi_\iota \rangle|^2.
\end{aligned}
$$

Hence

$$
\| f - \varphi \|_2^2 = \| f \|_2^2 - \sum |\langle f, \varphi_\iota \rangle|^2,
$$

which gives

$$
\sum |\langle f, \varphi_\iota \rangle|^2 \leq \| f \|_2^2.
$$

Taking the supremum with respect to all such sums gives the required result. □

If we consider the orthonormal family $\{e^{int}\}_{n \in \mathbb{Z}}$, then Bessel's inequality is written as

$$
\sum_{n=-\infty}^{\infty} |\hat{f}(n)|^2 \leq \frac{1}{2\pi} \int_{-\pi}^{\pi} |f(e^{it})|^2\, dt
$$

or equivalently, $\hat{f} \in \ell^2(\mathbb{Z})$ with

$$
\| \hat{f} \|_2 \leq \| f \|_2
$$

for each $f \in L^2(\mathbb{T})$. Therefore, the mapping

$$
\begin{aligned}
L^2(\mathbb{T}) &\longrightarrow \ell^2(\mathbb{Z}) \\
f &\longmapsto \hat{f}
\end{aligned}
$$

is well-defined and, by the uniqueness theorem (Corollary 2.13), is injective. Now, we show that it is also surjective.

Theorem 2.21 (Riesz–Fischer theorem) *Let $(a_n)_{n \in \mathbb{Z}} \in \ell^2(\mathbb{Z})$. Then there is an $f \in L^2(\mathbb{T})$ such that*

$$
\hat{f}(n) = a_n
$$

for all $n \in \mathbb{Z}$.

Proof. Let

$$
\chi_k(e^{it}) = e^{ikt}, \qquad (k \in \mathbb{Z}),
$$

and let

$$
f_n = \sum_{k=-n}^{n} a_k\, \chi_k.
$$

Hence

$$\hat{f}_n(k) = \langle f_n, \chi_k \rangle = \begin{cases} a_k & \text{if } n \ge |k|, \\ 0 & \text{if } n < |k|. \end{cases}$$

Let $m > n$. Then

$$\|f_m - f_n\|_2^2 = \sum_{|k|=n+1}^{m} |a_k|^2$$

and thus $(f_n)_{n \ge 1}$ is a Cauchy sequence in $L^2(\mathbb{T})$. Hence it is convergent, say to $f \in L^2(\mathbb{T})$. Therefore, for each $k \in \mathbb{Z}$, we have

$$\hat{f}(k) = \langle f, \chi_k \rangle = \lim_{n \to \infty} \langle f_n, \chi_k \rangle = a_k.$$

$\square$

The proof of the Riesz–Fischer theorem contains more than what was stated in the theorem. We saw that $\lim_{n \to \infty} \|f_n - f\|_2 = 0$, where $f_n = \sum_{k=-n}^{n} a_k \chi_k$. Hence,

$$\|f\|_2^2 = \lim_{n \to \infty} \|f_n\|_2^2 = \sum_{k=-\infty}^{\infty} |a_k|^2.$$

Bessel's inequality (Lemma 2.20) ensures that $\hat{f} \in \ell^2(\mathbb{Z})$, whenever $f \in L^2(\mathbb{T})$. Hence, given $f \in L^2(\mathbb{T})$, if we pick $a_k = \hat{f}(k)$, an appeal to the uniqueness theorem (Corollary 2.13) shows that

$$\lim_{n \to \infty} \|f_n - f\|_2 = 0, \tag{2.6}$$

where

$$f_n(e^{it}) = \sum_{k=-n}^{n} \hat{f}(k) e^{ikt}.$$

This result is an improvement of Corollary 2.16 when $p = 2$ (see also Exercise 2.1.3). Moreover,

$$\|f\|_2^2 = \sum_{k=-\infty}^{\infty} |\hat{f}(k)|^2.$$

This last identity is very important and we state it as a corollary.

Corollary 2.22 (Parseval's identity) *Let* $f \in L^2(\mathbb{T})$. *Then*

$$\frac{1}{2\pi} \int_{-\pi}^{\pi} |f(e^{it})|^2 \, dt = \sum_{n=-\infty}^{\infty} |\hat{f}(n)|^2.$$

Parseval's identity can be rewritten as

$$\|f\|_2 = \|\hat{f}\|_2.$$

Hence the Fourier transform

$$
\begin{array}{ccc}
L^2(\mathbb{T}) & \longrightarrow & \ell^2(\mathbb{Z}) \\
f & \longmapsto & \hat{f}
\end{array}
$$

is bijective and shows that the Hilbert spaces $L^2(\mathbb{T})$ and $\ell^2(\mathbb{Z})$ are isomorphically isometric.

Exercises

Exercise 2.7.1 [Polarization identity] Let $\mathcal{H}$ be a complex inner product space and let $x, y \in \mathcal{H}$. Show that

$$
4\langle x, y \rangle = \|x + y\|^2 - \|x - y\|^2 + i\|x + iy\|^2 - i\|x - iy\|^2.
$$

Exercise 2.7.2 Let $f, g \in L^2(\mathbb{T})$. Show that

$$
\frac{1}{2\pi} \int_{-\pi}^{\pi} f(e^{it}) \, \overline{g(e^{it})} \, dt = \sum_{n=-\infty}^{\infty} \hat{f}(n) \, \overline{\hat{g}(n)}.
$$

Hint: Use Parseval's identity and Exercise 2.7.1.

Chapter 3

Harmonic functions in the unit disc

3.1 Series representation of harmonic functions

Let U be a harmonic function on the disc $D_R = \{\, |z| < R \,\}$. In the proof of the following theorem, we will see that there is another harmonic function V such that $F = U + iV$ is analytic on D_R. Such a function V is called a *harmonic conjugate* of U. It is determined up to an additive constant and we usually normalize it so that $V(0) = 0$. Remember that

$$\operatorname{sgn}(n) = \left\{ \begin{array}{rcl} 1 & \text{if} & n > 0, \\ 0 & \text{if} & n = 0, \\ -1 & \text{if} & n < 0. \end{array} \right.$$

Theorem 3.1 *Let U be harmonic on the disc D_R. Then, for each $n \in \mathbb{Z}$, the quantity*

$$a_n = \frac{\rho^{-|n|}}{2\pi} \int_{-\pi}^{\pi} U(\rho\, e^{it})\, e^{-int}\, dt, \qquad (0 < \rho < R), \tag{3.1}$$

is independent of ρ, and we have

$$U(re^{i\theta}) = \sum_{n=-\infty}^{\infty} a_n\, r^{|n|}\, e^{in\theta}, \qquad (re^{i\theta} \in D_R). \tag{3.2}$$

The function

$$V(re^{i\theta}) = \sum_{n=-\infty}^{\infty} -i\, \operatorname{sgn}(n)\, a_n\, r^{|n|}\, e^{in\theta}, \qquad (re^{i\theta} \in D_R), \tag{3.3}$$

is the unique harmonic conjugate of U such that $V(0) = 0$. The series in (3.2) and (3.3) are absolutely and uniformly convergent on compact subsets of D_R.

Proof. Without loss of generality, assume that U is real. Let

$$G(z) = \frac{\partial U}{\partial x}(z) - i\frac{\partial U}{\partial y}(z).$$

Since U satisfies the Laplace equation (1.1), the real and imaginary parts of G satisfy the Cauchy–Riemann equations. Hence, G is analytic on D_R. Let

$$F(z) = U(0) + \int_0^z G(w)\,dw.$$

Since D_R is simply connected, F is well-defined (the value of the integral is independent of the path of integration from 0 to z in D_R) and we have $F' = G$. Let $\mathbb{U} = \Re F$ and note that $\mathbb{U}(0) = U(0)$ and $\Im F(0) = 0$. Now, on the one hand,

$$F'(z) = G(z) = \frac{\partial U}{\partial x}(z) - i\frac{\partial U}{\partial y}(z)$$

and, on the other hand, by the Cauchy–Riemann equations

$$F'(z) = \frac{\partial \mathbb{U}}{\partial x}(z) - i\frac{\partial \mathbb{U}}{\partial y}(z).$$

Hence
$$\frac{\partial (U - \mathbb{U})}{\partial x}(z) = 0 \qquad \text{and} \qquad \frac{\partial (U - \mathbb{U})}{\partial y}(z) = 0$$

on D_R, which along with $\mathbb{U}(0) = U(0)$ imply $U \equiv \mathbb{U}$. Thus, writing V for $\Im F$, we have

$$F = U + iV$$

with $V(0) = 0$. Since F is analytic on D_R it has the unique power series representation

$$F(re^{i\theta}) = \sum_{n=0}^{\infty} \alpha_n\, r^n\, e^{in\theta},$$

with

$$\sum_{n=0}^{\infty} |\alpha_n|\, r^n < \infty \qquad\qquad\qquad (3.4)$$

for all $r < R$. Hence

$$
\begin{aligned}
U(re^{i\theta}) &= \Re\left\{ F(re^{i\theta}) \right\} = \Re\left\{ \sum_{n=0}^{\infty} \alpha_n\, r^n\, e^{in\theta} \right\} \\[2mm]
&= \alpha_0 + \frac{1}{2}\sum_{n=1}^{\infty} \alpha_n\, r^n\, e^{in\theta} + \frac{1}{2}\sum_{n=1}^{\infty} \overline{\alpha}_n\, r^n\, e^{-in\theta} \\[2mm]
&= \sum_{n=-\infty}^{\infty} a_n\, r^{|n|}\, e^{in\theta},
\end{aligned}
$$

where

$$a_n = \begin{cases} \alpha_n/2 & \text{if } n > 0, \\ \alpha_0 & \text{if } n = 0, \\ \overline{\alpha}_{-n}/2 & \text{if } n < 0. \end{cases}$$

Condition (3.4) ensures the absolute and uniform convergence of

$$\sum_{n=-\infty}^{\infty} a_n r^{|n|} e^{in\theta}$$

on compact subsets of D_R. Moreover, for each $m \in \mathbb{Z}$ and $0 < r < R$,

$$
\begin{aligned}
\frac{r^{-|m|}}{2\pi} \int_{-\pi}^{\pi} U(re^{it}) \, e^{-imt} \, dt &= \frac{r^{-|m|}}{2\pi} \int_{-\pi}^{\pi} \left(\sum_{n=-\infty}^{\infty} a_n r^{|n|} e^{int} \right) e^{-imt} \, dt \\
&= r^{-|m|} \sum_{n=-\infty}^{\infty} a_n r^{|n|} \left(\frac{1}{2\pi} \int_{-\pi}^{\pi} e^{i(n-m)t} \, dt \right) \\
&= a_m.
\end{aligned}
$$

Finally, we have

$$
\begin{aligned}
V(re^{i\theta}) &= \Im\left\{ F(re^{i\theta}) \right\} = \Im\left\{ \sum_{n=0}^{\infty} a_n r^n e^{in\theta} \right\} \\
&= \frac{1}{2i} \sum_{n=1}^{\infty} a_n r^n e^{in\theta} - \frac{1}{2i} \sum_{n=1}^{\infty} \overline{a}_n r^n e^{-in\theta} \\
&= \sum_{n=-\infty}^{\infty} -i \operatorname{sgn}(n) \, a_n r^{|n|} e^{in\theta}.
\end{aligned}
$$

Condition (3.4) also implies the absolute and uniform convergence of this series on compact subsets of D_R. $\qquad \square$

Based on the content of the preceding theorem, the conjugate of any trigonometric series

$$S = \sum_{n=-\infty}^{\infty} a_n e^{in\theta}$$

is defined by

$$\tilde{S} = \sum_{n=-\infty}^{\infty} -i \operatorname{sgn}(n) \, a_n e^{in\theta}.$$

A special case of (3.1), corresponding to $m = 0$, will be used often. We mention it as a corollary.

Corollary 3.2 *Let U be harmonic on D_R. Then*

$$U(0) = \frac{1}{2\pi} \int_{-\pi}^{\pi} U(re^{i\theta}) \, d\theta$$

for all r, $0 \leq r < R$.

Proof. It is enough to note that $a_0 = U(0)$.   □

Exercises

Exercise 3.1.1   Let U be harmonic on D_R. Show that

$$\frac{1}{\pi r^2} \int_{-\pi}^{\pi} \int_0^r U(\rho\, e^{i\theta}) \, \rho \, d\rho \, d\theta = U(0)$$

for all r, $0 < r < R$.
Hint: Use Corollary 3.2.

Exercise 3.1.2   Let U be a real harmonic function on $\mathbb{C}$ and let $N \geq 0$.
Suppose that
$$U(re^{i\theta}) \leq c\, r^N + c' \tag{3.5}$$
for all $r \geq 0$ and all θ; c and c' are two positive constants. Show that

$$U(re^{i\theta}) = \sum_{n=-N}^{N} a_n \, r^{|n|} \, e^{in\theta},$$

where the coefficients a_n are given by (3.1).
Hint: By Corollary 3.2 and (3.1)

$$a_0 \pm r^{|n|} \Re a_n = \frac{1}{2\pi} \int_{-\pi}^{\pi} U(re^{it}) \, (1 \pm \cos(nt)) \, dt.$$

The advantage of this representation is that $(1 \pm \cos(nt)) \geq 0$.
Remark: We emphasize that in (3.5) we have U and not $|U|$. Not using the absolute values is crucial in some applications.

Exercise 3.1.3   Let Ω be a simply connected domain and let U be a harmonic function on Ω. Show that there is a harmonic function V on Ω such that $F = U + iV$ is analytic over Ω. Moreover, show that V is unique up to an additive constant.
Hint: See the first part of the proof of Theorem 3.1.

3.2 Hardy spaces on $\mathbb{D}$

The family of all complex harmonic functions on the open unit disc $\mathbb{D}$ is denoted by $h(\mathbb{D})$. Let $U \in h(\mathbb{D})$ and write

$$\| U \|_p = \sup_{0 \le r < 1} \| U_r \|_p = \sup_{0 \le r < 1} \left(\frac{1}{2\pi} \int_0^{2\pi} |U(r\, e^{i\theta})|^p \, d\theta \right)^{\frac{1}{p}},$$

if $p \in (0, \infty)$, and

$$\| U \|_\infty = \sup_{z \in \mathbb{D}} | U(z) |.$$

We define

$$h^p(\mathbb{D}) = \big\{ U \in h(\mathbb{D}) : \ \| U \|_p < \infty \big\},$$

where $p \in (0, \infty]$. It is straightforward to see that $h^p(\mathbb{D})$, $1 \le p \le \infty$, is a normed vector space and, by Hölder's inequality,

$$h^\infty(\mathbb{D}) \subset h^q(\mathbb{D}) \subset h^p(\mathbb{D})$$

if $0 < p < q < \infty$. We will see that $h^1(\mathbb{D})$ and $h^p(\mathbb{D})$, $1 < p \le \infty$, are Banach spaces respectively isomorphic to $\mathcal{M}(\mathbb{T})$ and $L^p(\mathbb{T})$, $1 < p \le \infty$.

A complex harmonic function F is simply of the form $F = U + iV$ where U and V are real harmonic functions and there is no other relation between U and V. However, if we assume that V is a harmonic conjugate of U, and thus F is analytic on $\mathbb{D}$, a whole new family of functions with profound properties emerges. Let us denote the family of all *analytic* functions on $\mathbb{D}$ by $H(\mathbb{D})$. Hence, $H(\mathbb{D}) \subset h(\mathbb{D})$. Then, parallel to our previous definitions, we consider the *Hardy classes* of analytic functions on the unit disc

$$H^p(\mathbb{D}) = \big\{ F \in H(\mathbb{D}) : \ \| F \|_p < \infty \big\}$$

for $0 < p \le \infty$. Clearly,

$$H^p(\mathbb{D}) \subset h^p(\mathbb{D}).$$

As a matter of fact, that is why we assumed that the elements of $h(\mathbb{D})$ are complex-valued harmonic functions. As a consequence, any representation theorem for $h^p(\mathbb{D})$ functions is also automatically valid for the elements of the smaller subclass $H^p(\mathbb{D})$. The Hardy space $H^p(\mathbb{D})$, $1 \le p \le \infty$, is a normed vector space and, by Hölder's inequality,

$$H^\infty(\mathbb{D}) \subset H^q(\mathbb{D}) \subset H^p(\mathbb{D})$$

if $0 < p < q < \infty$. We will see that $H^p(\mathbb{D})$, $1 \le p \le \infty$, is also a Banach space isomorphic to a closed subspace of $L^p(\mathbb{T})$ denoted by $H^p(\mathbb{T})$.

Using these new notations, Corollaries 2.11, 2.15 and 2.19 can be rewritten as follows.

Theorem 3.3 *Let* $1 \le p \le \infty$. *If* $u \in L^p(\mathbb{T})$, *then* $U = P * u \in h^p(\mathbb{D})$ *and* $\|U\|_p = \|u\|_p$. *If* $\mu \in \mathcal{M}(\mathbb{T})$, *then* $U = P * \mu \in h^1(\mathbb{D})$ *and* $\|U\|_1 = \|\mu\|$.

In the following sections, we study the converse of this theorem. More precisely, we start with a harmonic function in $h^p(\mathbb{D})$ and show that it can be represented as $P * u$ or $P * \mu$ with a suitable function u or measure μ.

Exercises

Exercise 3.2.1   Let $F = U + iV$ be analytic on $\mathbb{D}$, and let $0 < p \leq \infty$. Show that $F \in H^p(\mathbb{D})$ if and only if U and V are real harmonic functions in $h^p(\mathbb{D})$.

Exercise 3.2.2   Let $0 < p < q < \infty$. Show that

$$h^\infty(\mathbb{D}) \subsetneqq h^q(\mathbb{D}) \subsetneqq h^p(\mathbb{D})$$

and that

$$H^\infty(\mathbb{D}) \subsetneqq H^q(\mathbb{D}) \subsetneqq H^p(\mathbb{D}).$$

Exercise 3.2.3   Let $u \in L^2(\mathbb{T})$. Define

$$F(z) = \frac{1}{2\pi} \int_{-\pi}^{\pi} \frac{e^{it} + z}{e^{it} - z} \, u(e^{it}) \, dt, \qquad (z \in \mathbb{D}).$$

Show that $F \in H^2(\mathbb{D})$ and

$$\|F\|_2^2 = |\hat{u}(0)|^2 + 4 \sum_{n=1}^{\infty} |\hat{u}(n)|^2.$$

Deduce

$$\|F\|_2 \leq 2 \, \|u\|_2$$

and show that 2 is the best possible constant.
Hint: Use Exercise 1.3.3 and Parseval's identity (Corollary 2.22).

3.3   Poisson representation of $h^\infty(\mathbb{D})$ functions

In studying the boundary values of harmonic functions on the unit disc, the *best* possible assumption is to consider harmonic functions which are actually defined on discs larger than the unit disc. Hence, let us consider

$$h(\overline{\mathbb{D}}) = \{ U : \nabla^2 U = 0 \text{ on } |z| < R, \text{ for some } R > 1 \}.$$

The constant R is not universal and it depends on U. Putting such a strong assumption on the elements of $h(\overline{\mathbb{D}})$ makes it the *smallest* subclass of $h(\mathbb{D})$ in our discussion.

Lemma 3.4 *Let $U \in h(\overline{\mathbb{D}})$. Then*

$$U(re^{i\theta}) = \frac{1}{2\pi} \int_{-\pi}^{\pi} \frac{1-r^2}{1+r^2 - 2r\cos(\theta - t)} U(e^{it})\, dt, \qquad (re^{i\theta} \in \mathbb{D}). \quad (3.6)$$

Proof. Since $U \in h(\overline{\mathbb{D}})$, there exists $R > 1$ such that U is harmonic on D_R. Hence, by Theorem 3.1, for all $re^{i\theta} \in D_R$, we have

$$U(re^{i\theta}) = \sum_{n=-\infty}^{\infty} a_n\, r^{|n|}\, e^{in\theta},$$

where a_n are given by

$$a_n = \frac{1}{2\pi} \int_{-\pi}^{\pi} U(e^{it})\, e^{-int}\, dt, \qquad (n \in \mathbb{Z}).$$

In particular, for all $re^{i\theta} \in \mathbb{D}$, we obtain

$$U(re^{i\theta}) = \sum_{n=-\infty}^{\infty} \left(\frac{1}{2\pi} \int_{-\pi}^{\pi} U(e^{it}) e^{-int}\, dt \right) r^{|n|}\, e^{in\theta}.$$

Fix r and θ. Since U is bounded on $\mathbb{T}$, the absolute and uniform convergence of the series $\sum_{n=-\infty}^{\infty} r^{|n|}\, e^{in(\theta - t)}\, U(e^{it})$, as a function of t, enables us to change the order of summation and integration. Hence,

$$U(re^{i\theta}) = \frac{1}{2\pi} \int_{-\pi}^{\pi} \left(\sum_{n=-\infty}^{\infty} r^{|n|}\, e^{in(\theta - t)} \right) U(e^{it})\, dt, \qquad (re^{i\theta} \in \mathbb{D}).$$

But, as we saw in (1.5),

$$\sum_{n=-\infty}^{\infty} r^{|n|}\, e^{in(\theta - t)} = \frac{1-r^2}{1+r^2 - 2r\cos(\theta - t)}.$$

$\square$

In the following we show that the integral representation (3.6) is valid for some larger subclasses of $h(\mathbb{D})$. Since $h(\overline{\mathbb{D}}) \subset h^\infty(\mathbb{D})$, the following result is the first generalization of Lemma 3.4.

Theorem 3.5 (Fatou [6]) *Let $U \in h^\infty(\mathbb{D})$. Then there exists a unique $u \in L^\infty(\mathbb{T})$ such that*

$$U(re^{i\theta}) = \frac{1}{2\pi} \int_{-\pi}^{\pi} \frac{1-r^2}{1+r^2 - 2r\cos(\theta - t)} u(e^{it})\, dt, \qquad (re^{i\theta} \in \mathbb{D}),$$

and

$$\|U\|_\infty = \|u\|_\infty.$$

Remark: Compare with Corollary 2.19.

Proof. The uniqueness is a consequence of Corollary 2.12. The rest of the proof is based on the Poisson representation of harmonic functions in $h(\overline{\mathbb{D}})$ and the following two facts:

(i) $L^\infty(\mathbb{T})$ is the *dual* of $L^1(\mathbb{T})$;

(ii) $L^1(\mathbb{T})$ is a *separable* space (see Exercise 2.5.2).

Step 1: *Picking a family of bounded linear functionals on $L^1(\mathbb{T})$.*

Put

$$U_n(z) = U\left(\left(1 - \frac{1}{n}\right) z \right), \qquad (n \geq 2).$$

First of all, U_n is defined on the disc $\{|z| < n/(n-1)\}$. Hence, $U_n \in h(\overline{\mathbb{D}})$ and by Theorem 3.4,

$$U\left(\left(1 - \frac{1}{n}\right) z \right) = \frac{1}{2\pi} \int_{-\pi}^{\pi} \frac{1 - r^2}{1 + r^2 - 2r\cos(\theta - t)} U_n(e^{it}) \, dt \qquad (3.7)$$

for all $re^{i\theta} \in \mathbb{D}$. Let $n \to \infty$. The left side clearly tends to $U(re^{it})$. We show that the limit of the right side has an integral representation. Define

$$\Lambda_n : L^1(\mathbb{T}) \longrightarrow \mathbb{C}$$

by

$$\Lambda_n(f) = \frac{1}{2\pi} \int_{-\pi}^{\pi} f(e^{it}) U_n(e^{it}) \, dt, \qquad (n \geq 2).$$

Since

$$|\Lambda_n(f)| \leq \|U_n\|_\infty \|f\|_1 \leq \|U\|_\infty \|f\|_1, \qquad (3.8)$$

each Λ_n is a *bounded linear functional* on $L^1(\mathbb{T})$ with

$$\|\Lambda_n\| \leq \|U\|_\infty.$$

Step 2: *Extracting a convergent subsequence of Λ_n.*

At this point we can use the Banach–Alaoglu theorem and deduce that Λ_n has a convergent subsequence in the weak* topology. In other words, there is a bounded linear functional Λ on $L^1(\mathbb{T})$ and a subsequence Λ_{n_k} such that

$$\lim_{k \to \infty} \Lambda_{n_k}(f) = \Lambda(f)$$

for all $f \in L^1(\mathbb{T})$. However, we give a direct proof of this fact.

Take a countable dense subset of $L^1(\mathbb{T})$, say $\{f_1, f_2, \dots\}$. By (3.8), there is a subsequence of $\{\Lambda_n\}_{n \geq 2}$, say $\{\Lambda_{n_{1j}}\}_{j \geq 1}$, such that

$$\lim_{j \to \infty} \Lambda_{n_{1j}}(f_1)$$

exists. Again by (3.8), there is a subsequence of $\{\Lambda_{n_{1j}}\}_{j\geq1}$, say $\{\Lambda_{n_{2j}}\}_{j\geq1}$, such that

$$\lim_{j\to\infty} \Lambda_{n_{2j}}(f_2)$$

exists. Continuing this process, for any $i \geq 1$, we find a subsequence $\{\Lambda_{n_{ij}}\}_{j\geq1}$ of $\{\Lambda_{n_{(i-1)j}}\}_{j\geq1}$ such that $\lim_{j\to\infty} \Lambda_{n_{ij}}(f_i)$ exists. To apply Cantor's method, consider the diagonal subsequence $\{\Lambda_{n_{kk}}\}_{k\geq1}$. Since $\{\Lambda_{n_{kk}}\}_{k\geq1}$ is eventually a subsequence of $\{\Lambda_{n_{ij}}\}_{j\geq1}$, the limit

$$\lim_{k\to\infty} \Lambda_{n_{kk}}(f_i)$$

exists for any $i \geq 1$. Moreover, we show that this limit actually exists for all $f \in L^1(\mathbb{T})$. To do so, we use the fact that Λ_n's are uniformly bounded and f_i's are dense in $L^1(\mathbb{T})$. Fix $f \in L^1(\mathbb{T})$ and $\varepsilon > 0$. Hence, there is f_i such that $\|f - f_i\|_1 < \varepsilon$. By (3.8), we have

$$
\begin{aligned}
|\Lambda_{n_{kk}}(f) - \Lambda_{n_{ll}}(f)| &\leq |\Lambda_{n_{kk}}(f) - \Lambda_{n_{kk}}(f_i)| \\
&+ |\Lambda_{n_{kk}}(f_i) - \Lambda_{n_{ll}}(f_i)| \\
&+ |\Lambda_{n_{ll}}(f_i) - \Lambda_{n_{ll}}(f)| \\
&\leq 2\|U\|_\infty \|f - f_i\|_1 + |\Lambda_{n_{kk}}(f_i) - \Lambda_{n_{ll}}(f_i)| \\
&\leq 2\|U\|_\infty \varepsilon + |\Lambda_{n_{kk}}(f_i) - \Lambda_{n_{ll}}(f_i)|.
\end{aligned}
$$

Picking k, l large enough, we obtain

$$|\Lambda_{n_{kk}}(f) - \Lambda_{ll}(f)| \leq (2\|U\|_\infty + 1)\varepsilon.$$

Hence

$$\Lambda(f) = \lim_{k\to\infty} \Lambda_{n_{kk}}(f)$$

exists for all $f \in L^1(\mathbb{T})$ and, again by (3.8),

$$|\Lambda(f)| \leq \|U\|_\infty \|f\|_1.$$

In other words, Λ is a bounded linear functional on $L^1(\mathbb{T})$ with $\|\Lambda\| \leq \|U\|_\infty$.

Step 3: *Appealing to Riesz's theorem.*

For a fixed $z = re^{i\theta}$,

$$f_z(e^{it}) = P_r(e^{i(\theta-t)}) = \frac{1-r^2}{1+r^2-2r\cos(\theta-t)},$$

as a function of t, is in $L^1(\mathbb{T})$. Hence, by (3.7), we have

$$
\begin{aligned}
\Lambda(f_z) &= \lim_{k\to\infty} \Lambda_{n_{kk}}(f_z) \\
&= \lim_{k\to\infty} \frac{1}{2\pi} \int_{-\pi}^{\pi} P_r(e^{i(\theta-t)}) U_{n_{kk}}(e^{it})\, dt \\
&= \lim_{k\to\infty} U\left(\left(1 - \frac{1}{n_{kk}}\right)re^{i\theta}\right) = U(re^{i\theta}).
\end{aligned}
$$

On the other hand, by Riesz's theorem, there is a $u \in L^\infty(\mathbb{T})$ such that

$$\Lambda(f) = \frac{1}{2\pi} \int_{-\pi}^{\pi} f(e^{it})\, u(e^{it})\, dt$$

for all $f \in L^1(\mathbb{T})$. First of all, by choosing $f = f_z$, we obtain

$$U(re^{i\theta}) = \frac{1}{2\pi} \int_{-\pi}^{\pi} P_r(e^{i(\theta-t)})\, u(e^{it})\, dt.$$

Secondly, by Corollary 2.19, this identity implies $\|U\|_\infty = \|u\|_\infty$. $\square$

Exercises

Exercise 3.3.1   Let U be harmonic on a domain Ω and let $\overline{D(a,R)} \subset \Omega$. Show that

$$U(a+re^{i\theta}) = \frac{1}{2\pi} \int_{-\pi}^{\pi} \frac{R^2 - r^2}{R^2 + r^2 - 2Rr\cos(\theta-t)}\, U(a+Re^{it})\, dt, \qquad (0 \le r < R).$$

Hint: Use Lemma 3.4 and make a change of variable.

Exercise 3.3.2   Let $0 \le r_0 < 1$. Show that

$$U(re^{i\theta}) = \frac{r(1+r_0^2)\cos\theta - r_0(1+r^2)}{1+r_0^2\, r^2 - 2r_0\, r\cos\theta}$$

and

$$V(re^{i\theta}) = \frac{r(1-r_0^2)\sin\theta}{1+r_0^2\, r^2 - 2r_0\, r\cos\theta}$$

are bounded harmonic functions on $\mathbb{D}$. Find the Fourier series expansions of U_r and V_r.
Hint: Besides the direct verification of each fact, it might be easier to show that $U + iV$ is a bounded analytic function on $\mathbb{D}$.

Exercise 3.3.3   Find a bounded harmonic function on the unit disc such that its conjugate is not bounded.

Exercise 3.3.4   Let U be harmonic on $\mathbb{D}$. Show that there exists a unique $u \in \mathcal{C}(\mathbb{T})$ such that

$$U(re^{i\theta}) = \frac{1}{2\pi} \int_{-\pi}^{\pi} \frac{1-r^2}{1+r^2 - 2r\cos(\theta-t)}\, u(e^{it})\, dt, \qquad (re^{i\theta} \in \mathbb{D}),$$

if and only if the family $(U_r)_{0 \le r < 1}$ is Cauchy in $\mathcal{C}(\mathbb{T})$ as $r \to 1$.

Exercise 3.3.5 [Harnack's theorem] Let $(U_n)_{n \geq 1}$ be a sequence of harmonic functions on a domain Ω. Suppose that on each compact subset of Ω, U_n converges uniformly to U. Show that U is harmonic on Ω.
Hint: Use Corollary 2.3 and Exercise 3.3.1.

3.4 Poisson representation of $h^p(\mathbb{D})$ functions $(1 < p < \infty)$

The proof of Theorem 3.5 can be modified slightly to give a representation formula for $h^p(\mathbb{D})$, $1 < p < \infty$, functions. The modification is based on the following two facts:

(i) $L^p(\mathbb{T})$ is the *dual* of $L^q(\mathbb{T})$, where $1/p + 1/q = 1$;

(ii) $L^p(\mathbb{T})$ is a *separable* space.

Since $h(\overline{\mathbb{D}}) \subset h^\infty(\mathbb{D}) \subset h^p(\mathbb{D})$, the following result is a generalization of Theorems 3.4 and 3.5.

Theorem 3.6 *Let $U \in h^p(\mathbb{D})$, $1 < p < \infty$. Then there exists a unique $u \in L^p(\mathbb{T})$ such that*

$$U(re^{i\theta}) = \frac{1}{2\pi} \int_{-\pi}^{\pi} \frac{1 - r^2}{1 + r^2 - 2r\cos(\theta - t)} \, u(e^{it}) \, dt, \qquad (re^{i\theta} \in \mathbb{D}),$$

and

$$\|U\|_p = \|u\|_p.$$

Remark: Compare with Corollary 2.15.

If $U \in h^2(\mathbb{D})$ this result, along with Theorem 2.1, gives us a unique $u \in L^2(\mathbb{T})$ such that

$$U(re^{i\theta}) = \frac{1}{2\pi} \int_{-\pi}^{\pi} \frac{1 - r^2}{1 + r^2 - 2r\cos(\theta - t)} \, u(e^{it}) \, dt = \sum_{n=-\infty}^{\infty} \hat{u}(n) \, r^{|n|} \, e^{in\theta}$$

for all $re^{i\theta} \in \mathbb{D}$. Hence, by Parseval's identity (Corollary 2.22),

$$\frac{1}{2\pi} \int_{-\pi}^{\pi} |U(re^{i\theta})|^2 \, d\theta = \sum_{n=-\infty}^{\infty} |\hat{u}(n)|^2 \, r^{2|n|}.$$

Thanks to the uniform and bounded convergence of the series $\sum_n \hat{u}(n) \, r^{|n|} \, e^{in\theta}$ for each fixed $r < 1$, and the fact that $(e^{in\theta})_{n \in \mathbb{Z}}$ is an orthonormal family, the preceding identity can also be proved by direct computation. Now, by the monotone convergence theorem, and again by Parseval's identity (this time we really need it), if we let $r \to 1$, we obtain

$$\|U\|_2 = \|u\|_2 = \left(\sum_{n=-\infty}^{\infty} |\hat{u}_n|^2 \right)^{\frac{1}{2}}.$$

There is no such relation between a function and its Fourier coefficients in other L^p classes.

Exercises

Exercise 3.4.1 Show that

$$U(re^{i\theta}) = \log(1 + r^2 - 2r\cos\theta) \in h^2(\mathbb{D}).$$

Hint: Study the analytic function $f(z) = 2\log(1-z)$ to find the Fourier series expansion of U_r.

Exercise 3.4.2 Find V, the harmonic conjugate of

$$U(re^{i\theta}) = \log(1 + r^2 - 2r\cos\theta).$$

Do we have $V \in h^2(\mathbb{D})$?

3.5   Poisson representation of $h^1(\mathbb{D})$ functions

In the proof of Theorems 3.5 and 3.6 we used the fact that $L^p(\mathbb{T})$ is the dual of $L^q(\mathbb{T})$, whenever $1 < p \leq \infty$. But $L^1(\mathbb{T})$ is not the dual of any space. That is why, in Theorem 3.6, the assumption $p > 1$ is essential and the suggested proof for $1 < p < \infty$ does not work if $p = 1$. To overcome this difficulty, we consider $L^1(\mathbb{T})$ as a subset of $\mathcal{M}(\mathbb{T})$. By Riesz's theorem, $\mathcal{M}(\mathbb{T})$ is the dual of $C(\mathbb{T})$. Now we proceed as in the proof of Theorem 3.5. The only difference is that this time there is a unique Borel measure that represents our continuous linear functional on $C(\mathbb{T})$. Therefore, an element of $h^1(\mathbb{D})$ is represented by the Poisson integral of a measure (and not necessarily of an $L^1(\mathbb{T})$ function).

Since $h(\overline{\mathbb{D}}) \subset h^\infty(\mathbb{D}) \subset h^p(\mathbb{D}) \subset h^1(\mathbb{D})$, $1 < p < \infty$, the following result is the last step in the generalization of Theorems 3.4, 3.5 and 3.6.

Theorem 3.7 *Let $U \in h^1(\mathbb{D})$. Then there exists a unique $\mu \in \mathcal{M}(\mathbb{T})$ such that*

$$U(re^{i\theta}) = \int_{\mathbb{T}} \frac{1 - r^2}{1 + r^2 - 2r\cos(\theta - t)} \, d\mu(e^{it}), \qquad (re^{i\theta} \in \mathbb{D}),$$

and

$$\|U\|_1 = \|\mu\|.$$

Remark: Compare with Corollary 2.11.

Let U be a *positive harmonic function* on $\mathbb{D}$. As a convention in this context, positive means ≥ 0 (see Exercise 3.5.7). Then, by Corollary 3.2,

$$\frac{1}{2\pi} \int_{-\pi}^{\pi} |U(re^{i\theta})| \, d\theta = \frac{1}{2\pi} \int_{-\pi}^{\pi} U(re^{i\theta}) \, d\theta = U(0),$$

and thus $U \in h^1(\mathbb{D})$. Hence we can apply Theorem 3.7 and obtain a measure whose Poisson integral is U. But a closer look at its proof suggests an improvement. Here, the measures

$$d\mu_n(e^{it}) = U((1 - 1/n)e^{it}) \, dt/2\pi$$

are positive and thus their weak limit $d\mu$ has to be positive too. Therefore, we arrive at the following special case of Theorem 3.7.

Theorem 3.8 (Herglotz) *Let U be a positive harmonic function on $\mathbb{D}$. Then there exists a unique finite positive Borel measure μ on $\mathbb{T}$ such that*

$$U(re^{i\theta}) = \int_{\mathbb{T}} \frac{1 - r^2}{1 + r^2 - 2r \cos(\theta - t)} \, d\mu(e^{it}), \qquad (re^{i\theta} \in \mathbb{D}).$$

Herglotz's theorem provides an easy proof of Harnack's inequality. The main feature of Harnack's inequality is that the constants appearing in the lower and upper bounds do not depend on the function U.

Corollary 3.9 (Harnack's inequality) *Let U be a positive harmonic function on $\mathbb{D}$. Then, for each $re^{i\theta} \in \mathbb{D}$,*

$$\frac{1 - r}{1 + r} \, U(0) \leq U(re^{i\theta}) \leq \frac{1 + r}{1 - r} \, U(0).$$

Proof. By Theorem 3.8, U is the Poisson integral of a positive measure μ. Since, for all θ and t,

$$\frac{1 - r}{1 + r} \leq \frac{1 - r^2}{1 + r^2 - 2r \cos(\theta - t)} \leq \frac{1 + r}{1 - r},$$

and since $d\mu \geq 0$, we have

$$\frac{1 - r}{1 + r} \int_{\mathbb{T}} d\mu(e^{it}) \leq \int_{\mathbb{T}} \frac{1 - r^2}{1 + r^2 - 2r \cos(\theta - t)} \, d\mu(e^{it}) \leq \frac{1 + r}{1 - r} \int_{\mathbb{T}} d\mu(e^{it}).$$

But, again according to Theorem 3.8,

$$\int_{\mathbb{T}} d\mu(e^{it}) = U(0).$$

$\square$

Exercises

Exercise 3.5.1 Show that the Poisson kernel

$$P(re^{i\theta}) = \frac{1 - r^2}{1 + r^2 - 2r \cos \theta},$$

as a positive harmonic function on $\mathbb{D}$, is in $h^p(\mathbb{D})$, $0 < p \leq 1$. What is the measure promised in Theorem 3.8? Moreover, show that $P \notin h^p(\mathbb{D})$ for any $p > 1$.

Exercise 3.5.2   Let U be a real harmonic function on $\mathbb{D}$. Show that $U \in h^1(\mathbb{D})$ if and only if U is the difference of two positive harmonic functions.
Hint: Use Theorems 3.7 and 3.8 and the fact that each $\mu \in \mathcal{M}(\mathbb{T})$ can be decomposed as $\mu = \mu_1 - \mu_2$ where μ_1 and μ_2 are finite positive Borel measures.

Exercise 3.5.3   Let U be harmonic on $\mathbb{D}$. Show that there exists a unique $u \in L^1(\mathbb{T})$ such that

$$U(re^{i\theta}) = \frac{1}{2\pi} \int_{-\pi}^{\pi} \frac{1 - r^2}{1 + r^2 - 2r\cos(\theta - t)} u(e^{it}) \, dt, \qquad (re^{i\theta} \in \mathbb{D}),$$

if and only if the family $(U_r)_{0 \le r < 1}$ is Cauchy in $L^1(\mathbb{T})$ as $r \to 1$.

Exercise 3.5.4   [Generalized Harnack's inequality] Let U be a positive harmonic function on the disc $D(a, R) = \{z : |z - a| < R\}$. Show that, for all $0 \le r < R$ and θ,

$$\frac{R - r}{R + r} U(a) \le U(a + re^{i\theta}) \le \frac{R + r}{R - r} U(a).$$

Exercise 3.5.5   Let $\mathcal{U}$ be the collection of all positive harmonic functions U on $\mathbb{D}$ with $U(0) = 1$. Find

$$\sup_{U \in \mathcal{U}} U(1/2)$$

and

$$\inf_{U \in \mathcal{U}} U(1/2).$$

Hint: Use Harnack's inequality (Corollary 3.9).

Exercise 3.5.6   Let Ω be a domain in $\mathbb{C}$. Fix z and w in Ω. Show that there exists $\tau \ge 1$ such that, for *every* positive harmonic function U on Ω,

$$\tau^{-1} U(w) \le U(z) \le \tau U(w).$$

Remark: $\tau_\Omega(z, w)$, the *Harnack distance* between z and w, is by definition the smallest τ satisfying the last relation.

Exercise 3.5.7   Let U be a positive harmonic function on a domain Ω. Show that either U is identically zero on Ω or it never vanishes there.
Hint: Use Exercise 3.5.6.

Exercise 3.5.8   Let f be a conformal mapping between Ω and Ω'. Show that

$$\tau_\Omega(z,w) = \tau_{\Omega'}(f(z),f(w))$$

for all z, $w \in \Omega$.

Exercise 3.5.9   Show that

$$\tau_\mathbb{D}(0,r) = \frac{1+r}{1-r}, \qquad (0 \leq r < 1).$$

Then use the conformal mapping

$$\varphi(z) = e^{i\beta}\frac{z-\alpha}{1-\bar{\alpha}z}, \qquad (|\alpha| < 1,\ \beta \in \mathbb{R}),$$

(an *automorphism* of the unit disc) to show that

$$\tau_\mathbb{D}(z,w) = \frac{1 + \left|\frac{z-w}{1-z\bar{w}}\right|}{1 - \left|\frac{z-w}{1-z\bar{w}}\right|}$$

for all $z, w \in \mathbb{D}$.
Remark: The content of Section 7.1 might help.

Exercise 3.5.10   Let U be a positive harmonic function on $\mathbb{D}$. Show that

$$|\nabla U(0)| \leq 2\,U(0),$$

where

$$\nabla U = \partial U/\partial x + i\,\partial U/\partial y.$$

Hint: Use Harnack's inequality and the fact that $\nabla U(0)$ is the maximum directional derivative at zero, i.e.

$$|\nabla U(0)| = \sup_{-\pi \leq \theta \leq \pi} \left| \lim_{r \to 0} \frac{U(re^{i\theta}) - U(0)}{r} \right|.$$

Exercise 3.5.11   [Liouville] Use Harnack's inequality to show that every positive harmonic function on $\mathbb{C}$ is constant.

Exercise 3.5.12   Let $(U_n)_{n \geq 1}$ be a sequence of positive harmonic functions on a domain Ω, and let $z_0 \in \Omega$. Suppose that

$$\lim_{n \to \infty} U_n(z_0) = \infty.$$

Show that U_n converges uniformly to infinity on compact subsets of Ω.
Hint: Use Exercise 3.3.1.

Exercise 3.5.13 Let $(U_n)_{n \geq 1}$ be a sequence of positive harmonic functions on a domain Ω, and let $z_0 \in \Omega$. Suppose that

$$\lim_{n \to \infty} U_n(z_0) = 0.$$

Show that U_n converges uniformly to zero on compact subsets of Ω.
Hint: Use Exercise 3.3.1.

Exercise 3.5.14 [Harnack's theorem] Let $(U_n)_{n \geq 1}$ be an increasing sequence of harmonic functions on a domain Ω. Show that either U_n converges uniformly on compact subsets of Ω to a harmonic function, or it converges uniformly to infinity on each compact subset.
Hint: Use Exercises 3.3.5, 3.5.12 and 3.5.6.

Exercise 3.5.15 [Herglotz] Let f be an analytic function on the unit disc with values in the right half plane $\{\Re z > 0\}$. Suppose that $f(0) > 0$. Show that there exists a positive Borel measure on $\mathbb{T}$, say $\mu \in \mathcal{M}(\mathbb{T})$, such that

$$f(z) = \int_{\mathbb{T}} \frac{e^{it} + z}{e^{it} - z} \, d\mu(e^{it})$$

for all $z \in \mathbb{D}$.
Hint: Use Theorem 3.8 and Exercise 1.3.2.

Exercise 3.5.16 Let U be a positive harmonic function on $\mathbb{D}$. We know that there exists a positive Borel measure μ on $\mathbb{T}$ such that

$$U(re^{i\theta}) = \int_{\mathbb{T}} \frac{1 - r^2}{1 + r^2 - 2r\cos(\theta - t)} \, d\mu(e^{it}), \qquad (re^{i\theta} \in \mathbb{D}).$$

Show that

$$\|U\|_1 = U(0) = \|\mu\| = \mu(\mathbb{T}).$$

3.6   Radial limits of $h^p(\mathbb{D})$ functions $(1 \leq p \leq \infty)$

The following result is a generalization of Fatou's theorem, which is about the boundary values of bounded analytic functions in the unit disc. However, his theorem works in a more general setting.

Lemma 3.10 *Let $u \in L^1(\mathbb{T})$, and let*

$$U(re^{i\theta}) = \frac{1}{2\pi} \int_{-\pi}^{\pi} \frac{1 - r^2}{1 + r^2 - 2r\cos(\theta - t)} \, u(e^{it}) \, dt, \qquad (re^{i\theta} \in \mathbb{D}).$$

Then

$$\lim_{r \to 1} U(re^{i\theta}) = u(e^{i\theta}) \qquad\qquad (3.9)$$

for almost all $e^{i\theta} \in \mathbb{T}$.

Proof. According to a classical result of Lebesgue,

$$\lim_{t \to 0} \frac{1}{2t} \int_{\theta-t}^{\theta+t} u(e^{is}) \, ds = u(e^{i\theta})$$

for almost all $e^{i\theta} \in \mathbb{T}$. We prove that at such a point (3.9) holds. Without loss of generality, assume that $\theta = 0$. Put

$$\mathfrak{U}(x) = \int_{-\pi}^{x} u(e^{it}) \, dt, \qquad x \in [-\pi, \pi].$$

Then, doing integration by parts, we obtain

$$
\begin{aligned}
2\pi \, U(r) &= \int_{-\pi}^{\pi} \frac{1 - r^2}{1 + r^2 - 2r \cos t} \, u(e^{it}) \, dt \\
&= \frac{1 - r^2}{1 + r^2 - 2r \cos t} \, \mathfrak{U}(t) \bigg|_{t=-\pi}^{\pi} - \int_{-\pi}^{\pi} \frac{\partial}{\partial t} \left\{ \frac{1 - r^2}{1 + r^2 - 2r \cos t} \right\} \mathfrak{U}(t) \, dt \\
&= \frac{1 - r}{1 + r} \, \mathfrak{U}(\pi) + \int_{-\pi}^{\pi} \frac{(1 - r^2) \, 2r \, \sin t}{(1 + r^2 - 2r \cos t)^2} \, \mathfrak{U}(t) \, dt \\
&= \frac{1 - r}{1 + r} \, \mathfrak{U}(\pi) + \frac{2r}{1 + r} \int_{-\pi}^{\pi} \frac{(1 + r)^2 \, (1 - r) \, t \, \sin t}{(1 + r^2 - 2r \cos t)^2} \times \frac{\mathfrak{U}(t) - \mathfrak{U}(-t)}{2t} \, dt.
\end{aligned}
$$

But we know that

$$F_r(e^{it}) = \frac{(1 + r)^2 \, (1 - r) \, t \, \sin t}{(1 + r^2 - 2r \cos t)^2}$$

is a positive approximate identity on $\mathbb{T}$ and, by assumption,

$$\lim_{t \to 0} \frac{\mathfrak{U}(t) - \mathfrak{U}(-t)}{2t} = \lim_{t \to 0} \frac{1}{2t} \int_{-t}^{t} u(e^{is}) \, ds = u(1).$$

In other words, the function

$$
\Psi(e^{it}) = \begin{cases} \dfrac{\mathfrak{U}(t) - \mathfrak{U}(-t)}{2t} & \text{if} \quad 0 < |t| \leq \pi, \\[2mm] u(1) & \text{if} \quad t = 0 \end{cases}
$$

is continuous at $t = 0$, and moreover, $\Psi(e^{-it}) = \Psi(e^{it})$. Hence, by Theorem 2.5,

$$
\begin{aligned}
\lim_{r \to 1} U(r) &= \lim_{r \to 1} \frac{1}{2\pi} \left\{ \frac{1 - r}{1 + r} \, \mathfrak{U}(\pi) + \frac{2r}{1 + r} \int_{-\pi}^{\pi} F_r(e^{it}) \, \frac{\mathfrak{U}(t) - \mathfrak{U}(-t)}{2t} \, dt \right\} \\
&= \lim_{r \to 1} \frac{1}{2\pi} \int_{-\pi}^{\pi} F_r(e^{it}) \, \Psi(e^{it}) \, dt \\
&= \lim_{r \to 1} (F_r * \Psi)(1) = \Psi(1) = u(1).
\end{aligned}
$$

$\square$

A slight modification of the preceding lemma yields the following result about harmonic functions generated by singular measures. Some authors prefer to combine both results and present them as a single theorem, as we do later in this section.

Lemma 3.11 *Let $\sigma \in \mathcal{M}(\mathbb{T})$ be singular with respect to the Lebesgue measure, and let*

$$U(re^{i\theta}) = \int_{\mathbb{T}} \frac{1 - r^2}{1 + r^2 - 2r\cos(\theta - t)} \, d\sigma(e^{it}).$$

Then

$$\lim_{r \to 1} U(re^{i\theta}) = 0 \tag{3.10}$$

for almost all $e^{i\theta} \in \mathbb{T}$.

Proof. Since σ is singular with respect to the Lebesgue measure, we have

$$\lim_{t \to 0} \frac{\sigma\big(\{e^{is} : s \in (\theta - t, \theta + t)\}\big)}{2t} = 0$$

for almost all $e^{i\theta} \in \mathbb{T}$. We prove that at such a point (3.10) holds. Without loss of generality, assume that $\theta = 0$. First, we extract the Dirac measure (if any) at the point -1. Hence,

$$
\begin{aligned}
U(r) &= \int_{\mathbb{T}} \frac{1 - r^2}{1 + r^2 - 2r\cos t} \, d\sigma(e^{it}) \\
&= \frac{1 - r^2}{1 + r^2 - 2r\cos\pi} \, \sigma(\{-1\}) + \int_{\mathbb{T}\setminus\{-1\}} \frac{1 - r^2}{1 + r^2 - 2r\cos t} \, d\sigma(e^{it}).
\end{aligned}
$$

Since

$$\lim_{r \to 1} \frac{1 - r^2}{1 + r^2 - 2r\cos\pi} = 0,$$

without loss of generality we may assume that σ has no point mass at -1. Put

$$\mathfrak{U}(x) = \sigma\big(\{e^{is} : s \in (-\pi, x)\}\big), \qquad x \in (-\pi, \pi).$$

Then, doing integration by parts,

$$
\begin{aligned}
U(r) &= \int_{\mathbb{T}} \frac{1 - r^2}{1 + r^2 - 2r\cos t} \, d\sigma(e^{it}) \\
&= \left\{ \frac{1 - r^2}{1 + r^2 - 2r\cos t} \, \mathfrak{U}(t) \right\}\Big|_{t=-\pi}^{\pi} - \int_{-\pi}^{\pi} \frac{\partial}{\partial t}\left\{ \frac{1 - r^2}{1 + r^2 - 2r\cos t} \right\} \mathfrak{U}(t) \, dt \\
&= \frac{1 - r}{1 + r} \, \mathfrak{U}(\pi) + \int_{-\pi}^{\pi} \frac{(1 - r^2)\, 2r \sin t}{(1 + r^2 - 2r\cos t)^2} \, \mathfrak{U}(t) \, dt \\
&= \frac{1 - r}{1 + r} \, \mathfrak{U}(\pi) + \frac{2r}{1 + r} \int_{-\pi}^{\pi} \frac{(1 + r)^2 (1 - r)\, t \sin t}{(1 + r^2 - 2r\cos t)^2} \times \frac{\mathfrak{U}(t) - \mathfrak{U}(-t)}{2t} \, dt.
\end{aligned}
$$

Since

$$\lim_{t \to 0} \frac{\mathfrak{U}(t) - \mathfrak{U}(-t)}{2t} = \lim_{t \to 0} \frac{1}{2t} \int_{-t}^{t} d\sigma(e^{is}) = 0,$$

by Theorem 2.5,

$$\lim_{r \to 1} U(r) = \lim_{r \to 1} \int_{-\pi}^{\pi} F_r(e^{it}) \frac{\mathfrak{U}(t) - \mathfrak{U}(-t)}{2t} \, dt = 0.$$

$\square$

Let $\mu \in \mathcal{M}(\mathbb{T})$. By *Lebesgue's decomposition theorem* there are $u \in L^1(\mathbb{T})$ and a measure $\sigma \in \mathcal{M}(\mathbb{T})$, singular with respect to the Lebesgue measure, such that

$$d\mu(e^{it}) = u(e^{it}) \, dt/2\pi + d\sigma(e^{it}).$$

Moreover,

$$\mu'(e^{it}) = \lim_{\tau \to 0} \frac{\mu\big(\{ e^{is} : s \in (t - \tau, t + \tau) \} \big)}{2\tau}$$

exists and equals $u(e^{it})/2\pi$, for almost all $e^{it} \in \mathbb{T}$. Now, applying Lemmas 3.10 and 3.11, we obtain the following result about the radial limits of $h^1(\mathbb{D})$ functions.

Theorem 3.12 (Fatou) *Let $\mu \in \mathcal{M}(\mathbb{T})$ and let*

$$U(re^{i\theta}) = \int_{\mathbb{T}} \frac{1 - r^2}{1 + r^2 - 2r\cos(\theta - t)} \, d\mu(e^{it}), \qquad (re^{i\theta} \in \mathbb{D}).$$

Then

$$\lim_{r \to 1} U(re^{i\theta}) = 2\pi \, \mu'(e^{i\theta})$$

for almost all $e^{i\theta} \in \mathbb{T}$.

Exercises

Exercise 3.6.1   Let $u(e^{i\theta}) = \theta$, for $0 < \theta < 2\pi$, and let $U(re^{i\theta}) = (P_r * u)(e^{i\theta})$. Show that

$$U(re^{i\theta}) = \pi - 2\arctan\left(\frac{r\sin\theta}{1 - r\cos\theta} \right), \qquad (re^{i\theta} \in \mathbb{D}).$$

Find all possible values of $\lim_{n \to \infty} U(z_n)$, where z_n is a sequence in $\mathbb{D}$ converging to 1.

Exercise 3.6.2   Let $u \in L^1(\mathbb{T})$, and let $U(re^{i\theta}) = (P_r * u)(e^{i\theta})$. Suppose that $\lim_{\theta \to 0+} u(e^{i\theta}) = L^+$, $\lim_{\theta \to 2\pi-} u(e^{i\theta}) = L^-$ and that $L^- \le L^+$.

(a) Let z_n be a sequence in $\mathbb{D}$ such that $\lim_{n\to\infty} z_n = 1$ and $L = \lim_{n\to\infty} U(z_n)$ exists. Show that $L \in [L^-, L^+]$.

(b) Let $L \in [L^-, L^+]$. Show that there exists a sequence $z_n \in \mathbb{D}$ such that $\lim_{n\to\infty} z_n = 1$ and $\lim_{n\to\infty} U(z_n) = L$.

Hint: Consider

$$v(e^{i\theta}) = u(e^{i\theta}) - L^+ + \frac{L^+ - L^-}{2\pi}\theta, \qquad \theta \in (0, 2\pi),$$

and apply Exercise 3.6.1 and Corollary 2.6.

Exercise 3.6.3 Let μ be a real signed Borel measure in $\mathcal{M}(\mathbb{T})$. Let

$$U(re^{i\theta}) = \frac{1}{2\pi}\int_{\mathbb{T}} \frac{1-r^2}{1+r^2 - 2r\cos(\theta - t)}\, d\mu(e^{it}), \qquad (re^{i\theta} \in \mathbb{D}).$$

For each $e^{i\theta_0} \in \mathbb{T}$, define

$$\overline{\mu'}(e^{i\theta_0}) = \limsup_{\tau\to 0} \frac{\mu(\{e^{is} : s \in (\theta_0 - \tau, \theta_0 + \tau)\})}{2\tau},$$

$$\underline{\mu'}(e^{i\theta_0}) = \liminf_{\tau\to 0} \frac{\mu(\{e^{is} : s \in (\theta_0 - \tau, \theta_0 + \tau)\})}{2\tau}.$$

Show that

$$\underline{\mu'}(e^{i\theta_0}) \le \liminf_{r\to 1} U(re^{i\theta_0}) \le \limsup_{r\to 1} U(re^{i\theta_0}) \le \overline{\mu'}(e^{i\theta_0}).$$

Remark: This is a generalization of Theorem 3.12.

Exercise 3.6.4 Let μ be a real signed Borel measure in $\mathcal{M}(\mathbb{T})$. Let

$$U(re^{i\theta}) = \int_{\mathbb{T}} \frac{1-r^2}{1+r^2 - 2r\cos(\theta - t)}\, d\mu(e^{it}), \qquad (re^{i\theta} \in \mathbb{D}).$$

Let $e^{i\theta_0} \in \mathbb{T}$ be such that

$$\mu'(e^{i\theta_0}) = +\infty.$$

Show that

$$\lim_{r\to 1} U(re^{i\theta_0}) = +\infty.$$

Hint: Apply Exercise 3.6.3.

Exercise 3.6.5 Let $\sigma \in \mathcal{M}(\mathbb{T})$ be positive and singular with respect to the Lebesgue measure, and let

$$U(re^{i\theta}) = \int_{\mathbb{T}} \frac{1 - r^2}{1 + r^2 - 2r\cos(\theta - t)} \, d\sigma(e^{it}), \qquad (re^{i\theta} \in \mathbb{D}).$$

Show that

$$\lim_{r \to 1} U(re^{i\theta}) = +\infty \qquad (3.11)$$

for *almost all* $e^{i\theta} \in \mathbb{T}$ (almost all with respect to σ).
Hint: Use Exercise 3.6.4.
Remark 1: If $\sigma \neq 0$, then at least for one point (3.11) holds.
Remark 2: Compare with Lemma 3.11 in which *almost all* is with respect to the Lebesgue measure.

Exercise 3.6.6 Let U be a positive harmonic function on $\mathbb{D}$ such that

$$\lim_{r \to 1} U(re^{i\theta}) = 0$$

for all $e^{i\theta} \in \mathbb{T} \setminus \{1\}$. Show that

$$U(re^{i\theta}) = c \, \frac{1 - r^2}{1 + r^2 - 2r\cos\theta},$$

where c is a positive constant.
Hint: Use Theorem 3.8 and Exercise 3.6.5.

Exercise 3.6.7 Let $\mu \in \mathcal{M}(\mathbb{T})$, and let

$$U(re^{i\theta}) = \int_{\mathbb{T}} \frac{1 - r^2}{1 + r^2 - 2r\cos(\theta - t)} \, d\mu(e^{it}), \qquad (re^{i\theta} \in \mathbb{D}).$$

Suppose that

$$\lim_{r \to 1} U(re^{i\theta}) = 0$$

for *all* $e^{i\theta} \in \mathbb{T}$. Show that $U \equiv 0$.
Hint: Use Theorem 3.12 and Exercise 3.6.5.

Exercise 3.6.8 Let U be a harmonic function on the open unit disc and suppose that

$$\lim_{r \to 1} U(re^{i\theta}) = 0$$

for *all* $e^{i\theta} \in \mathbb{T}$. Can we deduce that $U \equiv 0$?
Hint: Consider

$$U(re^{i\theta}) = \frac{2r(1 - r^2)\sin\theta}{(1 + r^2 - 2r\cos\theta)^2} = -\frac{\partial P}{\partial \theta}(re^{i\theta}).$$

Remark: Compare with Exercise 3.6.7.

Exercise 3.6.9 Let $\mu \in \mathcal{M}(\mathbb{T})$ and let

$$U(re^{i\theta}) = \int_{\mathbb{T}} \frac{1 - r^2}{1 + r^2 - 2r\cos(\theta - t)} \, d\mu(e^{it}), \qquad (re^{i\theta} \in \mathbb{D}).$$

Show that, for almost all $\theta_0 \in \mathbb{T}$,

$$\lim_{\substack{z \to e^{i\theta_0} \\ z \in S_\alpha(\theta_0)}} U(z) = 2\pi\, \mu'(e^{i\theta_0}),$$

where $S_\alpha(\theta_0)$ is the Stoltz domain

$$S_\alpha(\theta_0) = \{\, z \in \mathbb{D} : |z - e^{i\theta_0}| \le C_\alpha\,(1 - |z|) \,\}.$$

(See Figure 3.1.)
Remark 1: $C_\alpha > 1$ is an arbitrary constant. Near the point $e^{i\theta_0}$, the boundaries
of $S_\alpha(\theta_0)$ are tangent to a triangular-shaped region with angle

$$2\alpha = 2\arccos(1/C_\alpha)$$

and vertex at $e^{i\theta_0}$.
Remark 2: We say that $2\pi\,\mu'(e^{i\theta_0})$ is the nontangential limit of U at $e^{i\theta_0}$.

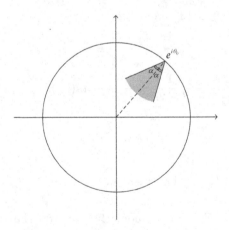

Fig. 3.1. The Stoltz domain $S_\alpha(\theta_0)$.

Exercise 3.6.10 Let μ be a positive Borel measure in $\mathcal{M}(\mathbb{T})$. Let

$$U(re^{i\theta}) = \int_{\mathbb{T}} \frac{1 - r^2}{1 + r^2 - 2r\cos(\theta - t)} \, d\mu(e^{it}), \qquad (re^{i\theta} \in \mathbb{D}).$$

Let $e^{i\theta_0} \in \mathbb{T}$ be such that

$$\mu'(e^{i\theta_0}) = +\infty.$$

Show that, for each $\alpha \geq 0$,

$$\lim_{\substack{z \to e^{i\theta_0} \\ z \in S_\alpha(\theta_0)}} U(z) = +\infty.$$

Remark: This is a generalization of a particular case of Exercise 3.6.4.

Exercise 3.6.11   Construct a real signed Borel measure $\mu \in \mathcal{M}(\mathbb{T})$ such that

$$\mu'(1) = +\infty,$$

but

$$\lim_{t \to 0} U(1 - te^{i\pi/4}) \neq +\infty,$$

where

$$U(re^{i\theta}) = \int_{\mathbb{T}} \frac{1 - r^2}{1 + r^2 - 2r\cos(\theta - t)} \, d\mu(e^{it}), \qquad (re^{i\theta} \in \mathbb{D}).$$

Remark: Compare with Exercises 3.6.4 and 3.6.10.

3.7   Series representation of the harmonic conjugate

Let $\mu \in \mathcal{M}(\mathbb{T})$, and let

$$U(re^{i\theta}) = \int_{\mathbb{T}} \frac{1 - r^2}{1 + r^2 - 2r\cos(\theta - t)} \, d\mu(e^{it}), \qquad (re^{i\theta} \in \mathbb{D}).$$

Then, by Theorem 2.1,

$$U(re^{i\theta}) = \sum_{n=-\infty}^{\infty} \hat{\mu}(n) \, r^{|n|} \, e^{in\theta}, \qquad (re^{i\theta} \in \mathbb{D}).$$

Hence, by Theorem 3.1, the harmonic conjugate of U is given by

$$V(re^{i\theta}) = \sum_{n=-\infty}^{\infty} -i\,\mathrm{sgn}(n)\,\hat{\mu}(n)\,r^{|n|}\,e^{in\theta}, \qquad (re^{i\theta} \in \mathbb{D}).$$

Thus,

$$
\begin{aligned}
V(re^{i\theta}) &= \sum_{n=-\infty}^{\infty} -i\,\mathrm{sgn}(n) \left(\int_{\mathbb{T}} e^{-int} \, d\mu(e^{it}) \right) r^{|n|}\, e^{in\theta} \\
&= \int_{\mathbb{T}} \left(\sum_{n=-\infty}^{\infty} -i\,\mathrm{sgn}(n)\, r^{|n|}\, e^{in(\theta - t)} \right) d\mu(e^{it}).
\end{aligned}
$$

But, a simple calculation shows that

$$Q_r(e^{it}) = \sum_{n=-\infty}^{\infty} -i \operatorname{sgn}(n)\, r^{|n|}\, e^{int} = \frac{2r \sin t}{1 + r^2 - 2r \cos t}. \qquad (3.12)$$

(See Figure 3.2. Figure 3.3 shows the spectrum of $\frac{1}{i} Q_r$.) This function is called the *conjugate Poisson kernel*. Therefore, the harmonic conjugate of U is given by the *conjugate Poisson integral*

$$V(re^{i\theta}) = \frac{1}{2\pi} \int_{-\pi}^{\pi} \frac{2r \sin(\theta - t)}{1 + r^2 - 2r \cos(\theta - t)}\, d\mu(e^{it}), \qquad (re^{i\theta} \in \mathbb{D}). \quad (3.13)$$

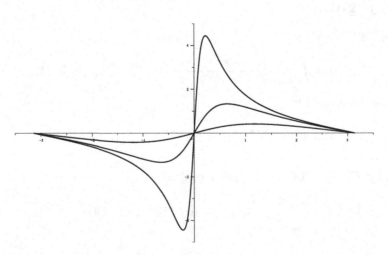

Fig. 3.2. The conjugate Poisson kernel $Q_r(e^{it})$ for $r = 0.2, 0.5, 0.8$.

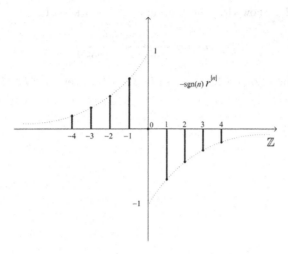

Fig. 3.3. The spectrum of $\frac{1}{i} Q_r$.

We state the preceding result as a theorem which reveals the relation between the harmonic conjugate V and the Fourier coefficients of μ when V is given by the conjugate Poisson integral of μ.

Theorem 3.13 *Let $\mu \in \mathcal{M}(\mathbb{T})$, and let*

$$V(re^{i\theta}) = \int_{\mathbb{T}} \frac{2r\sin(\theta - t)}{1 + r^2 - 2r\cos(\theta - t)} \, d\mu(e^{it}), \qquad (re^{i\theta} \in \mathbb{D}).$$

Then

$$V(re^{i\theta}) = \sum_{n=-\infty}^{\infty} -i\,sgn(n)\,\hat{\mu}(n)\,r^{|n|}\,e^{in\theta}, \qquad (re^{i\theta} \in \mathbb{D}).$$

The series is absolutely and uniformly convergent on compact subsets of $\mathbb{D}$. The function V is the unique harmonic conjugate of

$$U(re^{i\theta}) = \int_{\mathbb{T}} \frac{1 - r^2}{1 + r^2 - 2r\cos(\theta - t)} \, d\mu(e^{it}), \qquad (re^{i\theta} \in \mathbb{D}),$$

on $\mathbb{D}$ with $V(0) = 0$.

Exercises

Exercise 3.7.1 Let $\mu \in \mathcal{M}(\mathbb{T})$, and let $U = P * \mu$ and $V = Q * \mu$. Let $F = U + iV$. Show that

$$F(z) = \int_{\mathbb{T}} \frac{e^{it} + z}{e^{it} - z} \, d\mu(e^{it}), \qquad (z \in \mathbb{D}).$$

Exercise 3.7.2   Show that the conjugate Poisson kernel

$$Q(re^{i\theta}) = \frac{2r \sin \theta}{1 + r^2 - 2r \cos \theta},$$

as a harmonic function on $\mathbb{D}$, is not in $h^1(\mathbb{D})$. However, $Q \in h^p(\mathbb{D})$, $0 < p < 1$.

Chapter 4

Logarithmic convexity

4.1 Subharmonic functions

Let Ω be a topological space. A function

$$\Phi : \Omega \longmapsto [-\infty, \infty)$$

is *upper semicontinuous* on Ω if the set $\{z \in \Omega : \Phi(z) < c\}$ is open for every $c \in \mathbb{R}$. Note that $-\infty$ is included as a possible value for $\Phi(z)$. As a matter of fact, according to our definition, $\Phi \equiv -\infty$ is an upper semicontinuous function.

By a fundamental theorem of real analysis, a continuous function over a compact set is bounded and besides, it attains its maximum and its minimum. A similar, but certainly weaker, result holds for upper semicontinuous functions.

Lemma 4.1 *Let Ω be a topological space, and let $\Phi : \Omega \longmapsto [-\infty, \infty)$ be upper semicontinuous on Ω. Let K be a compact subset of Ω. Then there exists $z_0 \in K$ such that*

$$\Phi(z) \leq \Phi(z_0)$$

for all $z \in K$.

Proof. Clearly

$$K \subset \bigcup_{n=1}^{\infty} \{z \in \Omega : \Phi(z) < n\}$$

and, by definition, each set $\{z \in \Omega : \Phi(z) < n\}$ is an open subset of Ω. Therefore, K has a finite subcover, say

$$K \subset \bigcup_{n=1}^{N} \{z \in \Omega : \Phi(z) < n\},$$

which implies

$$M = \sup_{z \in K} \Phi(z) \leq N < \infty.$$

On the other hand, each set $K_n = \{z \in K : \Phi(z) \geq M - 1/n\}$ is compact, nonempty and $K_n \supset K_{n+1}$, for $n \geq 1$. Therefore, by the finite intersection property for compact sets,

$$\bigcap_{n=1}^{\infty} K_n = \{z \in K : \Phi(z) = M\}$$

is not empty. Any $z_0 \in \bigcap_{n=1}^{\infty} K_n$ is a global maximum point.   □

Upper semicontinuous functions can be pointwise approximated by continuous functions on compact sets. This result plays a key role in developing the theory of subharmonic functions. In our applications, the compact set is usually a circle.

Theorem 4.2 *Let Ω be an open subset of $\mathbb{C}$, and let $\Phi : \Omega \longmapsto [-\infty, \infty)$ be upper semicontinuous on Ω. Let K be a compact subset of Ω. Then there exist continuous functions $\Phi_n : K \longmapsto \mathbb{R}$, $n \geq 1$, such that*

$$\Phi_1 \geq \Phi_2 \geq \cdots \geq \Phi$$

on K and

$$\lim_{n \to \infty} \Phi_n(z) = \Phi(z)$$

for each $z \in K$.

Proof. If $\Phi \equiv -\infty$, simply take $\Phi_n \equiv -n$. Hence, assuming that $\Phi \not\equiv -\infty$, let

$$\Phi_n(z) = \sup_{w \in K} \left(\Phi(w) - n|w - z| \right), \qquad (z \in K).$$

According to Lemma 4.1, for each $n \geq 1$,

$$\Phi_n(z) \leq M = \sup_{w \in K} \Phi(w) < \infty, \qquad (z \in K).$$

Moreover, by the triangle inequality,

$$|\Phi_n(z) - \Phi_n(z')| \leq n|z - z'|$$

for all z, $z' \in K$, and clearly

$$\Phi_1(z) \geq \Phi_2(z) \geq \cdots \geq \Phi(z), \qquad (z \in K).$$

It remains to show that $\Phi_n(z)$ converges to $\Phi(z)$. Fix $z_0 \in K$. Given $\varepsilon > 0$, choose $R > 0$ such that

$$\Phi(z) < \Phi(z_0) + \varepsilon$$

whenever $|z - z_0| < R$. Hence

$$\Phi_n(z_0) \leq \max\{ \Phi(z_0) + \varepsilon, M - nR \}, \qquad (n \geq 1).$$

Therefore, for all $n > M/R$,

$$\Phi(z_0) \leq \Phi_n(z_0) \leq \Phi(z_0) + \varepsilon.$$

□

A function $\Phi : \Omega \longmapsto [-\infty, \infty)$ is *subharmonic* on Ω if it is upper semicontinuous and for each $z \in \Omega$ there exists $r_z > 0$ with

$$\overline{D(z, r_z)} \subset \Omega,$$

such that

$$\Phi(z) \le \frac{1}{2\pi} \int_0^{2\pi} \Phi(z + r\, e^{i\theta})\, d\theta \qquad (4.1)$$

whenever $r < r_z$. Our prototype of a subharmonic function is $\Phi = \log|F|$, where F is an analytic function on Ω. Indeed, Φ is a continuous function on Ω with values in $[-\infty, \infty)$. If $F(z) = 0$, then (4.1) is trivial and if $F(z) \ne 0$, then $\log|F|$ is harmonic in an open neighborhood of z and thus, by Corollary 3.2, equality holds in (4.1) for sufficiently small values of r. The subharmonicity is a local property. In other words, the property (4.1) should be verified in a disc around z whose radius may depend on z. Nevertheless, we will see that (4.1) holds as long as $\overline{D(z, r)} \subset \Omega$. The passage from a local to a global property is a *precious tool* in studying subharmonic functions.

Exercises

Exercise 4.1.1 Let Ω be an open subset of $\mathbb{C}$. Show that $\Phi : \Omega \longmapsto [-\infty, \infty)$ is upper semicontinuous if and only if

$$\limsup_{\substack{w \to z \\ w \ne z}} \Phi(w) \le \Phi(z)$$

for all $z \in \Omega$.
Hint: Remember that

$$\limsup_{\substack{w \to z \\ w \ne z}} \Phi(w) = \lim_{r \to 0} \left(\sup_{0 < |w-z| < r} \Phi(w) \right).$$

Exercise 4.1.2 Let Ω be an open subset of $\mathbb{C}$, let $z_0 \in \Omega$, and let $\Phi : \Omega \longmapsto [-\infty, \infty)$ be upper semicontinuous. Suppose that $\Phi(z_0) = -\infty$. Show that Φ is continuous at z_0, i.e.

$$\lim_{z \to z_0} \Phi(z) = \Phi(z_0).$$

Exercise 4.1.3 (Dini's theorem) Let K be a compact set. Let $\Phi_n : K \longmapsto \mathbb{R}$, $n \ge 1$, and $\Phi : K \longmapsto \mathbb{R}$ be continuous functions on K. Suppose that

$$\Phi_1(z) \ge \Phi_2(z) \ge \cdots \ge \Phi(z)$$

and that

$$\lim_{n \to \infty} \Phi_n(z) = \Phi(z)$$

for each $z \in K$. Show that Φ_n converges uniformly to Φ on K.

Remark: Here we assumed that Φ is continuous on K. Hence, in general, we cannot apply Dini's theorem to conclude that the sequence $(\Phi_n)_{n \geq 1}$ given in Theorem 4.2 converges uniformly to Φ on K.

Exercise 4.1.4   Let Ω be an open subset of $\mathbb{C}$ and let U be harmonic on Ω. Let $1 \leq p < \infty$. Show that $|U|^p$ is subharmonic on Ω.

Hint: Use Corollary 3.2 and Hölder's inequality. Jensen's inequality can be applied too.

Exercise 4.1.5   Let Ω be an open subset of $\mathbb{C}$, let $\Phi : \Omega \longrightarrow [-\infty, \infty)$ be subharmonic and let $\Psi : [-\infty, \infty) \longrightarrow [-\infty, \infty)$ be nondecreasing, continuous and convex on $(-\infty, \infty)$. Show that $\Psi \circ \Phi : \Omega \longrightarrow [-\infty, \infty)$ is subharmonic. In particular, $\exp(\Phi)$ and Φ^+ are subharmonic functions.

Hint: Use Jensen's inequality.

Remark: We assume $\exp(-\infty) = 0$.

Exercise 4.1.6   Let Ω be an open subset of $\mathbb{C}$ and let F be analytic on Ω. Let $0 < p < \infty$. Show that $|F|^p$ and $\log^+ |F|$ are subharmonic on Ω.

Hint: Use Exercise 4.1.5 and the fact that $\Phi = \log |F|^p$ is subharmonic.

Remark: The case $1 \leq p < \infty$ also follows from Exercise 4.1.4.

Exercise 4.1.7   Let U be harmonic on $\mathbb{D}$. Show that $U \in h^p(\mathbb{D})$, $1 \leq p < \infty$, if and only if the subharmonic function $|U|^p$ has a harmonic majorant on $\mathbb{D}$, i.e. there is a harmonic function $V \in h(\mathbb{D})$ such that

$$|U(z)|^p \leq V(z), \qquad (re^{i\theta} \in \mathbb{D}).$$

Remark: The function V is called a harmonic majorant of $|U|^p$. Compare with Exercise 11.4.1.

4.2   The maximum principle

If F is analytic on Ω and continuous on $\overline{\Omega}$, the maximum principle for analytic functions says that

$$\max_{z \in \overline{\Omega}} |F(z)| = \max_{\zeta \in \partial\Omega} |F(\zeta)|.$$

In this section we show that this fundamental result is also fulfilled by subharmonic functions.

Theorem 4.3 (Maximum principle for open sets) *Let Ω be a domain in $\mathbb{C}$, and let Φ be subharmonic on Ω. Suppose that there is a $z_0 \in \Omega$ such that*

$$\Phi(z) \leq \Phi(z_0)$$

for all $z \in \Omega$. Then Φ is constant.

Proof. Let $M = \Phi(z_0)$, and let

$$\Omega_1 = \{\, z \in \Omega : \Phi(z) < M \,\}$$

and

$$\Omega_2 = \{\, z \in \Omega : \Phi(z) = M \,\}.$$

Clearly $\Omega = \Omega_1 \cup \Omega_2$, $\Omega_1 \cap \Omega_2 = \emptyset$ and $\Omega_2 \neq \emptyset$. Hence, if we show that Ω_1 and Ω_2 are open subsets of Ω, then the connectivity of Ω forces $\Omega_1 = \emptyset$, and thus $\Phi \equiv M$.

Since Φ is upper semicontinuous, by definition, Ω_1 is open. On the other hand, for any $z \in \Omega_2$, there exists $r_z > 0$ with $\overline{D(z, r_z)} \subset \Omega$, such that

$$M = \Phi(z) \leq \frac{1}{2\pi} \int_0^{2\pi} \Phi(z + r\, e^{i\theta})\, d\theta,$$

whenever $r < r_z$. But, by assumption, we also have $\Phi(z + r\, e^{i\theta}) \leq M$ for all θ. If for some θ_0, $\Phi(z + r\, e^{i\theta_0}) < M$ holds, then, by upper semicontinuity, $\Phi(z + r\, e^{i\theta}) < M$ must hold on some open arc around θ_0, which would imply

$$\frac{1}{2\pi} \int_0^{2\pi} \Phi(z + r\, e^{i\theta})\, d\theta < M.$$

Therefore, $\Phi(z + r\, e^{i\theta}) = M$, for all θ and all $0 \leq r < r_z$. In other words, $D(z, r_z) \subset \Omega_2$. Hence, Ω_2 is also open. $\qquad\square$

In the following, $\partial\Omega$ denotes the boundary of Ω as a subset of $\mathbb{C}$. Hence, if Ω is unbounded, we do not assume that $\infty \in \partial\Omega$. Therefore, whenever a property has to be satisfied at boundary points of Ω and also at infinity, the latter requirement is explicitly expressed.

Corollary 4.4 *Let Ω be a domain in $\mathbb{C}$, and let Φ be subharmonic on Ω. Suppose that*

$$\limsup_{\substack{z \to \zeta \\ z \in \Omega}} \Phi(z) \leq 0$$

for all $\zeta \in \partial\Omega$. If Ω is unbounded, we also assume that

$$\limsup_{\substack{z \to \infty \\ z \in \Omega}} \Phi(z) \leq 0. \tag{4.2}$$

Then

$$\Phi(z) \leq 0, \qquad (z \in \Omega).$$

Proof. Let $M = \sup_{z \in \Omega} \Phi(z)$, and suppose that $M > 0$ (possibly $+\infty$). Then there is a sequence $(z_n)_{n \geq 1}$ in Ω such that $\lim_{n \to \infty} \Phi(z_n) = M$. Since we assumed that

$$\limsup_{\substack{z \to \zeta \\ z \in \Omega}} \Phi(z) \leq 0$$

for all $\zeta \in \partial\Omega$, and also for $\zeta = \infty$ if Ω is unbounded, the sequence $(z_n)_{n\geq 1}$ is bounded and all its accumulation points are inside Ω. But, if $z_0 \in \Omega$ and $(z_{n_k})_{k\geq 1}$ converges to z_0, then

$$M = \lim_{k\to\infty} \Phi(z_{n_k}) \leq \limsup_{\substack{z\to z_0 \\ z\in\Omega}} \Phi(z) \leq \Phi(z_0) \leq M.$$

Thus $\Phi(z_0) = M$, which means z_0 is a global maximum point. Therefore, by Theorem 4.3, $\Phi \equiv M > 0$, which contradicts our assumptions about the behavior of Φ as we approach the boundary points of Ω. Hence, $M \leq 0$. $\square$

The condition (4.2) cannot be relaxed for unbounded domains. For example, $\Phi(x + iy) = y$ is harmonic in the upper half plane $\mathbb{C}_+$ and

$$\limsup_{\substack{z\to t \\ z\in\mathbb{C}_+}} \Phi(z) = 0$$

for all $t \in \mathbb{R}$. However, $\Phi > 0$ on $\mathbb{C}_+$. Nevertheless, we can sometimes replace (4.2) by another condition. The following corollary is a result of this type.

Corollary 4.5 *Let Ω be a proper unbounded domain in $\mathbb{C}$, and let Φ be subharmonic and bounded above on Ω. Suppose that*

$$\limsup_{\substack{z\to\zeta \\ z\in\Omega}} \Phi(z) \leq 0$$

for all $\zeta \in \partial\Omega$. Then
$$\Phi(z) \leq 0, \qquad (z \in \Omega).$$

Proof. Without loss of generality, assume that $0 \in \partial\Omega$. Given $\varepsilon > 0$, there is $0 < R < 1/2$ such that $\Phi(z) \leq \varepsilon$ for all $z \in \Omega$ with $|z| \leq R$. Since

$$\frac{\log(|z|/R)}{\log(1/R)}$$

is a positive harmonic function on $\{\, |z| > R \,\}$,

$$\Psi(z) = \Phi(z) - \varepsilon - \varepsilon \, \frac{\log(|z|/R)}{\log(1/R)}$$

is subharmonic on $\Omega_R = \Omega \cap \{\, |z| > R \,\}$, and

$$\limsup_{\substack{z\to\zeta \\ z\in\Omega_R}} \Psi(z) \leq 0$$

for all $\zeta \in \partial\Omega_R$. The term $-\varepsilon$ in the definition of Ψ is added to ensure that this property holds even if $\zeta \in \partial\Omega_R$ and $|\zeta| = R$. Moreover, since Φ is bounded above on Ω,

$$\limsup_{\substack{z\to\infty \\ z\in\Omega_R}} \Psi(z) = -\infty \leq 0.$$

Hence, by Corollary 4.4, $\Psi(z) \leq 0$ for all $z \in \Omega_R$, which gives

$$\Phi(z) \leq \varepsilon + \varepsilon \, \frac{\log^+(|z|/R)}{\log(1/R)}$$

for all $z \in \Omega_R$. Again thanks to the extra ε that we added from the beginning, the last inequality is actually valid for *all* $z \in \Omega$. Let $\varepsilon \to 0$ (which may force $R \to 0$). Nevertheless, we get $\Phi(z) \leq 0$ for all $z \in \Omega$. $\qquad\square$

Exercises

Exercise 4.2.1 [Maximum principle for compact sets] Let Ω be a bounded domain in $\mathbb{C}$, and let Φ be upper semicontinuous on $\overline{\Omega}$ and subharmonic on Ω. Show that there is a $\zeta_0 \in \partial\Omega$ such that

$$\Phi(z) \leq \Phi(\zeta_0)$$

for all $z \in \overline{\Omega}$.
Hint: Apply Lemma 4.1 and Theorem 4.3.

Exercise 4.2.2 Let

$$\Phi(x + iy) = \begin{cases} 0 & \text{if} \quad x < 0, \\[2mm] x & \text{if} \quad x \geq 0. \end{cases}$$

Show that Φ is subharmonic on $\mathbb{C}$.
Hint: Use Exercise 4.3.1, part (a).
Remark: Every point $z = x + iy$, $x < 0$, is a local maximum. Is that a contradiction to the maximum principle?

Exercise 4.2.3 Let Ω be a bounded domain in $\mathbb{C}$, and let U be a real harmonic function on Ω. Suppose that

$$\lim_{\substack{z \to \zeta \\ z \in \Omega}} U(z) = 0$$

for all $\zeta \in \partial\Omega$. Show that $U \equiv 0$.
Hint: Apply Corollary 4.4 to U and $-U$.

Exercise 4.2.4 Let Ω be a proper unbounded domain in $\mathbb{C}$, and let F be a bounded analytic function on Ω. Suppose that

$$\limsup_{\substack{z \to \zeta \\ z \in \Omega}} |F(z)| \leq 1$$

for all $\zeta \in \partial\Omega$. Show that

$$|F(z)| \le 1$$

for all $z \in \Omega$.
Hint: Apply Corollary 4.5 to $\log|F|$.

Exercise 4.2.5 [Gauss's theorem] Let Ω be an open subset of $\mathbb{C}$, and let U be a real continuous function on Ω. Suppose that U satisfies the mean value property, i.e. if $\overline{D(z,r)} \subset \Omega$, then

$$U(z) = \frac{1}{2\pi} \int_{-\pi}^{\pi} U(z + re^{it})\, dt.$$

Show that U is harmonic on Ω.
Hint: The proof of Theorem 4.3 might help.

Exercise 4.2.6 [Schwarz's reflection principle] Let Ω be a domain in $\mathbb{C}$ symmetric with respect to the real axis. Suppose that F is analytic on $\Omega \cap \mathbb{C}_+$ and, for each $t \in \Omega \cap \mathbb{R}$,

$$\lim_{\substack{z \to t \\ z \in \Omega \cap \mathbb{C}_+}} \Im F(z) = 0.$$

Show that there is a unique analytic function G on Ω such that $G = F$ on $\Omega \cap \mathbb{C}_+$.
Hint: Put

$$V(z) = \begin{cases} \Im F(z) & \text{if } z \in \Omega \cap \mathbb{C}_+, \\ 0 & \text{if } z \in \Omega \cap \mathbb{R}, \\ -\Im F(\bar{z}) & \text{if } z \in \Omega \cap \mathbb{C}_-, \end{cases}$$

and apply Exercise 4.2.5 to show that V is harmonic on Ω.

4.3   A characterization of subharmonic functions

Corollary 4.4 enables us to provide an equivalent definition of subharmonic functions which justifies the title *subharmonic*.

Theorem 4.6 *Let Ω be an open subset of $\mathbb{C}$, and let $\Phi : \Omega \longmapsto [-\infty, \infty)$ be upper semicontinuous. Then the following are equivalent:*

(a) Φ *is subharmonic on Ω;*

(b) *for any bounded subdomain Ξ with $\overline{\Xi} \subset \Omega$, and for any harmonic function U on Ξ, if*

$$\limsup_{\substack{z \to \zeta \\ z \in \Xi}} \left(\Phi(z) - U(z) \right) \le 0$$

for every $\zeta \in \partial\Xi$, *then we have*

$$\Phi(z) \le U(z), \qquad (z \in \Xi);$$

(c) *if* $z \in \Omega$, *and* $\overline{D(z,R)} \subset \Omega$, *then*

$$\Phi(z + re^{i\theta}) \le \frac{1}{2\pi} \int_{-\pi}^{\pi} \frac{R^2 - r^2}{R^2 + r^2 - 2Rr\cos(\theta - t)} \, \Phi(z + Re^{it}) \, dt$$

for all $0 \le r < R$ *and all* θ.

Proof. $(a) \Longrightarrow (b)$: It follows from Corollary 4.4.

$(b) \Longrightarrow (c)$: By Theorem 4.2, there exist continuous functions Φ_n, $n \ge 1$, defined on the circle $C_R = \{\zeta : |\zeta - z| = R\}$, so that $\Phi_n(\zeta) \ge \Phi_{n+1}(\zeta)$ and $\lim_{n\to\infty} \Phi_n(\zeta) = \Phi(\zeta)$, for all $\zeta \in C_R$. By Corollary 2.3, if we also define

$$\Phi_n(z + re^{i\theta}) = \frac{1}{2\pi} \int_{-\pi}^{\pi} \frac{R^2 - r^2}{R^2 + r^2 - 2Rr\cos(\theta - t)} \, \Phi_n(z + Re^{it}) \, dt$$

for $z + re^{i\theta} \in D(z, R)$, then Φ_n is continuous on the closed disc $\overline{D(z,R)}$, and besides it is harmonic inside that disc. Since Φ is upper semicontinuous, for each $\zeta \in \partial D_R = C_R$, we have

$$\limsup_{\substack{w \to \zeta \\ w \in D_R}} \left(\Phi(w) - \Phi_n(w) \right) \le \Phi(\zeta) - \Phi_n(\zeta) \le 0.$$

Hence, by assumption (b), $\Phi(z + re^{i\theta}) \le \Phi_n(z + re^{i\theta})$, i.e.

$$\Phi(z + re^{i\theta}) \le \frac{1}{2\pi} \int_{-\pi}^{\pi} \frac{R^2 - r^2}{R^2 + r^2 - 2Rr\cos(\theta - t)} \, \Phi_n(z + Re^{it}) \, dt$$

for all $0 \le r < R$ and all θ. Let $n \to \infty$ and use the monotone convergence theorem to get (c).

$(c) \Longrightarrow (a)$: Put $r = 0$. $\qquad \square$

Exercises

Exercise 4.3.1 Let Φ_1 and Φ_2 be subharmonic functions on an open set Ω. Show that

(a) $\max\{\,\Phi_1, \Phi_2\,\}$ and

(b) $\alpha_1\,\Phi_1 + \alpha_2\,\Phi_2$, where $\alpha_1, \alpha_2 \ge 0$,

are also subharmonic on Ω. Is $\min\{\,\Phi_1, \Phi_2\,\}$ necessarily subharmonic on Ω?

Exercise 4.3.2   Let Ω be an open subset of $\mathbb{C}$, and let $\Phi : \Omega \longmapsto \mathbb{R}$ be twice continuously differentiable on Ω. Show that Φ is subharmonic on Ω if and only if $\nabla^2 \Phi \geq 0$ on Ω.
Hint: Use Theorem 4.6(b) with the upper semicontinuous function

$$
\Psi_\varepsilon(z) = \begin{cases} \Phi(z) - U(z) + \varepsilon |z|^2 & \text{if} \quad z \in \Xi, \\[2mm] \varepsilon |z|^2 & \text{if} \quad z \in \partial\Xi. \end{cases}
$$

Exercise 4.3.3   Let Ω_1, Ω_2 be open subsets of $\mathbb{C}$, let $F : \Omega_1 \longrightarrow \Omega_2$ be analytic, and let $\Phi : \Omega_2 \longmapsto [-\infty, \infty)$ be subharmonic on Ω_2. Show that $\Phi \circ F : \Omega_1 \longmapsto [-\infty, \infty)$ is subharmonic on Ω_1.
Hint: Use Exercise 1.1.4 and Theorem 4.6, part (b).

4.4   Various means of subharmonic functions

Given a *finite* family of subharmonic functions, by taking positive linear combinations or by taking their supremum, we can create new subharmonic functions. In this section, we show that under certain conditions the preceding two procedures still create subharmonic functions even if the family is not finite. This topic is thoroughly treated in [17].

In the following theorem we assume that T is any compact topological space. But in our applications, it is always the unit circle $\mathbb{T}$.

Theorem 4.7 *Let T be a compact topological space, and let Ω be an open subset of $\mathbb{C}$. Suppose that $\Phi : \Omega \times T \longmapsto [-\infty, \infty)$ has the following properties:*

(i) *Φ is upper semicontinuous on $\Omega \times T$;*

(ii) *for each fixed $t \in T$,*

$$
\Phi_t(z) = \Phi(z, t), \qquad (z \in \Omega),
$$

as a function of z, is subharmonic on Ω.

Let
$$
\Phi(z) = \sup_{t \in T} \Phi_t(z), \qquad (z \in \Omega).
$$

Then Φ is subharmonic on Ω.

Proof. First, let us show that Φ is upper semicontinuous. Fix $c \in \mathbb{R}$, and suppose that $\Phi(z_0) < c$. Then, according to the definition of Φ, we have $\Phi(z_0, t) < c$ for all $t \in T$. Since Φ is upper semicontinuous on $\Omega \times T$, for each $t \in T$, there exist $r_t > 0$ and an open set $V_t \subset T$ with $t \in V_t$ such that

$$
D(z_0, r_t) \times V_t \subset \{ (z, \tau) : \Phi(z, \tau) < c \}.
$$

Since $T = \bigcup_t V_t$, and T is a compact set, T has a finite subcover, say

$$T = V_{t_1} \cup \cdots \cup V_{t_n}$$

for some $t_1, t_2, \ldots, t_n \in T$. Let $r = \min\{\, r_{t_1}, r_{t_2}, \ldots, r_{t_n} \,\}$ and note that $r > 0$. Hence

$$D(z_0, r) \times V_{t_k} \subset \{\, (z, \tau) : \Phi(z, \tau) < c - \varepsilon \,\}$$

for $k = 1, 2, \ldots, n$. Therefore, taking the union over k, we obtain

$$D(z_0, r) \times T \subset \{\, (z, \tau) : \Phi(z, \tau) < c \,\},$$

which implies

$$D(z_0, r) \subset \{\, z : \Phi(z) < c \,\}.$$

Note that $\{z\} \times T$ is compact and thus, by Lemma 4.1,

$$\Phi(z) = \sup_{\tau \in T} \Phi(z, \tau) < c$$

provided that $\Phi(z, \tau) < c$ holds for all $\tau \in T$. Hence Φ is upper semicontinuous.

Secondly, we show that Φ is subharmonic. Fix $z_0 \in \Omega$, and assume that $\overline{D(z_0, R)} \subset \Omega$. Since Φ_t is subharmonic, Theorem 4.6 implies

$$\Phi_t(z_0) \le \frac{1}{2\pi} \int_{-\pi}^{\pi} \Phi_t(z_0 + re^{i\theta})\, d\theta$$

for all $r < R$. Appealing to Theorem 4.6 is a crucial step in the proof. Since otherwise, based on the original definition of subharmonicity, the last inequality would be valid for $r < R = R(t)$ and we have no control over $R(t)$. Hence,

$$\Phi_t(z_0) \le \frac{1}{2\pi} \int_{-\pi}^{\pi} \Phi(z_0 + re^{i\theta})\, d\theta$$

for all $r < R$. Now, once more, take the supremum over $t \in T$ to get the required result. □

A subharmonic function Ψ on D_R is called *radial* if

$$\Psi(z) = \Psi(|z|)$$

for all $z \in D_R$. Given a subharmonic function on a disc, the preceding result enables us to create a radial subharmonic function by taking its supremum over each circle.

Corollary 4.8 *Let Φ be subharmonic on the disc D_R. Let*

$$\Psi(z) = \sup_{e^{i\theta} \in \mathbb{T}} \Phi(|z|e^{i\theta}), \qquad (z \in D_R).$$

Then Ψ is a radial subharmonic function on D_R.

Proof. The function

$$
\begin{aligned}
D_R \times \mathbb{T} &\longrightarrow [-\infty, \infty) \\
(z, e^{i\theta}) &\longmapsto \Phi(ze^{i\theta})
\end{aligned}
$$

fulfils all the requirements of Theorem 4.7. Hence,

$$
\Psi(z) = \sup_{e^{i\theta} \in \mathbb{T}} \Phi(ze^{i\theta}) = \sup_{e^{i\theta} \in \mathbb{T}} \Phi(|z|e^{i\theta})
$$

is subharmonic on D_R. The last identity also shows that $\Psi(z) = \Psi(|z|)$.   □

In the following result we assume that $(T, \mathfrak{M}, \mu)$ is any measure space with μ a finite positive measure on T. But in our applications, T is either the unit circle $\mathbb{T}$ equipped with the normalized Lebesgue measure $d\theta/2\pi$, or is the unit disc $\mathbb{D}$ with the two-dimensional normalized Lebesgue measure $rdrd\theta/\pi$.

Theorem 4.9 *Let $(T, \mathfrak{M}, \mu)$ be a measure space, $\mu \geq 0$, $\mu(T) < \infty$, and let Ω be an open subset of $\mathbb{C}$. Suppose that $\Phi : \Omega \times T \longmapsto [-\infty, \infty)$ has the following properties:*

(i) Φ is measurable on $\Omega \times T$;

(ii) for each fixed $t \in T$, $\Phi(z, t)$, as a function of z, is subharmonic on Ω;

(iii) for each $z \in \Omega$, there exists $r_z > 0$ with $\overline{D(z, r_z)} \subset \Omega$ such that

$$
\sup_{\overline{D(z,r_z)} \times T} \Phi(w, t) < \infty.
$$

Let

$$
\Phi(z) = \int_T \Phi(z, t) \, d\mu(t), \qquad (z \in \Omega).
$$

Then Φ is subharmonic on Ω.

Proof. Without loss of generality, assume that μ is a probability measure. We start by showing that Φ is upper semicontinuous on Ω. Fix $z \in \Omega$, and let

$$
M_z = \sup_{\overline{D(z,r_z)} \times T} \Phi(w, t).
$$

First of all, assumption (iii) shows that

$$
\Phi(z) \leq M_z < \infty.
$$

Secondly, since $\Phi(z,t)$ is subharmonic on Ω, by Fatou's lemma,

$$
\begin{aligned}
M_z - \limsup_{w \to z} \Phi(w) &= \liminf_{w \to z} \left(M_z - \Phi(w) \right) \\
&= \liminf_{w \to z} \int_T \left(M_z - \Phi(w,t) \right) d\mu(t) \\
&\geq \int_T \liminf_{w \to z} \left(M_z - \Phi(w,t) \right) d\mu(t) \\
&= \int_T \left(M_z - \limsup_{w \to z} \Phi(w,t) \right) d\mu(t) \\
&\geq \int_T \left(M_z - \Phi(z,t) \right) d\mu(t) = M_z - \Phi(z).
\end{aligned}
$$

Hence, we have

$$
\limsup_{w \to z} \Phi(w) \leq \Phi(z), \qquad (z \in \Omega),
$$

which ensures that Φ is upper semicontinuous on Ω. It remains to show that Φ fulfils the submean inequality. Let $r < r_z$ and let

$$
I_z(r) = \frac{1}{2\pi} \int_{-\pi}^{\pi} \Phi(z + re^{i\theta}) \, d\theta.
$$

Then, by Fubini's theorem,

$$
\begin{aligned}
M_z - I_z(r) &= \frac{1}{2\pi} \int_{-\pi}^{\pi} \left(M_z - \Phi(z + re^{i\theta}) \right) d\theta \\
&= \frac{1}{2\pi} \int_{-\pi}^{\pi} \left\{ \int_T \left(M_z - \Phi(z + re^{i\theta}, t) \right) d\mu(t) \right\} d\theta \\
&= \int_T \left\{ \frac{1}{2\pi} \int_{-\pi}^{\pi} \left(M_z - \Phi(z + re^{i\theta}, t) \right) d\theta \right\} d\mu(t) \\
&\leq \int_T \left(M_z - \Phi(z,t) \right) d\mu(t) = M_z - \Phi(z).
\end{aligned}
$$

Note that the integrand is nonnegative and thus Fubini's theorem can be applied. Thus, for all $r < r_z$,

$$
\Phi(z) \leq \frac{1}{2\pi} \int_{-\pi}^{\pi} \Phi(z + re^{i\theta}) \, d\theta.
$$

□

Given a subharmonic function on a disc, the preceding result enables us to create a radial subharmonic function by taking its integral means over each circle.

Corollary 4.10 *Let Φ be subharmonic on the disc D_R, and let*

$$
\Psi(z) = \frac{1}{2\pi} \int_{-\pi}^{\pi} \Phi(ze^{i\theta}) \, d\theta, \qquad (z \in D_R).
$$

Then Ψ is a radial subharmonic function on D_R.

Proof. The function

$$
\begin{aligned}
D_R \times \mathbb{T} &\longrightarrow [-\infty, \infty) \\
(z, e^{i\theta}) &\longmapsto \Phi(ze^{i\theta})
\end{aligned}
$$

fulfils all the requirements of Theorem 4.9. Hence,

$$
\Psi(z) = \frac{1}{2\pi} \int_{-\pi}^{\pi} \Phi(ze^{i\theta})\, d\theta = \frac{1}{2\pi} \int_{-\pi}^{\pi} \Phi(|z|e^{i\theta})\, d\theta
$$

is subharmonic on D_R. The last identity also shows that $\Psi(z) = \Psi(|z|)$. $\square$

Exercises

Exercise 4.4.1   Let Φ be subharmonic on the disc D_R. Let

$$
\Psi(z) = \frac{1}{\pi} \int_{-\pi}^{\pi} \int_0^1 \Phi(z\rho e^{i\theta})\, \rho d\rho d\theta, \qquad (z \in D_R).
$$

Show that Ψ is a radial subharmonic function on D_R.
Hint: Consider the function

$$
\begin{aligned}
D_R \times \mathbb{D} &\longrightarrow [-\infty, \infty) \\
(z, \rho e^{i\theta}) &\longmapsto \Phi(z\rho e^{i\theta})
\end{aligned}
$$

and apply Theorem 4.9.
Remark: By a simple change of variable, we also have

$$
\Psi(z) = \frac{1}{\pi|z|^2} \int_{-\pi}^{\pi} \int_0^{|z|} \Phi(\rho e^{i\theta})\, \rho d\rho d\theta,
$$

if $z \neq 0$.

Exercise 4.4.2   Let Ω be a domain in $\mathbb{C}$ and define

$$
\Phi(z) = -\log \operatorname{dist}(z, \partial\Omega), \qquad (z \in \Omega),
$$

where

$$
\operatorname{dist}(z, \partial\Omega) = \inf_{\zeta \in \partial\Omega} |\zeta - z|.
$$

Show that Φ is subharmonic on Ω.
Hint: Use Theorem 4.7.

4.5 Radial subharmonic functions

Let Φ be subharmonic on D_R, and let

$$
\begin{aligned}
\mathcal{M}_\Phi(r) &= \sup_{|z|=r} \Phi(z), \\
\mathcal{C}_\Phi(r) &= \frac{1}{2\pi} \int_{-\pi}^{\pi} \Phi(re^{i\theta})\, d\theta
\end{aligned}
$$

for $0 \le r < R$. In Corollaries 4.8 and 4.10, we saw that $\Psi(z) = \mathcal{M}_\Phi(|z|)$ and $\Psi(z) = \mathcal{C}_\Phi(|z|)$ are radial subharmonic functions on D_R. The following result characterizes all radial subharmonic functions.

We say that $\Psi(r)$ is a convex function of $\log r$, if $\Psi(e^\rho)$ is a convex function of ρ. In other words, Ψ satisfies

$$
\Psi(r) \le \frac{\log r_2 - \log r}{\log r_2 - \log r_1}\, \Psi(r_1) + \frac{\log r - \log r_1}{\log r_2 - \log r_1}\, \Psi(r_2),
$$

whenever $r_1 < r < r_2$, and of course r, r_1, r_2 are in the domain of definition of Ψ.

Theorem 4.11 *Let Ψ be a radial subharmonic function on D_R. Then $\Psi(r)$, $0 < r < R$, is an increasing convex function of $\log r$, and $\lim_{r \to 0} \Psi(r) = \Psi(0)$.*

Proof. Let $0 < r_1 < r_2 < R$. Since Ψ is subharmonic and radial,

$$
\Psi(0) \le \frac{1}{2\pi} \int_{-\pi}^{\pi} \Psi(r_1 e^{i\theta})\, d\theta = \Psi(r_1). \tag{4.3}
$$

Also, by the maximum principle,

$$
\Psi(r_1) \le \sup_{|z|=r_2} \Psi(z) = \Psi(r_2).
$$

Thus Ψ is increasing. Moreover, by (4.3), and the fact that Ψ is upper semicontinuous on D_R, we have

$$
\Psi(0) \le \liminf_{r \to 0} \Psi(r) \le \limsup_{r \to 0} \Psi(r) \le \Psi(0).
$$

Therefore, Ψ is continuous at the origin.

To prove that $\Psi(r)$ is a convex function of $\log r$, fix r_1 and r_2, and let

$$
\Upsilon(z) = \Psi(z) - \frac{\log r_2 - \log |z|}{\log r_2 - \log r_1}\, \Psi(r_1) - \frac{\log |z| - \log r_1}{\log r_2 - \log r_1}\, \Psi(r_2)
$$

for

$$
z \in A(r_1, r_2) = \{\, z : r_1 < |z| < r_2 \,\}.
$$

Then Υ is a subharmonic function on the annulus $A(r_1, r_2)$, and

$$
\limsup_{\substack{z \to \zeta \\ z \in A(r_1, r_2)}} \Upsilon(z) \le 0
$$

for all $\zeta \in \partial A(r_1, r_2)$, where $\partial A(r_1, r_2)$ consists of two circles $\{|\zeta| = r_1\}$ and $\{|\zeta| = r_2\}$. Therefore, by the maximum principle (Corollary 4.4), $\Upsilon(z) \leq 0$ for all $z \in A(r_1, r_2)$. This fact is equivalent to

$$\Psi(r) \leq \frac{\log r_2 - \log r}{\log r_2 - \log r_1} \Psi(r_1) + \frac{\log r - \log r_1}{\log r_2 - \log r_1} \Psi(r_2),$$

whenever $r_1 < r < r_2$.   □

Corollary 4.12 *Let Φ be a subharmonic function on D_R. Then*

(a) $\mathcal{M}_\Phi(r)$ and $\mathcal{C}_\Phi(r)$ are increasing convex functions of $\log r$,

(b) $\lim_{r \to 0} \mathcal{M}_\Phi(r) = \lim_{r \to 0} \mathcal{C}_\Phi(r) = \Phi(0)$,

(c) $\Phi(0) \leq \mathcal{C}_\Phi(r) \leq \mathcal{M}_\Phi(r)$, for all $0 < r < R$.

Proof. The first two claims are direct consequences of Corollaries 4.8 and 4.10 and Theorem 4.11. Moreover, directly from the definition, $\mathcal{C}_\Phi(r) \leq \mathcal{M}_\Phi(r)$, and by (4.3), $\Phi(0) \leq \mathcal{C}_\Phi(r)$ for $0 < r < R$.   □

Let $u \in L^p(\mathbb{T})$, $1 \leq p \leq \infty$, and let $U = P * u$. According to Corollaries 2.11, 2.15 and 2.19, we have $U \in h^p(\mathbb{D})$ and

$$\|U\|_p = \sup_{0 \leq r < 1} \|U_r\|_p = \lim_{r \to 1} \|U_r\|_p = \|u\|_p.$$

However, based on Corollary 4.12, we can say more about the behavior of $\|U_r\|_p$ as a function of r.

Corollary 4.13 *Let $u \in L^p(\mathbb{T})$, $1 \leq p \leq \infty$, and let $U = P * u$. Then $\|U_r\|_p$ is an increasing convex function of $\log r$ and*

$$\lim_{r \to 1} \|U_r\|_p = \|U\|_p.$$

Let $F \in H^p(\mathbb{D})$ with $0 < p \leq \infty$. If $1 \leq p \leq \infty$ then the preceding result applies and thus $\|F_r\|_p$ is an increasing function of r with

$$\lim_{r \to 1} \|F_r\|_p = \|F\|_p = \|f\|_p,$$

where f represents the boundary values of F. If $0 < p < 1$, first of all, we have not yet shown that F has boundary values almost everywhere on $\mathbb{T}$. At this point, we do not have enough tools to prove this fact. However, since $|F|^p$ is a subharmonic function on $\mathbb{D}$, even if $0 < p < 1$, Corollary 4.12 applies and we obtain the following result.

Corollary 4.14 *Let $F \in H^p(\mathbb{D})$, $0 < p \leq \infty$. Then $\|F_r\|_p$ is an increasing convex function of $\log r$ and*

$$\lim_{r \to 1} \|F_r\|_p = \|F\|_p.$$

Exercises

Exercise 4.5.1   Let Φ be subharmonic on D_R and let

$$\mathcal{B}_\Phi(r) = \frac{1}{\pi r^2} \int_{-\pi}^{\pi} \int_0^r \Phi(\rho e^{i\theta}) \, \rho \, d\rho \, d\theta$$

for $0 < r < R$. Show that $\mathcal{B}_\Phi(r)$ is an increasing convex function of $\log r$,

$$\Phi(0) \leq \mathcal{B}_\Phi(r) < \mathcal{C}_\Phi(r) \leq \mathcal{M}_\Phi(r)$$

for all $0 < r < R$, and that

$$\lim_{r \to 0} \mathcal{B}_\Phi(r) = \Phi(0).$$

Exercise 4.5.2   Let Ψ be a radial function on D_R. Suppose that $\Psi(r)$, $0 < r < R$, is an increasing convex function of $\log r$ and that $\lim_{r \to 0} \Psi(r) = \Psi(0)$. Show that Ψ is subharmonic on D_R.

4.6   Hardy's convexity theorem

G. H. Hardy, in a famous paper [7], showed that $\log(\, \|F_r\|_p \,)$ is a convex function of $\log r$, whenever F is an analytic function on a disc. This result is considered as the starting point of the theory of Hardy spaces.

Theorem 4.15 (Hardy) *Let F be analytic on the disc D_R, and let $0 < p < \infty$. Then $\log(\|F_r\|_p)$ is an increasing convex function of $\log r$.*

Proof. First of all, note that

$$\|F_r\|_p^p = \frac{1}{2\pi} \int_{-\pi}^{\pi} |F(re^{i\theta})|^p \, d\theta = \mathcal{C}_{|F|^p}(r)$$

and thus, by Corollary 4.12, $\|F_r\|_p$ is an increasing convex function of $\log r$.
Fix $\lambda \in \mathbb{R}$. Since F is analytic on D_R,

$$\Psi(z) = |z|^\lambda \, |F(z)|^p$$

is subharmonic there, and furthermore

$$\mathcal{C}_\Psi(r) = r^\lambda \, \mathcal{C}_{|F|^p}(r). \tag{4.4}$$

Fix r, r_1 and r_2 with $0 < r_1 < r < r_2 < R$. By Corollary 4.12 and by (4.4), we have

$$\mathcal{C}_\Psi(r) \leq \frac{\log(r_2) - \log(r)}{\log(r_2) - \log(r_1)} \, \mathcal{C}_\Psi(r_1) + \frac{\log(r) - \log(r_1)}{\log(r_2) - \log(r_1)} \, \mathcal{C}_\Psi(r_2). \tag{4.5}$$

The right side is the arithmetic mean of

$$\alpha_1 = \mathcal{C}_\Psi(r_1) = r_1^\lambda \, \mathcal{C}_{|F|^p}(r_1) \qquad \text{and} \qquad \alpha_2 = \mathcal{C}_\Psi(r_2) = r_2^\lambda \, \mathcal{C}_{|F|^p}(r_2)$$

with weights

$$m_1 = \frac{\log(r_2) - \log(r)}{\log(r_2) - \log(r_1)} \qquad \text{and} \qquad m_2 = \frac{\log(r) - \log(r_1)}{\log(r_2) - \log(r_1)}.$$

Note that $r = r_1^{m_1} \, r_2^{m_2}$. Hence, by (4.4), we can rewrite (4.5) as

$$r^\lambda \, \mathcal{C}_{|F|^p}(r) \leq m_1 \, \alpha_1 + m_2 \, \alpha_2. \tag{4.6}$$

The arithmetic–geometric inequality says that

$$\alpha_1^{m_1} \, \alpha_2^{m_2} \leq m_1 \, \alpha_1 + m_2 \, \alpha_2$$

and that equality holds if and only if $\alpha_1 = \alpha_2$. Thus we choose λ such that $\alpha_1 = \alpha_2$, i.e.

$$r_1^\lambda \, \mathcal{C}_{|F|^p}(r_1) = r_2^\lambda \, \mathcal{C}_{|F|^p}(r_2).$$

Hence, with this particular choice of λ, we have

$$\begin{aligned} m_1 \, \alpha_1 + m_2 \, \alpha_2 &= \alpha_1^{m_1} \, \alpha_2^{m_2} \\ &= \left(r_1^\lambda \, \mathcal{C}_{|F|^p}(r_1) \right)^{m_1} \times \left(r_2^\lambda \, \mathcal{C}_{|F|^p}(r_2) \right)^{m_2} \\ &= r^\lambda \left(\mathcal{C}_{|F|^p}(r_1) \right)^{m_1} \times \left(\mathcal{C}_{|F|^p}(r_2) \right)^{m_2}. \end{aligned}$$

Therefore, by (4.6), we obtain

$$\mathcal{C}_{|F|^p}(r) \leq \left(\mathcal{C}_{|F|^p}(r_1) \right)^{m_1} \times \left(\mathcal{C}_{|F|^p}(r_2) \right)^{m_2}.$$

Since $\mathcal{C}_{|F|^p}(r) = \|F_r\|_p^p$, taking the logarithm of both sides gives

$$\begin{aligned} \log(\|F_r\|_p) \ \leq \ & \frac{\log(r_2) - \log(r)}{\log(r_2) - \log(r_1)} \, \log(\|F_{r_2}\|_p) \\ + \ & \frac{\log(r) - \log(r_1)}{\log(r_2) - \log(r_1)} \, \log(\|F_{r_1}\|_p), \end{aligned}$$

which is the required result. $\qquad\qquad\qquad\qquad\qquad\qquad\qquad\qquad\qquad\quad \square$

Hardy's convexity theorem is valid even for $p = \infty$. In this case it is called *Hadamard's three-circle theorem* [4, page 137]. If we use $\mathcal{M}_{|F|}$ instead of $\mathcal{C}_{|F|^p}$ in the preceding proof, we obtain a proof of Hadamard's theorem. However, we will also discuss a variation of this theorem in Section 8.2.

Exercises

Exercise 4.6.1 Let F be analytic on the disc D_R, and let $0 < p < \infty$. Define

$$\|F\|_{p,r} = \left\{ \frac{1}{\pi r^2} \int_{-\pi}^{\pi} \int_0^r |F(\rho e^{i\theta})|^p \, \rho d\rho d\theta \right\}^{\frac{1}{p}}, \qquad (0 < r < R).$$

Show that $\log(\|F\|_{p,r})$ is an increasing convex function of $\log r$.
Hint: Repeat the proof of Theorem 4.15, replacing $\mathcal{C}_{|F|^p}$ by $\mathcal{B}_{|F|^p}$.

Exercise 4.6.2 [Hardy [7]] Let F be an entire function, and let $0 < p \leq \infty$. Show that $\|F_r\|_p = O(r^n)$, as $r \to \infty$, if and only if F is a polynomial of degree at most n. Otherwise, $\|F_r\|_p$ tends to infinity more rapidly than any power of r.

Exercise 4.6.3 Let F be analytic on the annular domain $A(R_1, R_2)$, and let $0 < p < \infty$. Show that $\log(\|F_r\|_p)$, $R_1 < r < R_2$, is a convex function of $\log r$.
Remark: In this case, $\log(\|F_r\|_p)$ is not necessarily increasing.

Exercise 4.6.4 [Littlewood's subordination theorem] Let $\varphi : \mathbb{D} \longrightarrow \mathbb{D}$ be analytic with $|\varphi(z)| \leq |z|$ for all $z \in \mathbb{D}$. Let F and G be analytic on $\mathbb{D}$. We say that F is *subordinate* to G if $F(z) = G(\varphi(z))$ for all $z \in \mathbb{D}$.
    Let F be subordinate to G. Show that

$$\mathcal{M}_{|F|}(r) \leq \mathcal{M}_{|G|}(r), \qquad (0 \leq r < 1),$$

and, for $0 < p < \infty$,

$$\mathcal{C}_{|F|^p}(r) \leq \mathcal{C}_{|G|^p}(r), \qquad (0 \leq r < 1).$$

4.7 A complete characterization of $h^p(\mathbb{D})$ spaces

As far as the representation of harmonic functions on the open unit disc is concerned, our job is done. We summarize the results of the preceding sections to exhibit a complete characterization of $h^p(\mathbb{D})$ spaces. We only highlight the main points.

Let U be a harmonic function on the open unit disc $\mathbb{D}$.

Case 1: $p = 1$.

$U \in h^1(\mathbb{D})$ if and only if there exists $\mu \in \mathcal{M}(\mathbb{T})$ such that

$$U(re^{i\theta}) = \int_{\mathbb{T}} \frac{1 - r^2}{1 + r^2 - 2r \cos(\theta - t)} \, d\mu(e^{it}), \qquad (re^{i\theta} \in \mathbb{D}).$$

The measure μ is unique and

$$U(re^{i\theta}) = \sum_{n=-\infty}^{\infty} \hat{\mu}(n)\, r^{|n|}\, e^{in\theta}, \qquad (re^{i\theta} \in \mathbb{D}).$$

The family of measures $(U(re^{it})\, dt/2\pi)_{0 \le r < 1}$ converges to $d\mu(e^{it})$ in the weak* topology, as $r \to 1^-$, i.e.

$$\lim_{r \to 1} \int_{\mathbb{T}} \varphi(e^{it})\, U(re^{it})\, \frac{dt}{2\pi} = \int_{\mathbb{T}} \varphi(e^{it})\, d\mu(e^{it})$$

for all $\varphi \in \mathcal{C}(\mathbb{T})$, and we have

$$\|U\|_1 = \|\mu\| = |\mu|(\mathbb{T}).$$

Case 2: $1 < p < \infty$.

$U \in h^p(\mathbb{D})$, $1 < p < \infty$, if and only if there exists $u \in L^p(\mathbb{T})$ such that

$$U(re^{i\theta}) = \frac{1}{2\pi} \int_{-\pi}^{\pi} \frac{1 - r^2}{1 + r^2 - 2r\cos(\theta - t)}\, u(e^{it})\, dt, \qquad (re^{i\theta} \in \mathbb{D}).$$

The function u is unique and

$$U(re^{i\theta}) = \sum_{n=-\infty}^{\infty} \hat{u}(n)\, r^{|n|}\, e^{in\theta}, \qquad (re^{i\theta} \in \mathbb{D}).$$

Moreover,

$$\lim_{r \to 1} \|U_r - u\|_p = 0,$$

and thus

$$\|U\|_p = \|u\|_p.$$

Case 3: $p = \infty$.

$U \in h^\infty(\mathbb{D})$ if and only if there exists $u \in L^\infty(\mathbb{T})$ such that

$$U(re^{i\theta}) = \frac{1}{2\pi} \int_{-\pi}^{\pi} \frac{1 - r^2}{1 + r^2 - 2r\cos(\theta - t)}\, u(e^{it})\, dt, \qquad (re^{i\theta} \in \mathbb{D}).$$

The function u is unique and

$$U(re^{i\theta}) = \sum_{n=-\infty}^{\infty} \hat{u}(n)\, r^{|n|}\, e^{in\theta}, \qquad (re^{i\theta} \in \mathbb{D}).$$

The family $(U_r)_{0 \le r < 1}$ converges to u in the weak* topology, as $r \to 1$, i.e.

$$\lim_{r \to 1^-} \int_{-\pi}^{\pi} \varphi(e^{it})\, U(re^{it})\, dt = \int_{-\pi}^{\pi} \varphi(e^{it})\, u(e^{it})\, dt$$

for all $\varphi \in L^1(\mathbb{T})$. Hence,

$$\|U\|_\infty = \|u\|_\infty.$$

In all cases, the family $\|U_r\|_p$, $0 \le r < 1$, as a function of r, is increasing and a convex of $\log r$.

The first case shows that $h^1(\mathbb{D})$ and $\mathcal{M}(\mathbb{T})$ are isometrically isomorphic Banach spaces. Similarly, by the other two cases, $h^p(\mathbb{D})$ and $L^p(\mathbb{T})$, $1 < p \le \infty$, are isometrically isomorphic.

Exercises

Exercise 4.7.1 We saw that $h^1(\mathbb{D})$ and $\mathcal{M}(\mathbb{T})$ are isometrically isomorphic Banach spaces. Consider $L^1(\mathbb{T})$ as a subspace of $\mathcal{M}(\mathbb{T})$. What is the image of $L^1(\mathbb{T})$ in $h^1(\mathbb{D})$ under this correspondence? What is the image of positive measures?
Hint: Use Exercise 3.5.3.

Exercise 4.7.2 We know that $h^\infty(\mathbb{D})$ and $L^\infty(\mathbb{T})$ are isometrically isomorphic Banach spaces. Consider $\mathcal{C}(\mathbb{T})$ as a subspace of $L^\infty(\mathbb{T})$. What is the image of $\mathcal{C}(\mathbb{T})$ in $h^\infty(\mathbb{D})$ under this correspondence?
Hint: Use Exercise 3.3.4.

Chapter 5

Analytic functions in the unit disc

5.1 Representation of $H^p(\mathbb{D})$ functions $(1 < p \leq \infty)$

Since we assumed that the elements of $h^p(\mathbb{D})$ are complex-valued harmonic functions, we necessarily have $H^p(\mathbb{D}) \subset h^p(\mathbb{D})$. Hence, the facts we gathered in Section 4.7 are also valid for the analytic elements of $h^p(\mathbb{D})$, i.e. for elements of $H^p(\mathbb{T})$. As a matter of fact, we can say more in this case. If $F \in H^p(\mathbb{D})$, $1 < p \leq \infty$, then there is $f \in L^p(\mathbb{T})$ such that

$$F(z) = \sum_{n=-\infty}^{\infty} \hat{f}(n)\, r^{|n|}\, e^{in\theta} = \sum_{n=0}^{\infty} \hat{f}(n)\, z^n + \sum_{n=-\infty}^{-1} \hat{f}(n) \bar{z}^{-n}, \qquad (z = re^{i\theta} \in \mathbb{D}).$$

According to (3.1) in Theorem 3.1, we have

$$\hat{f}(n) = \frac{r^{-|n|}}{2\pi} \int_{-\pi}^{\pi} F(re^{it})\, e^{-int}\, dt, \qquad (n \in \mathbb{Z}, \, 0 < r < 1).$$

Hence, for $n \leq -1$,

$$\hat{f}(n) = \frac{r^{-|n|}}{2\pi i} \int_{|z|=r} F(z)\, z^{|n+1|}\, dz, \qquad (0 < r < 1),$$

which implies

$$\hat{f}(-1) = \hat{f}(-2) = \cdots = 0. \qquad (5.1)$$

On the other hand, if $f \in L^p(\mathbb{T})$, $1 < p \leq \infty$, satisfying (5.1) is given, and we define

$$F(z) = \frac{1}{2\pi} \int_{-\pi}^{\pi} \frac{1 - r^2}{1 + r^2 - 2r\cos(\theta - t)}\, f(e^{it})\, dt, \qquad (z = re^{i\theta} \in \mathbb{D}),$$

103

then, in the first place, $F \in h^p(\mathbb{D})$. Secondly,

$$F(z) = \sum_{n=-\infty}^{\infty} \hat{f}(n) \, r^{|n|} \, e^{in\theta} = \sum_{n=0}^{\infty} \hat{f}(n) \, z^n$$

and thus it represents an analytic function. Hence, $F \in H^p(\mathbb{D})$. This observation leads us to define the Hardy spaces on the unit circle by

$$H^p(\mathbb{T}) = \{ f \in L^p(\mathbb{T}) : \hat{f}(-1) = \hat{f}(-2) = \cdots = 0 \}, \qquad (1 \le p \le \infty). \quad (5.2)$$

In the following, the content of Section 4.7 is rewritten for analytic functions.

Theorem 5.1 *Let F be analytic on the open unit disc. Then $F \in H^p(\mathbb{D})$, $1 < p < \infty$, if and only if there exists $f \in H^p(\mathbb{T})$ such that*

$$F(re^{i\theta}) = \frac{1}{2\pi} \int_{-\pi}^{\pi} \frac{1 - r^2}{1 + r^2 - 2r\cos(\theta - t)} \, f(e^{it}) \, dt, \qquad (re^{i\theta} \in \mathbb{D}).$$

The function f is unique and

$$F(z) = \sum_{n=0}^{\infty} \hat{f}(n) \, z^n, \qquad (z \in \mathbb{D}).$$

Moreover,

$$\lim_{r \to 1} \| F_r - f \|_p = 0$$

and

$$\| F \|_p = \| f \|_p.$$

We have $F \in H^2(\mathbb{D})$ if and only if $f \in H^2(\mathbb{T})$, and in this case

$$\| F \|_2 = \| f \|_2 = \left(\sum_{n=0}^{\infty} |\hat{f}(n)|^2 \right)^{\frac{1}{2}}.$$

Theorem 5.2 *Let F be analytic on the open unit disc. Then $F \in H^\infty(\mathbb{D})$ if and only if there exists $f \in H^\infty(\mathbb{T})$ such that*

$$F(re^{i\theta}) = \frac{1}{2\pi} \int_{-\pi}^{\pi} \frac{1 - r^2}{1 + r^2 - 2r\cos(\theta - t)} \, f(e^{it}) \, dt, \qquad (re^{i\theta} \in \mathbb{D}).$$

The function f is unique and

$$F(z) = \sum_{n=0}^{\infty} \hat{f}(n) \, z^n, \qquad (z \in \mathbb{D}).$$

Moreover, F_r converges to f in the weak* topology, as $r \to 1$, i.e.

$$\lim_{r \to 1} \int_{-\pi}^{\pi} \varphi(e^{it}) \, F(re^{it}) \, dt = \int_{-\pi}^{\pi} \varphi(e^{it}) \, f(e^{it}) \, dt$$

for all $\varphi \in L^1(\mathbb{T})$, and

$$\| F \|_\infty = \| f \|_\infty.$$

The restriction of the Fourier transform to $H^p(\mathbb{T})$ classes

$$H^p(\mathbb{T}) \longrightarrow c_0(\mathbb{Z}^+)$$
$$f \longmapsto \hat{f}$$

is not surjective. However, based on the contents of Section 2.7 and Theorem 5.1, we can say that the maps

$$H^2(\mathbb{D}) \longrightarrow H^2(\mathbb{T}) \longrightarrow \ell^2(\mathbb{Z}^+)$$
$$F \longmapsto f \longmapsto \hat{f}$$

are bijective and preserve the inner product.

The preceding two theorems show that $H^p(\mathbb{D})$ and $H^p(\mathbb{T})$, $1 < p \leq \infty$, are isometrically isomorphic. One may naturally wonder why $H^1(\mathbb{D})$ was not treated in the same manner. Based on what we know about $h^1(\mathbb{D})$, if we follow the same procedure, we conclude that $H^1(\mathbb{D})$ is isometrically isomorphic to the subclass

$$\{\mu \in \mathcal{M}(\mathbb{T}) : \hat{\mu}(-1) = \hat{\mu}(-2) = \cdots = 0\},$$

and every such measure creates a unique element of $H^1(\mathbb{D})$ given by

$$F(z) = \int_{\mathbb{T}} \frac{1 - r^2}{1 + r^2 - 2r\cos(\theta - t)}\, d\mu(e^{it}) = \sum_{n=0}^{\infty} \hat{\mu}(n)\, z^n, \qquad (z = re^{i\theta} \in \mathbb{D}).$$

This assertion is absolutely true and can be viewed as a characterization of $H^1(\mathbb{D})$. However, it does not reveal the whole truth. According to a celebrated result of F. and M. Riesz, such a measure is necessarily absolutely continuous with respect to the Lebesgue measure. Therefore, Theorem 5.1 is valid even for $p = 1$. In the rest of this chapter, we develop some tools to prove the F. and M. Riesz theorem and thus complete the characterization of $H^1(\mathbb{D})$ functions.

Exercises

Exercise 5.1.1 Show that $H^p(\mathbb{T})$, $1 \leq p \leq \infty$, is a closed subspace of $L^p(\mathbb{T})$. Moreover, show that $H^\infty(\mathbb{D})$ is weak* closed in $L^\infty(\mathbb{T})$.

Exercise 5.1.2 Let $(a_n)_{n \geq 0} \in \ell^2(\mathbb{Z}_+)$, and let

$$F(z) = \sum_{n=0}^{\infty} a_n z^n, \qquad (z \in \mathbb{D}).$$

Show that $F \in H^2(\mathbb{D})$ and

$$\|F\|_2 = \left(\sum_{n=0}^{\infty} |a_n|^2 \right)^{\frac{1}{2}}.$$

Hint: Apply Theorem 5.1.

Exercise 5.1.3 We know that $H^\infty(\mathbb{D})$ and $H^\infty(\mathbb{T})$ are isometrically isomorphic Banach spaces. Let $\mathcal{A}(\mathbb{T})$ denote the closure of analytic polynomials in $H^\infty(\mathbb{T})$. What is the image of $\mathcal{A}(\mathbb{T})$ in $H^\infty(\mathbb{D})$ under this correspondence? *Hint*: Exercise 3.3.4 might help.

5.2 The Hilbert transform on $\mathbb{T}$

Let F be analytic on a disc D_R with $R > 1$ and suppose that $F(0) = 0$. Fix $e^{i\theta} \in \mathbb{T}$. Let $0 < \varepsilon < R - 1$ and let Γ_ε be the curve shown in Figure 5.1. The radius of the small circle is $r_\varepsilon = 2\sin(\varepsilon/2)$.

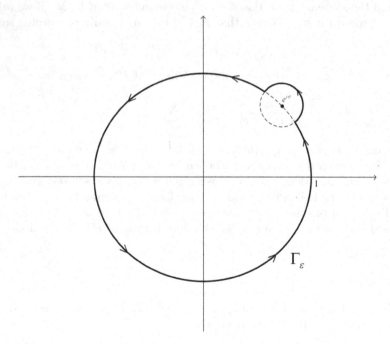

Fig. 5.1. The curve Γ_ε.

As Figure 5.2 shows, the curve $\Gamma_\varepsilon \cap \mathbb{T}$ is parameterized by

$$\zeta = e^{it}, \qquad \varepsilon \leq |\theta - t| \leq \pi,$$

and the rest of Γ_ε, according to the triangle Δ_ε, is given by

$$\zeta = e^{i\theta} + r_\varepsilon e^{it}, \qquad -\frac{\pi + \varepsilon}{2} \leq t \leq \frac{\pi + \varepsilon}{2}.$$

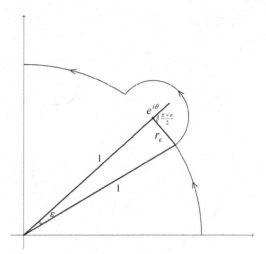

Fig. 5.2. The triangle Δ_ε.

Hence, by Cauchy's theorem,

$$
\begin{aligned}
F(e^{i\theta}) &= \frac{1}{2\pi i} \int_{\Gamma_\varepsilon} \frac{F(\zeta)}{\zeta - e^{i\theta}}\, d\zeta \\
&= \frac{1}{2\pi i} \int_{\varepsilon \le |t-\theta| \le \pi} \frac{F(e^{it})}{e^{it} - e^{i\theta}}\, ie^{it}\, dt + \frac{1}{2\pi} \int_{-\frac{\pi+\varepsilon}{2}}^{\frac{\pi+\varepsilon}{2}} F(e^{i\theta} + r_\varepsilon e^{it})\, dt.
\end{aligned}
$$

Since $F(e^{i\theta} + r_\varepsilon e^{it}) \to F(e^{i\theta})$, as $\varepsilon \to 0$, we obtain

$$
F(e^{i\theta}) = \lim_{\varepsilon \to 0} \frac{1}{\pi} \int_{\varepsilon \le |\theta - t| \le \pi} \frac{F(e^{it})}{1 - e^{i(\theta - t)}}\, dt. \tag{5.3}
$$

By a similar argument we have

$$
\begin{aligned}
0 &= \frac{1}{2\pi i} \int_{\Gamma_\varepsilon'} \frac{e^{i\theta} F(\zeta)}{\zeta(\zeta - e^{i\theta})}\, d\zeta \\
&= \frac{1}{2\pi i} \int_{\varepsilon \le |\theta - t| \le \pi} \frac{e^{i\theta} F(e^{it})}{e^{it}(e^{it} - e^{i\theta})}\, ie^{it}\, dt - \frac{1}{2\pi} \int_{\frac{\pi+\varepsilon}{2}}^{\frac{3\pi-\varepsilon}{2}} \frac{e^{i\theta} F(e^{i\theta} + r_\varepsilon e^{-it})}{e^{i\theta} + r_\varepsilon e^{-it}}\, dt,
\end{aligned}
$$

where Γ_ε' is the curve shown in Figure 5.3. Remember that $F(0) = 0$.

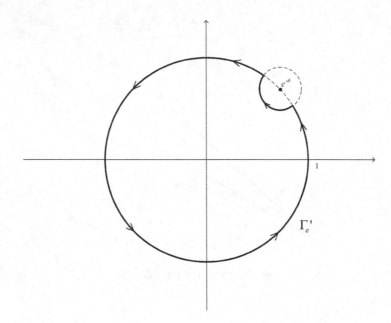

Fig. 5.3. The curve Γ'_ε.

Hence,

$$F(e^{i\theta}) = \lim_{\varepsilon \to 0} \frac{1}{\pi} \int_{\varepsilon \le |\theta - t| \le \pi} \frac{e^{i(\theta - t)} F(e^{it})}{1 - e^{i(\theta - t)}} \, dt. \qquad (5.4)$$

Adding (5.4) to (5.3) gives

$$F(e^{i\theta}) = \lim_{\varepsilon \to 0} \frac{i}{\pi} \int_{\varepsilon \le |\theta - t| \le \pi} \frac{F(e^{it})}{2 \tan(\frac{\theta - t}{2})} \, dt. \qquad (5.5)$$

The advantage of (5.5) over (5.4) and (5.3) is that the kernel appearing in it is a real function. Let us write $u = \Re F$ and $v = \Im F$ on the unit circle. Hence, (5.5) is equivalent to

$$v(e^{i\theta}) = \lim_{\varepsilon \to 0} \frac{1}{\pi} \int_{\varepsilon \le |\theta - t| \le \pi} \frac{u(e^{it})}{2 \tan(\frac{\theta - t}{2})} \, dt$$

and

$$u(e^{i\theta}) = - \lim_{\varepsilon \to 0} \frac{1}{\pi} \int_{\varepsilon \le |\theta - t| \le \pi} \frac{v(e^{it})}{2 \tan(\frac{\theta - t}{2})} \, dt.$$

The preceding argument highlights the importance of the transform

$$\lim_{\varepsilon \to 0} \frac{1}{\pi} \int_{\varepsilon \le |\theta - t| \le \pi} \frac{\phi(e^{it})}{2 \tan(\frac{\theta - t}{2})} \, dt$$

whenever ϕ is the real or imaginary part of a function analytic on a domain containing the closed unit disc. However, this transform is well-defined on other classes of functions and has far-reaching and profound properties. To have a meaningful Lebesgue integral we need at least to assume that $\phi \in L^1(\mathbb{T})$. Then the *Hilbert transform* of ϕ at the point $e^{i\theta} \in \mathbb{T}$ is defined by

$$
\begin{aligned}
\mathcal{H}\phi(e^{i\theta}) = \tilde{\phi}(e^{i\theta}) &= \lim_{\varepsilon \to 0} \frac{1}{\pi} \int_{\varepsilon \le |\theta - t| \le \pi} \frac{\phi(e^{it})}{2\tan(\frac{\theta-t}{2})}\, dt \\
&= \lim_{\varepsilon \to 0} \frac{1}{\pi} \int_{\varepsilon < |t| < \pi} \frac{\phi(e^{i(\theta-t)})}{2\tan(t/2)}\, dt \qquad (5.6)
\end{aligned}
$$

wherever the limit exists. The notation

$$
\tilde{\phi}(e^{i\theta}) = \frac{1}{\pi} \int_{-\pi}^{\pi} \frac{\phi(e^{i(\theta-t)})}{2\tan(t/2)}\, dt
$$

has exactly the same meaning as (5.6). Since around the origin $2\tan(t/2)$ behaves asymptotically like t, the transform

$$
H\phi(e^{i\theta}) = \lim_{\varepsilon \to 0} \frac{1}{\pi} \int_{\varepsilon < |t| < \pi} \frac{\phi(e^{i(\theta-t)})}{t}\, dt \qquad (5.7)
$$

has close connections to $\tilde{\phi}$. Indeed, some authors call $\tilde{\phi}$ the conjugate transform of u and keep the title "Hilbert transform" for $H\phi$. However, since

$$
\left| \frac{1}{2\tan(t/2)} - \frac{1}{t} \right| \le \frac{1}{\pi}, \qquad (0 < |t| \le \pi), \qquad (5.8)
$$

both transforms have the same behavior. More precisely, if one of them exists at $e^{i\theta} \in \mathbb{T}$, the other one exists too and we have

$$
|\mathcal{H}\phi(e^{i\theta}) - H\phi(e^{i\theta})| \le \frac{2}{\pi}. \qquad (5.9)
$$

Since $2\tan(t/2)$ is an odd function, a more appropriate way to write $\tilde{\phi}$ is

$$
\tilde{\phi}(e^{i\theta}) = \lim_{\varepsilon \to 0} \frac{1}{\pi} \int_{\varepsilon}^{\pi} \frac{\phi(e^{i(\theta-t)}) - \phi(e^{i(\theta+t)})}{2\tan(t/2)}\, dt. \qquad (5.10)
$$

Of course, if

$$
\frac{1}{\pi} \int_{0}^{\pi} \frac{|\phi(e^{i(\theta-t)}) - \phi(e^{i(\theta+t)})|}{2\tan(t/2)}\, dt < \infty, \qquad (5.11)
$$

then the Hilbert transform exists and is given by

$$
\tilde{\phi}(e^{i\theta}) = \frac{1}{\pi} \int_{0}^{\pi} \frac{\phi(c^{i(\theta-t)}) - \phi(e^{i(\theta+t)})}{2\tan(t/2)}\, dt. \qquad (5.12)
$$

However, even if (5.11) does not hold, we will show that the limit (5.10) exists almost everywhere on $\mathbb{T}$. To achieve this goal, we are led naturally to study the boundary behavior of the harmonic conjugate function.

Exercises

Exercise 5.2.1 Show that there is a $\phi \in \mathcal{C}(\mathbb{T})$ such that

$$\frac{1}{\pi} \int_0^\pi \frac{|\phi(e^{i(\theta-t)}) - \phi(e^{i(\theta+t)})|}{2\tan(t/2)}\, dt = \infty$$

for *all* $e^{i\theta} \in \mathbb{T}$.
Hint: Use the Baire category theorem [13, page 25].

Exercise 5.2.2 Verify the following Hilbert transform table:

$\phi(e^{i\theta})$	$\tilde{\phi}(e^{i\theta})$
1	0
$\cos\theta$	$\sin\theta$
$\sin\theta$	$-\cos\theta$

5.3   Radial limits of the conjugate function

In Chapter 3 we saw that, under certain conditions, a harmonic function can be represented as

$$U(re^{i\theta}) = \frac{1}{2\pi} \int_{-\pi}^\pi \frac{1 - r^2}{1 + r^2 - 2r\cos(\theta - t)}\, d\mu(e^{it}), \qquad (re^{i\theta} \in \mathbb{D}),$$

where $\mu \in \mathcal{M}(\mathbb{T})$. It is easy to verify that the harmonic conjugate of U is given by

$$V(re^{i\theta}) = \frac{1}{2\pi} \int_{-\pi}^\pi \frac{2r\sin(\theta - t)}{1 + r^2 - 2r\cos(\theta - t)}\, d\mu(e^{it}), \qquad (re^{i\theta} \in \mathbb{D}).$$

To verify this fact, simply note that

$$(U + iV)(z) = \frac{1}{2\pi} \int_{-\pi}^{\pi} \frac{e^{it} + z}{e^{it} - z} \, d\mu(e^{it}), \qquad (z \in \mathbb{D}),$$

which is an analytic function on the unit disc. According to Fatou's theorem, we know that $\lim_{r \to 1} U(re^{i\theta})$ exists for almost all $e^{i\theta} \in \mathbb{T}$. Hence, we naturally ask if the conjugate function behaves nicely too and $\lim_{r \to 1} V(re^{i\theta})$ exists for almost all $e^{i\theta} \in \mathbb{T}$. Since the conjugate Poisson kernel is not an approximate identity, we are not able to apply the techniques developed in Chapter 3. Nevertheless, the result is true. Let us start with a special case.

If μ is absolutely continuous with respect to the Lebesgue measure, then the harmonic conjugate of U is given by

$$V(re^{i\theta}) = \frac{1}{2\pi} \int_{-\pi}^{\pi} \frac{2r \sin(\theta - t)}{1 + r^2 - 2r \cos(\theta - t)} \, u(e^{it}) \, dt, \qquad (re^{i\theta} \in \mathbb{D}),$$

where $u \in L^1(\mathbb{T})$. Since the conjugate Poisson kernel is an odd function, the preceding formula for $V(re^{i\theta})$ can be rewritten as

$$V(re^{i\theta}) = \frac{1}{\pi} \int_0^{\pi} \frac{r \sin t}{1 + r^2 - 2r \cos t} \left(u(e^{i(\theta - t)}) - u(e^{i(\theta + t)}) \right) dt. \qquad (5.13)$$

This version is more appropriate in studying the boundary values of V. To save space, on some occasions we will write $V = Q * u$. To evaluate $\lim_{r \to 1} V(re^{i\theta})$, let us formally change the order of the integral and limit in (5.13) to obtain

$$\lim_{r \to 1} V(re^{i\theta}) = \frac{1}{\pi} \int_0^{\pi} \frac{u(e^{i(\theta - t)}) - u(e^{i(\theta + t)})}{2 \tan(t/2)} \, dt. \qquad (5.14)$$

We used the trigonometric identity

$$\frac{\sin t}{2(1 - \cos t)} = \frac{1}{2 \tan(t/2)}, \qquad (5.15)$$

which will appear again several times in our discussion. The formula (5.14) also highlights the relation between $\lim_{r \to 1} V(re^{i\theta})$ and $\tilde{u}(e^{i\theta})$. But if

$$\frac{1}{\pi} \int_0^{\pi} \frac{|u(e^{i(\theta - t)}) - u(e^{i(\theta + t)})|}{2 \tan(t/2)} \, dt = \infty, \qquad (5.16)$$

the integral in (5.14) is meaningless. In the first place, we show that if the integral in (5.16) is finite, then the radial limit $\lim_{r \to 1} V(re^{i\theta})$ and $\tilde{u}(e^{i\theta})$ both exist and they are equal. Secondly, we will see that, even if (5.16) may hold, the radial limits and the Hilbert transform still exist and are equal almost everywhere on $\mathbb{T}$. We start with the first case. The general case will be treated at the end of Section 5.3.

Theorem 5.3 *Let $u \in L^1(\mathbb{T})$ and let*

$$V(re^{i\theta}) = \frac{1}{2\pi} \int_{-\pi}^{\pi} \frac{2r \sin(\theta - t)}{1 + r^2 - 2r \cos(\theta - t)} \, u(e^{it}) \, dt, \qquad (re^{i\theta} \in \mathbb{D}).$$

Fix $e^{i\theta} \in \mathbb{T}$ and suppose that

$$\frac{1}{\pi} \int_0^{\pi} \frac{|u(e^{i(\theta-t)}) - u(e^{i(\theta+t)})|}{2\tan(t/2)} \, dt < \infty. \tag{5.17}$$

Then $\lim_{r\to 1} V(re^{i\theta})$ and $\tilde{u}(e^{i\theta})$ exist and

$$\lim_{r\to 1} V(re^{i\theta}) = \tilde{u}(e^{i\theta}) = \frac{1}{\pi} \int_0^{\pi} \frac{u(e^{i(\theta-t)}) - u(e^{i(\theta+t)})}{2\tan(t/2)} \, dt. \tag{5.18}$$

Proof. The fact that $\tilde{u}(e^{i\theta})$ exists and is given by

$$\tilde{u}(e^{i\theta}) = \frac{1}{\pi} \int_0^{\pi} \frac{u(e^{i(\theta-t)}) - u(e^{i(\theta+t)})}{2\tan(t/2)} \, dt$$

is a direct consequence of the definition of $\tilde{u}(e^{i\theta})$.

We need to estimate

$$\Delta_r = \left| V(re^{i\theta}) - \tilde{u}(e^{i\theta}) \right|.$$

Hence, by (5.13) and (5.15),

$$\begin{aligned}
\Delta_r &\leq \frac{1}{\pi} \int_0^{\pi} \left| \frac{r \sin t}{1 + r^2 - 2r \cos t} - \frac{\sin t}{2(1 - \cos t)} \right| |u(e^{i(\theta-t)}) - u(e^{i(\theta+t)})| \, dt \\
&= \frac{1}{\pi} \int_0^{\pi} \frac{(1-r)^2 \sin t}{2(1 - \cos t)(1 + r^2 - 2r \cos t)} |u(e^{i(\theta-t)}) - u(e^{i(\theta+t)})| \, dt \\
&= \frac{1}{\pi} \int_0^{\pi} \frac{(1-r)^2}{(1-r)^2 + 4r \sin^2(t/2)} \frac{|u(e^{i(\theta-t)}) - u(e^{i(\theta+t)})|}{2\tan(t/2)} \, dt.
\end{aligned}$$

Since

$$\frac{(1-r)^2}{(1-r)^2 + 4r \sin^2(t/2)} \leq 1$$

and

$$\lim_{r\to 1} \frac{(1-r)^2}{(1-r)^2 + 4r \sin^2(t/2)} = 0$$

for all $t \in (0, \pi]$, by the dominated convergence theorem, the last integral tends to zero as $r \to 1$. $\square$

We may make some stronger, but more familiar, assumptions on u at the point $e^{i\theta}$ to ensure that (5.17) holds. Two such conditions are mentioned below.

Corollary 5.4 *Let* $u \in L^1(\mathbb{T})$ *and let* $V = Q * u$. *Fix* $e^{i\theta} \in \mathbb{T}$ *and suppose that, for some* $\alpha > 0$,
$$u(e^{i(\theta-t)}) - u(e^{i(\theta+t)})$$
is Lip_α *at* $t = 0$, *i.e. for* $|t| \leq t_0$,
$$|u(e^{i(\theta-t)}) - u(e^{i(\theta+t)})| \leq C \, |t|^\alpha.$$
Then $\lim_{r \to 1} V(re^{i\theta})$ *and* $\tilde{u}(e^{i\theta})$ *exist and* (5.18) *holds.*

Corollary 5.5 *Let* $u \in L^1(\mathbb{T})$ *and let* $V = Q * u$. *Fix* $e^{i\theta} \in \mathbb{T}$ *and suppose that*
$$u(e^{i(\theta-t)}) - u(e^{i(\theta+t)}),$$
as a function of $t \in \mathbb{R}$, *is differentiable at* $t = 0$. *Then* $\lim_{r \to 1} V(re^{i\theta})$ *and* $\tilde{u}(e^{i\theta})$ *exist and* (5.18) *holds.*

5.4 The Hilbert transform of $C^1(\mathbb{T})$ functions

If the conjugate function $V = Q * u$ is produced by a continuously differentiable function u, then, by Corollary 5.5, we know that $\tilde{u}(e^{i\theta})$ and $\lim_{r \to 1} V(re^{i\theta})$ exist at *all* points of $\mathbb{T}$ and they are equal. We show that $\tilde{u}$ is continuous on $\mathbb{T}$.

Theorem 5.6 *Let* $u \in C^1(\mathbb{T})$, *and let*
$$V(re^{i\theta}) = \frac{1}{2\pi} \int_{-\pi}^{\pi} \frac{2r \sin(\theta - t)}{1 + r^2 - 2r \cos(\theta - t)} \, u(e^{it}) \, dt, \qquad (re^{i\theta} \in \mathbb{D}).$$

Then, as $r \to 1$, V_r *converges uniformly to* $\tilde{u}$ *on* $\mathbb{T}$.

Proof. Let
$$\Delta_r(e^{i\theta}) = |V(re^{i\theta}) - \tilde{u}(e^{i\theta})|.$$
Hence, as we saw in the proof of Theorem 5.3,
$$\Delta_r(e^{i\theta}) \leq \frac{1}{\pi} \int_0^{\pi} \frac{(1-r)^2}{1 + r^2 - 2r \cos t} \frac{|u(e^{i(\theta-t)}) - u(e^{i(\theta+t)})|}{2 \tan(t/2)} \, dt.$$

According to the mean value theorem, we have
$$|u(e^{i(\theta-t)}) - u(e^{i(\theta+t)})| \leq 2|t| \, \|u'\|_\infty.$$

Hence, there is an absolute constant C such that
$$\frac{|u(e^{i(\theta-t)}) - u(e^{i(\theta+t)})|}{2 \tan(t/2)} \leq C$$

for all $t \in [0, \pi]$. Therefore
$$\Delta_r(e^{i\theta})| \leq \frac{C}{\pi} \int_0^{\pi} \frac{(1-r)^2}{1 + r^2 - 2r \cos t} \, dt = C \frac{1-r}{1+r},$$

which gives
$$\|V_r - \tilde{u}\|_\infty \leq C\,(1-r).$$

$\square$

The uniform limit of a sequence of continuous functions is continuous. Hence, Theorem 5.6 implies immediately that $\tilde{u}$ is continuous on $\mathbb{T}$.

Corollary 5.7 *Let $u \in \mathcal{C}^1(\mathbb{T})$. Then $\tilde{u} \in \mathcal{C}(\mathbb{T})$.*

Theorem 5.6 and its corollary can be generalized for differentiable functions on an open arc $I \subset \mathbb{T}$. To prove this result we study a very special case in the first step. Note that if I is an open arc on $\mathbb{T}$ and an integrable function u is identically zero on I, then Corollary 5.5 ensures that $\tilde{u}(e^{i\theta})$ and $\lim_{r\to 1} V(re^{i\theta})$ exist at all $e^{i\theta} \in I$ and they are equal.

Lemma 5.8 *Let $u \in L^1(\mathbb{T})$, and let*

$$V(re^{i\theta}) = \frac{1}{2\pi} \int_{-\pi}^{\pi} \frac{2r\,\sin(\theta - t)}{1 + r^2 - 2r\cos(\theta - t)}\, u(e^{it})\, dt, \qquad (re^{i\theta} \in \mathbb{D}).$$

Let I be an open arc on $\mathbb{T}$ and suppose that $u(e^{it}) = 0$ for all $e^{it} \in I$. Let J be an open arc such that $\bar{J} \subset I$. Then, as $r \to 1$, V_r converges uniformly to $\tilde{u}$ on J.

Proof. Without loss of generality assume that $I = \{e^{it} : -\alpha < t < \alpha\}$ and $J = \{e^{it} : -\beta < t < \beta\}$ with $0 < \beta < \alpha < \pi$, since otherwise we can work with $V_r(e^{i(\theta + \theta_0)})$ where θ_0 is an appropriate rotation. Let

$$\Delta_r(e^{i\theta}) = |\,V(re^{i\theta}) - \tilde{u}(e^{i\theta})\,|, \qquad (e^{i\theta} \in J).$$

As in the proof of Theorem 5.6,

$$\Delta_r(e^{i\theta}) \leq \frac{1}{\pi} \int_0^{\pi} \frac{(1-r)^2 \sin t}{2(1 - \cos t)(1 + r^2 - 2r\cos t)}\, |u(e^{i(\theta - t)}) - u(e^{i(\theta + t)})|\, dt.$$

But, for all $t \in [0, \alpha - \beta]$ and all $e^{i\theta} \in J$, we have

$$u(e^{i(\theta - t)}) = u(e^{i(\theta + t)}) = 0.$$

Hence, for each $e^{i\theta} \in J$,

$$\begin{aligned}
\Delta_r(e^{i\theta}) &\leq \frac{1}{\pi} \int_{\alpha - \beta}^{\pi} \frac{(1-r)^2 \sin t}{2(1 - \cos t)(1 + r^2 - 2r\cos t)}\, |u(e^{i(\theta - t)}) - u(e^{i(\theta + t)})|\, dt \\
&\leq \frac{\|u\|_1}{(1 - \cos(\alpha - \beta))^2}\, (1 - r)^2,
\end{aligned}$$

which is equivalent to

$$\|V_r - \tilde{u}\|_{L^{\infty}(J)} \leq \frac{\|u\|_1}{(1 - \cos(\alpha - \beta))^2}\, (1 - r)^2.$$

$\square$

Theorem 5.9 *Let* $u \in L^1(\mathbb{T})$, *and let*

$$V(re^{i\theta}) = \frac{1}{2\pi} \int_{-\pi}^{\pi} \frac{2r \sin(\theta - t)}{1 + r^2 - 2r \cos(\theta - t)} \, u(e^{it}) \, dt, \qquad (re^{i\theta} \in \mathbb{D}).$$

Let I *be an open arc on* $\mathbb{T}$ *and suppose that* $u \in C^1(I)$. *Let* J *be an open arc such that* $\bar{J} \subset I$. *Then, as* $r \to 1$, V_r *converges uniformly to* $\tilde{u}$ *on* J. *Moreover,* $\tilde{u} \in C(I)$.

Proof. Let K be an open arc such that

$$\bar{J} \subset K \subset \overline{K} \subset I.$$

Then there exists a $C^\infty(\mathbb{T})$ function φ such that $\varphi \equiv 1$ on K and $\varphi \equiv 0$ on $\mathbb{T} \setminus I$. In this particular case, since I, J and K are open arcs, one can even provide an explicit formula for φ.

Now, consider $u_1 = u\varphi$ and $u_2 = u(1 - \varphi)$ on $\mathbb{T}$ and, using the conjugate Poisson kernel, extend them to the open unit disc, say $V_1 = Q * u_1$ and $V_2 = Q * u_2$.

The first observation is that $u_1 \in C^1(\mathbb{T})$. Hence, by Theorem 5.6, $\tilde{u}_1(e^{i\theta})$ and $\lim_{r \to 1} V_1(re^{i\theta})$ exist at all $e^{i\theta} \in \mathbb{T}$, they are equal and, as $r \to 1$, $V_{1,r}$ converges uniformly to $\tilde{u}_1$ on $\mathbb{T}$. Secondly, $u_2 \in L^1(\mathbb{T})$ and $u_2 \equiv 0$ on K. Hence, by Lemma 5.8, $\tilde{u}_2(e^{i\theta})$ and $\lim_{r \to 1} V_2(re^{i\theta})$ exist at all $e^{i\theta} \in K$, and as $r \to 1$, $V_{2,r}$ converges uniformly to $\tilde{u}_2$ on J. Since, on $\mathbb{D}$

$$V = V_1 + V_2,$$

and on I

$$\tilde{u} = \tilde{u}_1 + \tilde{u}_2,$$

we immediately see that, as $r \to 1$, $V_r = V_{1,r} + V_{2,r}$ converges uniformly to $\tilde{u}$ on J. The uniform convergence ensures that $\tilde{u}$ is continuous on J. Since J is an arbitrary open arc in I, with the only restriction $\bar{J} \subset I$, then $\tilde{u}$ is continuous at all points of I. $\qquad\qquad\square$

Exercises

Exercise 5.4.1 Let $u \in L^1(\mathbb{T})$ and suppose that $u(e^{it}) = 0$ for all $e^{it} \in I$, where I is an open arc on $\mathbb{T}$. Show that there is a harmonic function V on the domain

$$(\mathbb{C} \setminus \mathbb{T}) \cup I$$

whose values on the unit disc $\mathbb{D}$ are given by the conjugate Poisson integral $V = Q * u$.

Remark 1: This is a generalization of Lemma 5.8.
Remark 2: A similar result holds for the harmonic function $U = P * u$.

Exercise 5.4.2   Let $u \in L^1(\mathbb{T})$, and let $V = Q * u$. Let I be a *closed* arc on $\mathbb{T}$ and suppose that $u \in \mathcal{C}^1(I)$. Show that $\tilde{u} \in \mathcal{C}(I)$.

Exercise 5.4.3   Let $u \in \mathcal{C}^2(\mathbb{T})$. Show that $\tilde{u} \in \mathcal{C}^1(\mathbb{T})$ and that

$$\frac{d\tilde{u}}{d\theta}(e^{i\theta}) = -\frac{1}{4\pi} \int_0^\pi \frac{u(e^{i(\theta-t)}) + u(e^{i(\theta+t)}) - 2u(e^{i\theta})}{\sin^2(t/2)}\, dt$$

for all $e^{i\theta} \in \mathbb{T}$.

5.5   Analytic measures on $\mathbb{T}$

A measure $\mu \in \mathcal{M}(\mathbb{T})$ is called *analytic* if

$$\hat{\mu}(n) = \int_{\mathbb{T}} e^{-int}\, d\mu(e^{it}) = 0$$

for all $n \leq -1$. In other words, the negative part of the spectrum of μ vanishes completely. These measures appeared at the end of Section 5.1 in studying the Hardy space $H^1(\mathbb{D})$. According to F. and M. Riesz's theorem, such a measure is absolutely continuous with respect to the Lebesgue measure. To show this result we start with a construction due to Fatou.

Let E be a closed subset of $\mathbb{T}$ with $|E| = 0$, where $|E|$ stands for the Lebesgue measure of E. Fatou constructed a function F on the closed unit disc satisfying the following properties:

(i) F is continuous on $\overline{\mathbb{D}}$;

(ii) F is analytic on $\mathbb{D}$;

(iii) $|F(z)| < 1$, for all $z \in \mathbb{D}$;

(iv) $|F(\zeta)| < 1$, for all $\zeta \in \mathbb{T} \setminus E$;

(v) $F(\zeta) = 1$, for all $\zeta \in E$.

Such a function is called a *Fatou function* for E. Certainly F is not unique. For example, F^2 satisfies all preceding conditions if F does so. Without loss of generality, assume $1 \in E$. Since E is closed, we have

$$\mathbb{T} \setminus E = \bigcup_n I_n,$$

where I_n are some open arcs on $\mathbb{T}$, say

$$I_n = \{\, e^{i\theta} : 0 < \alpha_n < \theta < \beta_n < 2\pi \,\}.$$

The condition $|E| = 0$ is equivalent to

$$\sum_n (\beta_n - \alpha_n) = 2\pi.$$

Hence, there is a sequence $\eta_n > 0$ such that

$$\lim_{n \to \infty} \eta_n = \infty,$$

but still $\sum_n \eta_n (\beta_n - \alpha_n) < \infty$. Define

$$u(e^{it}) = \begin{cases} +\infty & \text{if} \quad e^{it} \in E, \\[2mm] \dfrac{\eta_n \ell_n}{\sqrt{\ell_n^2 - (t - \gamma_n)^2}} & \text{if} \quad e^{it} \in I_n, \end{cases}$$

where $\ell_n = (\beta_n - \alpha_n)/2$ and $\gamma_n = (\beta_n + \alpha_n)/2$. This positive function has two important properties. First of all, u is *integrable* on $\mathbb{T}$:

$$\begin{aligned} \int_{\mathbb{T}} |u(e^{it})|\, dt &= \int_{\mathbb{T}} u(e^{it})\, dt \\ &= \int_E u(e^{it})\, dt + \sum_n \int_{I_n} u(e^{it})\, dt \\ &= \sum_n \int_{\alpha_n}^{\beta_n} \frac{\eta_n \ell_n}{\sqrt{\ell_n^2 - (t - \gamma_n)^2}}\, dt = \frac{\pi}{2} \sum_n \eta_n (\beta_n - \alpha_n) < \infty. \end{aligned}$$

Secondly, u, as a function from $\mathbb{T}$ to $[0, +\infty]$, is *continuous*. As a matter of fact, u is infinitely differentiable on each I_n and tends to infinity when we approach α_n or β_n from within I_n. Moreover, $u(e^{it}) \geq \eta_n$ on I_n and η_n tends to infinity as n grows. That is why η_n entered our discussion. Hence, as $e^{it} \in \mathbb{T}$ approaches any $e^{it_0} \in E$, in any manner, $u(e^{it})$ tends to infinity, i.e.

$$\lim_{t \to t_0} u(e^{it}) = +\infty \tag{5.19}$$

for all $e^{it_0} \in E$. Define

$$U(re^{i\theta}) = \begin{cases} \dfrac{1}{2\pi} \displaystyle\int_{-\pi}^{\pi} \dfrac{1 - r^2}{1 + r^2 - 2r\cos(\theta - t)}\, u(e^{it})\, dt & \text{if} \quad 0 \leq r < 1, \\[4mm] u(e^{i\theta}) & \text{if} \quad r = 1. \end{cases}$$

By Corollary 2.9, U is continuous on $\overline{\mathbb{D}}$ with values in $[0, +\infty]$, and is harmonic on $\mathbb{D}$. In particular, for each $e^{it_0} \in E$,

$$\lim_{\substack{z \to e^{it_0} \\ z \in \mathbb{D}}} U(z) = +\infty. \tag{5.20}$$

Since u is continuously differentiable on each I_n, by Theorem 5.9, the harmonic conjugate

$$V(re^{i\theta}) = \frac{1}{2\pi} \int_{-\pi}^{\pi} \frac{2r \sin(\theta - t)}{1 + r^2 - 2r \cos(\theta - t)} u(e^{it}) \, dt, \qquad (re^{i\theta} \in \mathbb{D}),$$

has continuous extension up to I_n, and its boundary values on I_n are given by the Hilbert transform $\tilde{u}(e^{i\theta})$, $e^{i\theta} \in I_n$. Note that we did not define $\tilde{u}$ on E and moreover, for the rest of the discussion, we do not need to verify if $\tilde{u}(e^{i\theta})$ exists or not whenever $e^{i\theta} \in E$. Finally, let

$$F : \overline{\mathbb{D}} \longmapsto \mathbb{C}$$

be given by

$$F(z) = \begin{cases} \dfrac{U(z) + iV(z)}{1 + U(z) + iV(z)} & \text{if} \quad z \in \mathbb{D}, \\[2ex] \dfrac{u(z) + i\tilde{u}(z)}{1 + u(z) + i\tilde{u}(z)} & \text{if} \quad z \in \mathbb{T} \setminus E, \\[2ex] 1 & \text{if} \quad z \in E. \end{cases}$$

Clearly, F is analytic on $\mathbb{D}$ and continuous at all points of $\mathbb{T} \setminus E$. Since $u > 0$ and $U > 0$, we also have $|F(z)| < 1$ for all $z \in \mathbb{D}$ and all $z \in \mathbb{T} \setminus E$. Moreover, by (5.19) and (5.20), F is continuous at each point of E. Hence, F is continuous on $\overline{\mathbb{D}}$.

Now, we have all the necessary tools to prove the celebrated theorem of the Riesz brothers. This *fundamental* result is actually a genuine application of Fatou's functions for compact subsets of $\mathbb{T}$.

Theorem 5.10 (F. and M. Riesz) *Let $\mu \in \mathcal{M}(\mathbb{T})$ be an analytic measure. Then μ is absolutely continuous with respect to the Lebesgue measure.*

Proof. Put $d\nu(e^{it}) = e^{it} \, d\mu(e^{it})$. Hence $d\mu(e^{it}) = e^{-it} \, d\nu(e^{it})$, and thus it is enough to show that ν is absolutely continuous with respect to the Lebesgue measure. The advantage of ν is that

$$\hat{\nu}(n) = 0, \qquad \text{for all } n \leq 0.$$

(The identity also holds for $n = 0$.) Thus, for any analytic function G on the unit disc, we have

$$\int_{\mathbb{T}} G(re^{it}) \, d\nu(e^{it}) = \int_{\mathbb{T}} \left(\sum_{n=0}^{\infty} a_n r^n e^{int} \right) d\nu(e^{it})$$

$$= \sum_{n=0}^{\infty} a_n r^n \int_{\mathbb{T}} e^{int} \, d\nu(e^{it})$$

$$= \sum_{n=0}^{\infty} a_n r^n \, \hat{\nu}(n) = 0, \qquad (0 \leq r < 1).$$

Suppose $E \subset \mathbb{T}$ with $|E| = 0$. Let K be a compact subset of E and let F be a Fatou function for K. Hence, according to the preceding observation,

$$\int_{\mathbb{T}} \left(F(re^{it}) \right)^{\ell} d\nu(e^{it}) = 0, \qquad (0 \leq r < 1),$$

for all integers $\ell \geq 1$. First, let $r \to 1$. Since F is bounded on $\overline{\mathbb{D}}$, we obtain

$$\int_{\mathbb{T}} \left(F(e^{it}) \right)^{\ell} d\nu(e^{it}) = 0,$$

which is equivalent to

$$\nu(K) + \int_{\mathbb{T}\backslash K} \left(F(e^{it}) \right)^{\ell} d\nu(e^{it}) = 0.$$

Secondly, let $\ell \to \infty$. But F was constructed such that $|F(e^{it})| < 1$ for all $e^{it} \in \mathbb{T} \backslash K$. Thus, $\nu(K) = 0$. By regularity, $\nu(E) = 0$. $\qquad \square$

In the proof of the theorem, we used the original definition of absolute continuity. The measure μ is absolutely continuous with respect to ν provided that $\nu(E) = 0$ implies $\mu(E) = 0$. However, in application, we will use the following characterization of absolute continuity: μ is absolutely continuous with respect to ν if and only if there is an $f \in L^1(\nu)$ such that

$$d\mu = f \, d\nu.$$

Exercises

Exercise 5.5.1 Let $a_n \geq 0$ and $\sum_n a_n < \infty$. Show that there are $\eta_n \geq 0$ such that $\lim_{n \to \infty} \eta_n = \infty$ and still $\sum_n \eta_n a_n < \infty$.
Hint: Since the series $\sum_n a_n$ is convergent, there are $n_1 < n_2 < \cdots$ such that $\sum_{n > n_k} a_n < 2^{-k}$.

Exercise 5.5.2 Let $\mu \in \mathcal{M}(\mathbb{T})$. Suppose that μ has a one-sided spectrum, i.e.

$$\hat{\mu}(n) = \int_{\mathbb{T}} e^{-int} \, d\mu(e^{it}) = 0$$

either for $n \geq N$ or for $n \leq -N$, where N is an integer. Show that μ is absolutely continuous with respect to the Lebesgue measure.

Exercise 5.5.3 Let $\mu \in \mathcal{M}(\mathbb{T})$. Suppose that μ is analytic and also singular with respect to the Lebesgue measure. Show that $\mu = 0$.

Exercise 5.5.4 Construct a nonzero measure $\sigma \in \mathcal{M}(\mathbb{T})$, singular with respect to the Lebesgue measure, such that

$$\hat{\sigma}(2n+1) = 0, \quad \text{for all } n \in \mathbb{Z}.$$

Hint: Put appropriate Dirac measures at 1 and -1.

5.6   Representations of $H^1(\mathbb{D})$ functions

At this point we have developed all the necessary tools to show that the representations of Section 5.1 for $H^p(\mathbb{D})$, $1 < p < \infty$, classes are valid even if $p = 1$. This fact constitutes the fundamental difference between $h^1(\mathbb{D})$ and $H^1(\mathbb{D})$.

Let $F \in H^1(\mathbb{D}) \subset h^1(\mathbb{D})$. We saw that, by Theorems 2.1 and 3.7,

$$F(re^{i\theta}) = \int_{-\pi}^{\pi} \frac{1-r^2}{1+r^2-2r\cos(\theta-t)} \, d\mu(e^{it}) = \sum_{n=-\infty}^{\infty} \hat{\mu}(n) \, r^{|n|} \, e^{in\theta}, \quad (re^{i\theta} \in \mathbb{D}),$$

where $\mu \in \mathcal{M}(\mathbb{T})$. Moreover, the measures $F(re^{i\theta}) \, d\theta/2\pi$ converge in the weak* topology to μ as $r \to 1$. Hence, for any $n \leq -1$,

$$\hat{\mu}(n) = \int_{\mathbb{T}} e^{-in\theta} \, d\mu(e^{i\theta}) = \lim_{r \to 1} \frac{1}{2\pi} \int_{-\pi}^{\pi} F(re^{i\theta}) \, e^{-in\theta} \, d\theta.$$

But, since F is analytic,

$$\int_{-\pi}^{\pi} F(re^{i\theta}) \, e^{-in\theta} \, d\theta = r^{-n-1} \int_{|\zeta|=r} F(\zeta) \, \zeta^{-(n+1)} \, d\zeta = 0,$$

which implies $\hat{\mu}(n) = 0$ for all $n \leq -1$. Therefore, by the theorem of F. and M. Riesz, there exists a function $f \in L^1(\mathbb{T})$ such that

$$d\mu(e^{it}) = f(e^{it}) \, dt/2\pi$$

and

$$\hat{f}(n) = \hat{\mu}(n) = 0$$

for all $n \leq -1$. In other words, $f \in H^1(\mathbb{T})$.

On the other hand if $f \in H^1(\mathbb{T})$, then, by Theorem 2.1 and Corollary 2.15, the function

$$
\begin{aligned}
F(re^{i\theta}) &= \frac{1}{2\pi} \int_{-\pi}^{\pi} \frac{1-r^2}{1+r^2-2r\cos(\theta-t)} \, f(e^{it}) \, dt \\
&= \sum_{n=-\infty}^{\infty} \hat{f}(n) \, r^{|n|} \, e^{in\theta} = \sum_{n=0}^{\infty} \hat{f}(n) \, z^n, \qquad (z = re^{i\theta} \in \mathbb{D}),
\end{aligned}
$$

also represents an element of $H^1(\mathbb{D})$ and F_r converges to f in the $L^1(\mathbb{T})$ norm as $r \to 1$. We state the preceding results as a theorem.

Theorem 5.11 *Let F be analytic on the unit disc $\mathbb{D}$. Then $F \in H^1(\mathbb{D})$ if and only if there exists a unique $f \in H^1(\mathbb{T})$ such that*

$$F(z) = \frac{1}{2\pi} \int_{-\pi}^{\pi} \frac{1 - r^2}{1 + r^2 - 2r\cos(\theta - t)} \, f(e^{it}) \, dt = \sum_{n=0}^{\infty} \hat{f}(n) \, z^n \qquad (5.21)$$

for all $z = re^{i\theta} \in \mathbb{D}$. The series is uniformly convergent on compact subsets of $\mathbb{D}$. Moreover,

$$\lim_{r \to 1} \|F_r - f\|_1 = 0$$

and

$$\|F\|_1 = \|f\|_1.$$

Using the Cauchy integral formula and the norm convergence of F_r to f, we obtain another representation theorem for functions in $H^1(\mathbb{D})$.

Corollary 5.12 *Let $F \in H^1(\mathbb{D})$ and denote its boundary values by $f \in H^1(\mathbb{T})$. Then*

$$F(z) = \frac{1}{2\pi} \int_{-\pi}^{\pi} \frac{f(e^{it})}{1 - e^{-it}z} \, dt \qquad (5.22)$$

and

$$\int_{-\pi}^{\pi} \frac{\bar{z}\, e^{it}}{\bar{z}\, e^{it} - 1} \, f(e^{it}) \, dt = 0 \qquad (5.23)$$

for all $z \in \mathbb{D}$.

Proof. Fix $z \in \mathbb{D}$. By the Cauchy integral formula, if $\frac{1+|z|}{2} < r < 1$,

$$F(z) = \frac{1}{2\pi i} \int_{\{|\zeta| = r\}} \frac{F(\zeta)}{\zeta - z} \, d\zeta = \frac{1}{2\pi} \int_{-\pi}^{\pi} \frac{r e^{it}}{r e^{it} - z} \, F(re^{it}) \, dt.$$

Let $r \to 1$. Since

$$\left| \frac{r e^{it}}{r e^{it} - z} \right| \leq \frac{1}{r - |z|} \leq \frac{2}{1 - |z|}$$

and, by Theorem 5.11, F_r converges to f in the $L^1(\mathbb{T})$ norm, we have

$$F(z) = \lim_{r \to 1} \frac{1}{2\pi} \int_{-\pi}^{\pi} \frac{r e^{it}}{r e^{it} - z} \, F(re^{it}) \, dt = \frac{1}{2\pi} \int_{-\pi}^{\pi} \frac{e^{it}}{e^{it} - z} \, f(e^{it}) \, dt.$$

The second formula has a similar proof if we start with

$$\frac{1}{2\pi i} \int_{\{|\zeta| = r\}} \frac{\bar{z}\, F(\zeta)}{\bar{z}\, \zeta - 1} \, d\zeta = 0.$$

$\square$

Note that since

$$\frac{1-r^2}{1+r^2-2r\cos(\theta-t)} = \frac{e^{it}}{e^{it}-z} - \frac{\bar{z}e^{it}}{\bar{z}\,e^{it}-1}, \qquad (z=re^{i\theta}),$$

the Poisson integral formula (5.21) can also be deduced from (5.22) and (5.23).

Exercises

Exercise 5.6.1   Let $f_n \in H^1(\mathbb{T})$, $n \geq 1$, and let $f \in L^1(\mathbb{T})$. Suppose that

$$\lim_{n\to\infty} \|f_n - f\|_1 = 0.$$

Show that $f \in H^1(\mathbb{T})$.

Exercise 5.6.2   Let $f, g \in H^2(\mathbb{T})$. Show that $fg \in H^1(\mathbb{T})$.

Exercise 5.6.3   Let $F \in H^1(\mathbb{D})$ and denote its boundary values by $f \in H^1(\mathbb{T})$. Show that

$$F(z) = \frac{i}{2\pi}\int_{-\pi}^{\pi} \frac{2r\sin(\theta-t)}{1+r^2-2r\cos(\theta-t)}\, f(e^{it})\,dt + F(0), \qquad (z=re^{i\theta}\in\mathbb{D}).$$

Hint: Use Corollary 5.12 and the identity

$$\frac{i2r\sin(\theta-t)}{1+r^2-2r\cos(\theta-t)} = \frac{e^{it}}{e^{it}-z} + \frac{\bar{z}e^{it}}{\bar{z}\,e^{it}-1} - 1, \qquad (z=re^{i\theta}).$$

Exercise 5.6.4   Let $F \in H^1(\mathbb{D})$ and denote its boundary values by $f \in H^1(\mathbb{T})$. Show that

$$F(z) = \frac{1}{2\pi}\int_{-\pi}^{\pi} \frac{e^{it}+z}{e^{it}-z}\, \Re f(e^{it})\,dt + i\Im F(0), \qquad (z\in\mathbb{D}).$$

Hint: By (5.21),

$$\Re F(z) = \frac{1}{2\pi}\int_{-\pi}^{\pi} \frac{1-r^2}{1+r^2-2r\cos(\theta-t)}\, \Re f(e^{it})\,dt, \qquad (z=re^{i\theta}\in\mathbb{D}).$$

Moreover,

$$\frac{1-r^2}{1+r^2-2r\cos(\theta-t)} = \Re\left(\frac{e^{it}+z}{e^{it}-z}\right), \qquad (z=re^{i\theta}\in\mathbb{D}).$$

Exercise 5.6.5   Let $F \in H^1(\mathbb{D})$ and denote its boundary values by $f \in H^1(\mathbb{T})$. Show that

$$F(z) = \frac{i}{2\pi}\int_{-\pi}^{\pi} \frac{e^{it}+z}{e^{it}-z}\, \Im f(e^{it})\,dt + \Re F(0), \qquad (z\in\mathbb{D}).$$

Hint: Replace F by iF in Exercise 5.6.4.

5.7 The uniqueness theorem and its applications

An element $f \in H^1(\mathbb{T})$ is defined almost everywhere on $\mathbb{T}$. Hence we are free to change its values on a set of Lebesgue measure zero. Therefore, the level set

$$\{e^{i\theta} : f(e^{i\theta}) = 0\}$$

is also defined modulus a set of measure zero. Nevertheless, the Lebesgue measure of this set is well-defined. Amazingly, the size of this set is either 0 or 2π. Different versions of the following theorem are stated by Fatou, F. Riesz, M. Riesz and Szegö.

Theorem 5.13 (Uniqueness theorem) *Let $F \in H^1(\mathbb{D})$ with boundary values $f \in H^1(\mathbb{T})$. Let $E \subset \mathbb{T}$ with $|E| > 0$. Suppose that $f(e^{i\theta}) = 0$ for almost all $e^{i\theta} \in E$. Then $F \equiv 0$ on $\mathbb{D}$, or equivalently $f = 0$ almost everywhere on $\mathbb{T}$.*

Proof. If $|E| = 2\pi$, then $f = 0$ almost everywhere on $\mathbb{T}$ and thus by the Poisson integral representation given in Theorem 5.11, $F \equiv 0$ on $\mathbb{D}$.

Hence suppose that $0 < |E| < 2\pi$. We also suppose that $F \not\equiv 0$ and then we obtain a contradiction. Without loss of generality assume that $F(0) \neq 0$, since otherwise we can work with $F(z)/z^m$, where m is the order of the zero of F at the origin. Let

$$u(e^{it}) = \begin{cases} \dfrac{1}{|E|} & \text{if} \quad e^{it} \in E, \\ \dfrac{-1}{2\pi - |E|} & \text{if} \quad e^{it} \in \mathbb{T} \setminus E \end{cases}$$

and, for $re^{i\theta} \in \mathbb{D}$, let

$$U(re^{i\theta}) = \frac{1}{2\pi} \int_{-\pi}^{\pi} \frac{1-r^2}{1+r^2-2r\cos(\theta-t)} u(e^{it})\, dt,$$

$$V(re^{i\theta}) = \frac{1}{2\pi} \int_{-\pi}^{\pi} \frac{2r\sin(\theta-t)}{1+r^2-2r\cos(\theta-t)} u(e^{it})\, dt.$$

Clearly,

$$\frac{-1}{2\pi - |E|} \leq U(re^{i\theta}) \leq \frac{1}{|E|}$$

for all $re^{i\theta} \in \mathbb{D}$. Therefore,

$$\exp(U + iV) \in H^\infty(\mathbb{D}),$$

which implies

$$G_N = F\, e^{N(U+iV)} \in H^1(\mathbb{D})$$

for all integers $N \geq 1$. Moreover, $U(0) = V(0) = 0$, which gives $G_N(0) = F(0)$. By Lemma 3.10,

$$\lim_{r \to 1} U(re^{i\theta}) = \frac{-1}{2\pi - |E|} \tag{5.24}$$

for almost all $e^{i\theta} \in \mathbb{T} \setminus E$. Let g_N denote the boundary values of G_N. According to (5.24),

$$|g_N(e^{it})| = |f(e^{it})| \, \exp\left(-\frac{N}{2\pi - |E|}\right)$$

for almost all $e^{it} \in \mathbb{T} \setminus E$, and by our main assumption

$$g_N(e^{it}) = 0$$

for almost all $e^{it} \in E$. Therefore, again by the Poisson integral representation given in Theorem 5.11,

$$F(0) = G_N(0) = \frac{1}{2\pi} \int_{\mathbb{T}} g_N(e^{it}) \, dt = \frac{1}{2\pi} \int_{\mathbb{T}\setminus E} g_N(e^{it}) \, dt.$$

Hence, for all $N \geq 1$,

$$|F(0)| \leq \frac{1}{2\pi} \int_{\mathbb{T}\setminus E} |g_N(e^{it})| \, dt = \frac{1}{2\pi} \exp\left(-\frac{N}{2\pi - |E|}\right) \int_{\mathbb{T}\setminus E} |f(e^{it})| \, dt.$$

Let $N \to \infty$ to get $F(0) = 0$, which is a contradiction. $\square$

This result is not valid for the elements of $h^1(\mathbb{D})$. For example, the Poisson kernel

$$P(re^{i\theta}) = \frac{1 - r^2}{1 + r^2 - 2r\cos\theta}, \qquad (re^{i\theta} \in \mathbb{D}),$$

is in $h^1(\mathbb{D})$ and

$$\lim_{r\to 1} P(re^{i\theta}) = 0$$

for all $e^{i\theta} \in \mathbb{T} \setminus \{1\}$. However, $P \not\equiv 0$.

We know that elements of $H^\infty(\mathbb{D})$ have radial boundary values almost everywhere on the unit circle. This result, along with the uniqueness theorem for $H^1(\mathbb{D})$ functions, enables us to show that analytic functions with values in a half plane also have radial boundary values.

Lemma 5.14 *Let F be analytic on the unit disc and suppose that $\Re F(z) > 0$ for all $z \in \mathbb{D}$. Then*

$$\lim_{r\to 1} F(re^{i\theta})$$

exists and is finite for almost all $e^{i\theta} \in \mathbb{T}$.

Proof. We use a conformal mapping between the right half plane and the unit disc to obtain a function in $H^\infty(\mathbb{D})$. Put

$$G(z) = \frac{1 - F(z)}{1 + F(z)}, \qquad (z \in \mathbb{D}).$$

Hence

$$|G(z)| < 1 \tag{5.25}$$

for all $z \in \mathbb{D}$ and thus, by Theorem 3.10,

$$g(e^{i\theta}) = \lim_{r \to 1} G(re^{i\theta})$$

exists for almost all $e^{i\theta} \in \mathbb{T}$. Moreover, $g(e^{i\theta}) = -1$ at most on a set of Lebesgue measure zero. Since otherwise, according to Theorem 5.13, we necessarily would have $G \equiv -1$ on $\mathbb{D}$, which contradicts (5.25). Therefore,

$$\lim_{r \to 1} F(re^{i\theta}) = \lim_{r \to 1} \frac{1 - G(re^{i\theta})}{1 + G(re^{i\theta})} = \frac{1 - g(e^{i\theta})}{1 + g(e^{i\theta})}$$

also exists and is finite for almost all $e^{i\theta} \in \mathbb{T}$. $\qquad \square$

We are now in a position to prove that the conjugate function also has radial limits almost everywhere on $\mathbb{T}$. We emphasize that

$$L^p(\mathbb{T}) \subset L^1(\mathbb{T}) \subset \mathcal{M}(\mathbb{T})$$

if $1 \leq p \leq \infty$, and thus the following result applies to the conjugate of elements of all $h^p(\mathbb{D})$ classes.

Theorem 5.15 *Let $\mu \in \mathcal{M}(\mathbb{T})$, and let*

$$V(re^{i\theta}) = \frac{1}{2\pi} \int_{\mathbb{T}} \frac{2r \sin(\theta - t)}{1 + r^2 - 2r\cos(\theta - t)} \, d\mu(e^{it}), \qquad (re^{i\theta} \in \mathbb{D}).$$

Then

$$\lim_{r \to 1} V(re^{i\theta})$$

exists and is finite for almost all $e^{i\theta} \in \mathbb{T}$.

Proof. Without loss of generality, assume that μ is a positive Borel measure on $\mathbb{T}$. Since otherwise, using Hahn's decomposition theorem, we can write $\mu = (\mu_1 - \mu_2) + i(\mu_3 - \mu_4)$, where $\mu_k \geq 0$, and then consider each μ_k separately. Let

$$F(z) = \frac{1}{2\pi} \int_{\mathbb{T}} \frac{e^{it} + z}{e^{it} - z} \, d\mu(e^{it}), \qquad (z \in \mathbb{D}). \qquad (5.26)$$

Hence

$$\Re F(z) = \frac{1}{2\pi} \int_{\mathbb{T}} \frac{1 - r^2}{1 + r^2 - 2r\cos(\theta - t)} \, d\mu(e^{it}), \qquad (z = re^{i\theta} \in \mathbb{D}),$$

and, according to our assumption on μ, we have $\Re F(z) > 0$ for all $z \in \mathbb{D}$. Thus F is an analytic function mapping the unit disc into the right half plane. Therefore, by Lemma 5.14,

$$f(e^{i\theta}) = \lim_{r \to 1} F(re^{i\theta})$$

exists and is finite for almost all $e^{i\theta} \in \mathbb{T}$. In particular, considering the imaginary parts of both sides of (5.26), we see that $\lim_{r \to 1} V(re^{i\theta})$ exists and is finite for almost all $e^{i\theta} \in \mathbb{T}$. $\qquad \square$

In the following lemma, we show that the difference of two functions of r tends to zero almost everywhere as $r \to 1$. According to Theorem 5.15, we know that the first function has a finite limit almost everywhere. Hence, the same conclusion holds for the second one. Therefore, we will be able to conclude that the Hilbert transform of an integrable function exists almost everywhere.

Lemma 5.16 *Let $u \in L^1(\mathbb{T})$, and let*

$$V(re^{i\theta}) = \frac{1}{2\pi} \int_{-\pi}^{\pi} \frac{2r \sin(\theta - t)}{1 + r^2 - 2r \cos(\theta - t)} u(e^{it})\, dt, \qquad (re^{i\theta} \in \mathbb{D}).$$

Then, for almost all $e^{i\theta} \in \mathbb{T}$,

$$\lim_{r \to 1} \left\{ V(re^{i\theta}) - \frac{1}{\pi} \int_{1-r<|t|<\pi} \frac{u(e^{i(\theta-t)})}{2 \tan(t/2)}\, dt \right\} = 0.$$

Proof. We show that the theorem holds at points $e^{i\theta} \in \mathbb{T}$ where

$$\lim_{\delta \to 0} \frac{1}{\delta} \int_0^{\delta} |\, u(e^{i(\theta-t)}) - u(e^{i(\theta+t)})\,|\, dt = 0.$$

According to a well-known theorem of Lebesgue, u fulfils this property at almost all points of $\mathbb{T}$.

First write

$$V(re^{i\theta}) = \frac{1}{\pi} \left(\int_0^{1-r} + \int_{1-r}^{\pi} \right) \frac{r \sin t}{1 + r^2 - 2r \cos t} \left(u(e^{i(\theta-t)}) - u(e^{i(\theta+t)}) \right) dt.$$

Hence

$$V(re^{i\theta}) - \frac{1}{\pi} \int_{1-r}^{\pi} \frac{u(e^{i(\theta-t)}) - u(e^{i(\theta+t)})}{2 \tan(t/2)}\, dt = \mathcal{I}_1 + \mathcal{I}_2,$$

where

$$\mathcal{I}_1 = \int_0^{1-r} \frac{r \sin t}{1 + r^2 - 2r \cos t} \left(u(e^{i(\theta-t)}) - u(e^{i(\theta+t)}) \right) dt$$

and

$$\mathcal{I}_2 = \frac{1}{\pi} \int_{1-r}^{\pi} \left(\frac{r \sin t}{1 + r^2 - 2r \cos t} - \frac{\sin t}{2(1 - \cos t)} \right) \left(u(e^{i(\theta-t)}) - u(e^{i(\theta+t)}) \right) dt.$$

For the first integral, based on our assumption at $e^{i\theta}$, we have

$$|\mathcal{I}_1| \leq \int_0^{1-r} \frac{r \sin t}{1 + r^2 - 2r \cos t} \left| u(e^{i(\theta-t)}) - u(e^{i(\theta+t)}) \right| dt$$

$$\leq \int_0^{1-r} \frac{t}{(1-r)^2} \left| u(e^{i(\theta-t)}) - u(e^{i(\theta+t)}) \right| dt$$

$$\leq \frac{1}{1-r} \int_0^{1-r} \left| u(e^{i(\theta-t)}) - u(e^{i(\theta+t)}) \right| dt = o(1), \qquad (\text{as } r \to 1).$$

On the other hand,

$$
\begin{aligned}
|\mathcal{I}_2| &\leq \frac{1}{\pi} \int_{1-r}^{\pi} \left| \frac{r \sin t}{1 + r^2 - 2r \cos t} - \frac{\sin t}{2(1 - \cos t)} \right| \, |u(e^{i(\theta-t)}) - u(e^{i(\theta+t)})| \, dt \\
&= \frac{1}{\pi} \int_{1-r}^{\pi} \frac{(1-r)^2 \sin t}{2(1 - \cos t)(1 + r^2 - 2r \cos t)} \, |u(e^{i(\theta-t)}) - u(e^{i(\theta+t)})| \, dt \\
&= \frac{1}{\pi} \int_{1-r}^{\pi} \frac{(1-r)^2 \cos(t/2)}{2 \sin(t/2)((1-r)^2 + 4r \sin^2(t/2))} \, |u(e^{i(\theta-t)}) - u(e^{i(\theta+t)})| \, dt \\
&\leq \frac{1}{\pi} \int_{1-r}^{\pi} \frac{(1-r)^2}{8r \sin^3(t/2)} \, |u(e^{i(\theta-t)}) - u(e^{i(\theta+t)})| \, dt \\
&\leq \frac{(1-r)^2 \pi^2}{8r} \int_{1-r}^{\pi} \frac{|u(e^{i(\theta-t)}) - u(e^{i(\theta+t)})|}{t^3} \, dt.
\end{aligned}
$$

To show that $\mathcal{I}_2 \to 0$, define $F : \mathbb{R} \longrightarrow \mathbb{R}$ by

$$
F(x) = \begin{cases} \dfrac{1}{x} \displaystyle\int_0^x |u(e^{i(\theta-t)}) - u(e^{i(\theta+t)})| \, dt & \text{if} \quad x \neq 0, \\[2ex] 0 & \text{if} \quad x = 0. \end{cases}
$$

The function F is *continuous* at zero and *bounded* on $\mathbb{R}$. Therefore, doing integration by parts, we obtain

$$
\begin{aligned}
|\mathcal{I}_2| &\leq \frac{(1-r)^2 \pi^2}{8r} \int_{1-r}^{\pi} \frac{|u(e^{i(\theta-t)}) - u(e^{i(\theta+t)})|}{t^3} \, dt \\
&= \frac{(1-r)^2 \pi^2}{8r} \int_{1-r}^{\pi} \frac{d(\, t F(t)\,)}{t^3} \\
&= \frac{(1-r)^2 \pi^2}{8r} \left(\frac{t F(t)}{t^3} \right) \Bigg|_{t=1-r}^{\pi} + \frac{(1-r)^2 \pi^2}{8r} \int_{1-r}^{\pi} \frac{3t F(t)}{t^4} \, dt \\
&= \frac{(1-r)^2 F(\pi)}{8r} - \frac{\pi^2 F(1-r)}{8r} + \frac{3\pi^2}{8r} \int_{1}^{\pi/(1-r)} \frac{F((1-r)\tau)}{\tau^3} \, d\tau \\
&\leq o(1) + \frac{3\pi^2}{8r} \int_{1}^{\infty} \frac{F((1-r)\tau)}{\tau^3} \, d\tau.
\end{aligned}
$$

By the dominated convergence theorem, the last integral is also $o(1)$ as $r \to 1$. Hence,

$$
V(re^{i\theta}) - \frac{1}{\pi} \int_{1-r}^{\pi} \frac{u(e^{i(\theta-t)}) - u(e^{i(\theta+t)})}{2 \tan(t/2)} \, dt = o(1), \qquad (\text{as } r \to 1).
$$

$\square$

The following result is a direct consequence of Theorem 5.15 and Lemma 5.16.

Theorem 5.17 *Let $u \in L^1(\mathbb{T})$, and let*

$$V(re^{i\theta}) = \frac{1}{2\pi} \int_{-\pi}^{\pi} \frac{2r \sin(\theta - t)}{1 + r^2 - 2r \cos(\theta - t)} u(e^{it}) \, dt, \qquad (re^{i\theta} \in \mathbb{D}).$$

Then both limits

$$\tilde{u}(e^{i\theta}) = \lim_{\varepsilon \to 0} \frac{1}{\pi} \int_{\varepsilon < |t| < \pi} \frac{u(e^{i(\theta - t)})}{2 \tan(t/2)} \, dt$$

and

$$\lim_{r \to 1} V(re^{i\theta})$$

exist, are finite, and besides

$$\lim_{r \to 1} V(re^{i\theta}) = \tilde{u}(e^{i\theta})$$

for almost all $e^{i\theta} \in \mathbb{T}$.

Corollary 5.18 *Let $u \in L^1(\mathbb{T})$, and let*

$$F(z) = \frac{1}{2\pi} \int_{-\pi}^{\pi} \frac{e^{it} + z}{e^{it} - z} u(e^{it}) \, dt, \qquad (z \in \mathbb{D}).$$

Then

$$\lim_{r \to 1} F(re^{i\theta}) = u(e^{i\theta}) + i\, \tilde{u}(e^{i\theta})$$

for almost all $e^{i\theta} \in \mathbb{T}$.

Remark: The assumption $u \in L^1(\mathbb{T})$ is not enough to ensure that $F \in H^1(\mathbb{D})$.

Proof. It is enough to note that

$$\frac{1 - r^2}{1 + r^2 - 2r \cos(\theta - t)} = \Re\left(\frac{e^{it} + z}{e^{it} - z} \right)$$

and that

$$\frac{2r \sin(\theta - t)}{1 + r^2 - 2r \cos(\theta - t)} = \Im\left(\frac{e^{it} + z}{e^{it} - z} \right),$$

where $z = re^{i\theta} \in \mathbb{D}$. Thus

$$F(re^{i\theta}) = (P_r * u)(e^{i\theta}) + i(Q_r * u)(e^{i\theta}).$$

The required result follows immediately from Lemma 3.10 and Theorem 5.17.

$\square$

Exercises

Exercise 5.7.1   Let $f, g \in H^1(\mathbb{T})$. Let $E \subset \mathbb{T}$ with $|E| > 0$. Suppose that $f(e^{i\theta}) = g(e^{i\theta})$ for almost all $e^{i\theta} \in E$. Show that $f = g$ almost everywhere on $\mathbb{T}$.

Exercise 5.7.2   Let $\mu \in \mathcal{M}(\mathbb{T})$ be an analytic measure. Prove that either $\mu \equiv 0$ or its support is all of $\mathbb{T}$.

Exercise 5.7.3   Let $u, v \in L^1(\mathbb{T})$ be real and such that $f = u + iv \in H^1(\mathbb{T})$. Suppose that

$$\int_{-\pi}^{\pi} u(e^{it}) \, dt = \int_{-\pi}^{\pi} v(e^{it}) \, dt = 0.$$

Show that

$$\tilde{u} = v \qquad \text{and} \qquad \tilde{v} = -u,$$

or equivalently

$$\tilde{f} = -if.$$

Hint: Use Corollary 5.18.

Exercise 5.7.4   Let $f \in H^1(\mathbb{T})$, and let $u = \Re f$. Suppose that

$$\int_{-\pi}^{\pi} u(e^{it}) \, dt = 0.$$

Show that

$$\tilde{\tilde{u}} = -u.$$

Hint: Use Exercise 5.7.3.

Exercise 5.7.5   Let $F \in H^1(\mathbb{D})$ and denote its boundary values by $f \in H^1(\mathbb{T})$. Show that

$$F(z) = \frac{i}{2\pi} \int_{-\pi}^{\pi} \frac{e^{it} + z}{e^{it} - z} \, \widetilde{\Re f}(e^{it}) \, dt + F(0), \qquad (z \in \mathbb{D}).$$

Hint: Use Exercises 5.6.5 and 5.7.3.

Chapter 6

Norm inequalities for the conjugate function

6.1   Kolmogorov's theorems

Let $u \in L^1(\mathbb{T})$, and let $U = P * u$. Then, by Corollary 2.15, $U \in h^1(\mathbb{D})$ and, by Fatou's theorem,

$$\lim_{r \to 1} U(re^{i\theta}) = u(e^{i\theta})$$

for almost all $e^{i\theta} \in \mathbb{T}$. For the conjugate function $V = Q * u$, according to Theorem 5.17, we also know that

$$\lim_{r \to 1} V(re^{i\theta}) = \tilde{u}(e^{i\theta})$$

for almost all $e^{i\theta} \in \mathbb{T}$. However, the mere assumption $u \in L^1(\mathbb{T})$ is not enough to ensure that $\tilde{u} \in L^1(\mathbb{T})$ and $V \in h^1(\mathbb{D})$. The best result in this direction is due to Kolmogorov, showing that $\tilde{u}$ is in weak-$L^1(\mathbb{T})$ and that $V \in h^p(\mathbb{D})$ for all $0 < p < 1$.

Theorem 6.1 (Kolmogorov) *Let $u \in L^1(\mathbb{T})$. Then, for each $\lambda > 0$,*

$$m_{\tilde{u}}(\lambda) \leq \frac{64\pi \, \|u\|_1}{4\|u\|_1 + \lambda}.$$

Proof. (Carleson) Fix $\lambda > 0$. First suppose that $u \geq 0$ and $u \not\equiv 0$. Therefore, the analytic function

$$F(z) = U(z) + iV(z) = \frac{1}{2\pi} \int_{-\pi}^{\pi} \frac{e^{it} + z}{e^{it} - z} \, u(e^{it}) \, dt, \qquad (z \in \mathbb{D}),$$

maps the unit disc into the right half plane and

$$F(0) = \frac{1}{2\pi} \int_{-\pi}^{\pi} u(e^{it}) \, dt = \|u\|_1.$$

131

The conformal mapping

$$z \longmapsto \frac{2z}{z+\lambda}$$

maps the right half plane into the disc $\{|z - 1| \leq 1\}$ (See Figure 6.1). In particular, it maps $\{|z| \geq \lambda, \, \Re z \geq 0\}$ into $\{|z - 1| \leq 1, \, \Re z \geq 1\}$.

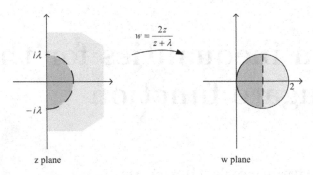

z plane   w plane

Fig. 6.1. The conformal mapping $z \longmapsto \frac{2z}{z+\lambda}$.

Therefore,

$$G(z) = \frac{2F(z)}{F(z) + \lambda}, \qquad (z \in \mathbb{D}),$$

is in $H^\infty(\mathbb{D})$ and, by Lemma 3.10 and Theorem 5.17, its boundary values are given by

$$g(e^{i\theta}) = \lim_{r \to 1} G(re^{i\theta}) = \frac{2(u(e^{i\theta}) + i\,\tilde{u}(e^{i\theta}))}{u(e^{i\theta}) + i\,\tilde{u}(e^{i\theta}) + \lambda}$$

for almost all $e^{i\theta} \in \mathbb{T}$. Hence, in the first place,

$$\Re\{g\} = 1 + \frac{u^2 + \tilde{u}^2 - \lambda^2}{(u+\lambda)^2 + \tilde{u}^2} \geq 1$$

provided that $\tilde{u} > \lambda$. Secondly, by Theorem 3.5,

$$G(0) = \frac{1}{2\pi} \int_{-\pi}^{\pi} g(e^{it}) \, dt,$$

which implies

$$\frac{2\|u\|_1}{\|u\|_1 + \lambda} = \frac{1}{2\pi} \int_{-\pi}^{\pi} \Re\{g(e^{it})\} \, dt.$$

Therefore,

$$\frac{4\pi\|u\|_1}{\|u\|_1 + \lambda} \geq \int_{\{|\tilde{u}|>\lambda\}} \Re\{g(e^{it})\} \, dt \geq \int_{\{|\tilde{u}|>\lambda\}} dt = m_{\tilde{u}}(\lambda).$$

For an arbitrary real $u \in L^1(\mathbb{T})$, write $u = u_1 - u_2$, where $u_k \geq 0$ and $u_1 u_2 \equiv 0$. Thus $\|u\|_1 = \|u_1\|_1 + \|u_2\|_1$ and $\tilde{u} = \tilde{u}_1 - \tilde{u}_2$. In this case, the inclusion

$$\{e^{i\theta} : |\tilde{u}(e^{i\theta})| > \lambda\} \subset \{e^{i\theta} : |\tilde{u}_1(e^{i\theta})| > \lambda/2\} \cup \{e^{i\theta} : |\tilde{u}_2(e^{i\theta})| > \lambda/2\}$$

implies

$$
\begin{aligned}
m_{\tilde{u}}(\lambda) &\leq m_{\tilde{u}_1}(\lambda/2) + m_{\tilde{u}_2}(\lambda/2) \\
&\leq \frac{4\pi \|u_1\|_1}{\|u_1\|_1 + \lambda/2} + \frac{4\pi \|u_2\|_1}{\|u_2\|_1 + \lambda/2} \\
&\leq \frac{8\pi (\|u_1\|_1 + \|u_2\|_1)}{(\|u_1\|_1 + \|u_2\|_1) + \lambda/2} = \frac{16\pi \|u\|_1}{2\|u\|_1 + \lambda}.
\end{aligned}
$$

For an arbitrary $u \in L^1(\mathbb{T})$, write $u = u_1 + i\,u_2$ where u_1 and u_2 are real functions in $L^1(\mathbb{T})$. Hence,

$$\|u_1\|_1 \leq \|u\|_1 \qquad \text{and} \qquad \|u_2\|_1 \leq \|u\|_1.$$

Since $\tilde{u} = \tilde{u}_1 + i\,\tilde{u}_2$, again the inclusion

$$\{e^{i\theta} : |\tilde{u}(e^{i\theta})| > \lambda\} \subset \{e^{i\theta} : |\tilde{u}_1(e^{i\theta})| > \lambda/2\} \cup \{e^{i\theta} : |\tilde{u}_2(e^{i\theta})| > \lambda/2\}$$

gives

$$m_{\tilde{u}}(\lambda) \leq m_{\tilde{u}_1}(\lambda/2) + m_{\tilde{u}_2}(\lambda/2),$$

and thus

$$m_{\tilde{u}}(\lambda) \leq \frac{16\pi \|u_1\|_1}{2\|u_1\|_1 + \lambda/2} + \frac{16\pi \|u_2\|_1}{2\|u_2\|_1 + \lambda/2} \leq \frac{64\pi \|u\|_1}{4\|u\|_1 + \lambda}.$$

$\square$

Let $u \in L^1(\mathbb{T})$, and let $V = Q * u$. The condition $u \in L^1(\mathbb{T})$ is not enough to conclude that $V \in h^1(\mathbb{D})$. Since otherwise, by Fatou's lemma, we would have

$$\|\tilde{u}\|_1 \leq \liminf_{r \to 1} \|V_r\|_1 < \infty$$

which is not true in the general case. However, we can show that $V \in h^p(\mathbb{D})$ for all $0 < p < 1$.

Corollary 6.2 (Kolmogorov) *Let $u \in L^1(\mathbb{T})$, and let*

$$V(re^{i\theta}) = \frac{1}{2\pi} \int_{-\pi}^{\pi} \frac{2r \sin(\theta - t)}{1 + r^2 - 2r \cos(\theta - t)} u(e^{it})\, dt, \qquad (re^{i\theta} \in \mathbb{D}).$$

Then, for each $0 < p < 1$, $V \in h^p(\mathbb{D})$, $\tilde{u} \in L^p(\mathbb{T})$ and

$$\|V\|_p = \|\tilde{u}\|_p \leq c_p \|u\|_1,$$

where c_p is a constant just depending on p.

Proof. Fix $0 < p < 1$. Let

$$U(re^{i\theta}) = \frac{1}{2\pi} \int_{-\pi}^{\pi} \frac{1 - r^2}{1 + r^2 - 2r\cos(\theta - t)} u(e^{it})\, dt, \qquad (re^{i\theta} \in \mathbb{D}).$$

Clearly $U \in h^1(\mathbb{D}) \supset h^p(\mathbb{D})$, and $F = U + iV$ is an analytic function on $\mathbb{D}$. Based on the content of Section 5.2, for each fixed $0 < r < 1$, the function $V_r(e^{i\theta})$ is the Hilbert transform of $U_r(e^{i\theta})$. Hence, by Theorem 6.1 and Corollary 4.13,

$$
\begin{aligned}
\|V_r\|_p^p &= \frac{1}{2\pi} \int_{-\pi}^{\pi} |V_r(e^{i\theta})|^p\, d\theta \\
&= \frac{1}{2\pi} \int_0^{\infty} p\,\lambda^{p-1}\, m_{V_r}(\lambda)\, d\lambda \\
&\leq \int_0^{\infty} p\,\lambda^{p-1}\, \frac{32\,\|U_r\|_1}{4\,\|U_r\|_1 + \lambda}\, d\lambda \\
&\leq \int_0^{\infty} p\,\lambda^{p-1}\, \frac{32\,\|u\|_1}{4\,\|u\|_1 + \lambda}\, d\lambda.
\end{aligned}
$$

See also Section A.5. Making the change of variable $\lambda = \|u\|_1\, s$ gives us

$$\|V_r\|_p \leq \left\{ 32p \int_0^{\infty} \frac{s^{p-1}}{4 + s}\, ds \right\}^{1/p} \|u\|_1.$$

Hence $V \in h^p(D)$ and $\|V\|_p \leq c_p \|u\|_1$. Since

$$\lim_{r \to 1} V(re^{i\theta}) = \tilde{u}(e^{i\theta})$$

for almost all $e^{i\theta} \in \mathbb{T}$, an application of Fatou's lemma implies that

$$\|\tilde{u}\|_p \leq \|V\|_p$$

and thus $\tilde{u} \in L^p(\mathbb{T})$. To prove that equality holds we need more tools, which will be discussed in Section 7.5. However, we do not use this fact. $\qquad \square$

A function $F = U + iV$ is in $H^p(\mathbb{D})$ if and only if $U, V \in h^p(D)$. If $u \in L^1(\mathbb{T})$ is given and we define

$$F(z) = U(z) + iV(z) = \frac{1}{2\pi} \int_{-\pi}^{\pi} \frac{e^{it} + z}{e^{it} - z} u(e^{it})\, dt, \qquad (z \in \mathbb{D}),$$

then, by Corollary 2.15, $U = P * u \in h^1(\mathbb{D})$, which implies $U \in h^p(\mathbb{D})$ for all $0 < p < 1$. On the other hand, Kolmogorov's theorem ensures that $V \in h^p(\mathbb{D})$ for all $0 < p < 1$. Hence, we obtain the following result.

Corollary 6.3 (Kolmogorov) *Let $u \in L^1(\mathbb{T})$, and let*

$$F(z) = \frac{1}{2\pi} \int_{-\pi}^{\pi} \frac{e^{it} + z}{e^{it} - z} u(e^{it})\, dt, \qquad (z \in \mathbb{D}).$$

Then $F \in H^p(\mathbb{D})$, for all $0 < p < 1$, and

$$\| F \|_p \le c_p \| u \|_1,$$

where c_p is a constant just depending on p.

Exercises

Exercise 6.1.1 Construct $u \in L^1(\mathbb{T})$ such that $\tilde{u} \notin L^1(\mathbb{T})$.

Exercise 6.1.2 Construct $u \in L^1(\mathbb{T})$ such that $F \notin H^1(\mathbb{D})$, where

$$F(z) = \frac{1}{2\pi} \int_{-\pi}^{\pi} \frac{e^{it} + z}{e^{it} - z} \, u(e^{it}) \, dt, \qquad (z \in \mathbb{D}).$$

Hint: Use Exercise 6.1.1.

6.2   Harmonic conjugate of $h^2(\mathbb{D})$ functions

The following deep result is a special case of M. Riesz's theorem (Theorem 6.6). However, we provide an extra proof for this particular, but important, case. We call your attention to the number of results used in its proof.

Theorem 6.4 *Let $u \in L^2(\mathbb{T})$, and let*

$$V(re^{i\theta}) = \frac{1}{2\pi} \int_{-\pi}^{\pi} \frac{2r \sin(\theta - t)}{1 + r^2 - 2r \cos(\theta - t)} \, u(e^{it}) \, dt, \qquad (re^{i\theta} \in \mathbb{D}).$$

Then $V \in h^2(\mathbb{D})$, $\tilde{u} \in L^2(\mathbb{T})$ and

$$\| V \|_2 = \| \tilde{u} \|_2 \le \| u \|_2.$$

Moreover,

$$V(re^{i\theta}) = \frac{1}{2\pi} \int_{-\pi}^{\pi} \frac{1 - r^2}{1 + r^2 - 2r \cos(\theta - t)} \, \tilde{u}(e^{it}) \, dt, \qquad (re^{i\theta} \in \mathbb{D}).$$

Proof. Since $u \in L^2(\mathbb{T})$, by Parseval's identity (Corollary 2.22),

$$\sum_{n=-\infty}^{\infty} |\hat{u}(n)|^2 = \| u \|_2^2 < \infty. \tag{6.1}$$

On the other hand, by Theorem 3.13,

$$V(re^{i\theta}) = \sum_{n=-\infty}^{\infty} (-i \operatorname{sgn} n) \, \hat{u}(n) \, r^{|n|} \, e^{in\theta}.$$

Hence, for $0 \leq r < 1$,

$$\|V_r\|_2^2 = \frac{1}{2\pi} \int_{-\pi}^{\pi} |V(re^{i\theta})|^2 \, dt = \sum_{\substack{n=-\infty \\ n \neq 0}}^{\infty} |\hat{u}(n)|^2 \, r^{2|n|}.$$

Thus, by the (discrete version) of the monotone convergence theorem,

$$\|V\|_2^2 = \sum_{\substack{n=-\infty \\ n \neq 0}}^{\infty} |\hat{u}(n)|^2.$$

Comparing with (6.1), we obtain

$$\|V\|_2^2 = \|u\|_2^2 - |\hat{u}(0)|^2.$$

Thus $V \in h^2(\mathbb{D})$ and $\|V\|_2 \leq \|u\|_2$.

Knowing that $V \in h^2(\mathbb{D})$, by Theorem 3.6, there exists $v \in L^2(\mathbb{T})$ such that

$$V(re^{i\theta}) = \frac{1}{2\pi} \int_{-\pi}^{\pi} \frac{1-r^2}{1+r^2-2r\cos(\theta-t)} \, v(e^{it}) \, dt, \qquad (re^{i\theta} \in \mathbb{D}),$$

and

$$\|V\|_2 = \|v\|_2.$$

According to Lemma 3.10, the radial limit $\lim_{r \to 1} V(re^{i\theta})$ exists and is equal to $v(e^{i\theta})$ for almost all $e^{i\theta} \in \mathbb{T}$. On the other hand, by Theorem 5.17,

$$\lim_{r \to 1} V(re^{i\theta}) = \tilde{u}(e^{i\theta})$$

for almost all $e^{i\theta} \in \mathbb{T}$. Hence $\tilde{u} = v$. $\square$

6.3   M. Riesz's theorem

To prove M. Riesz's theorem we consider two cases, $1 < p < 2$ and $2 < p < \infty$, and for the second case we need the following lemma.

Lemma 6.5 *Let U_1 and U_2 be harmonic functions on $\mathbb{D}$. Let V_1 and V_2 denote their harmonic conjugates satisfying $V_1(0) = V_2(0) = 0$. Then, for each $0 \leq r < 1$,*

$$\int_{-\pi}^{\pi} U_1(re^{i\theta}) \, V_2(re^{i\theta}) \, d\theta = - \int_{-\pi}^{\pi} V_1(re^{i\theta}) \, U_2(re^{i\theta}) \, d\theta.$$

Proof. The function

$$F(z) = \big(U_1(z) + iV_1(z)\big)\big(U_2(z) + iV_2(z)\big)$$

is analytic on $\mathbb{D}$. Hence, its imaginary part

$$U_1(z) \, V_2(z) + V_1(z) \, U_2(z)$$

is harmonic on $\mathbb{D}$ and

$$U_1(0)\, V_2(0) + V_1(0)\, U_2(0) = 0.$$

Therefore, by Corollary 3.2,

$$\frac{1}{2\pi} \int_{-\pi}^{\pi} \left(U_1(re^{i\theta})\, V_2(re^{i\theta}) + V_1(re^{i\theta})\, U_2(re^{i\theta}) \right) d\theta = 0.$$

$\square$

The following celebrated result is due to Marcel Riesz. According to this result an analytic function $F = U + iV$ is in $H^p(\mathbb{D})$, $1 < p < \infty$, *if and only if* its real part U is in $h^p(\mathbb{D})$. It has several interesting proofs. The one given here is by P. Stein. In the proof, two elementary identities for the Laplacian

$$\nabla^2 = \frac{\partial^2}{\partial x^2} + \frac{\partial^2}{\partial y^2}$$

of analytic functions are used. These identities are simple consequences of the Cauchy–Riemann equations. Moreover, we need Green's formula

$$\int_{-\pi}^{\pi} \frac{\partial \varphi}{\partial r}(re^{i\theta})\, r d\theta = \int_{0}^{r} \int_{-\pi}^{\pi} \nabla^2 \, \varphi(\rho e^{i\theta})\, \rho d\rho d\theta,$$

where φ is a twice continuously differentiable function on the unit disc $\mathbb{D}$.

Theorem 6.6 (M. Riesz) *Let $u \in L^p(\mathbb{T})$, $1 < p < \infty$, and let*

$$V(re^{i\theta}) = \frac{1}{2\pi} \int_{-\pi}^{\pi} \frac{2r\,\sin(\theta - t)}{1 + r^2 - 2r\cos(\theta - t)}\, u(e^{it})\, dt, \qquad (re^{i\theta} \in \mathbb{D}).$$

Then $V \in h^p(\mathbb{D})$, $\tilde{u} \in L^p(\mathbb{T})$ and

$$\|V\|_p = \|\tilde{u}\|_p \le c_p \,\| u \|_p,$$

where c_p is a constant just depending on p. Moreover,

$$V(re^{i\theta}) = \frac{1}{2\pi} \int_{-\pi}^{\pi} \frac{1 - r^2}{1 + r^2 - 2r\cos(\theta - t)}\, \tilde{u}(e^{it})\, dt, \qquad (re^{i\theta} \in \mathbb{D}).$$

Proof. The case $p = 2$ was established in Theorem 6.4 (as a matter of fact, a slight modification of the following proof also works for $p = 2$).

Case 1: Suppose that $1 < p < 2$, $u \ge 0$, $u \not\equiv 0$.

Let

$$U(re^{i\theta}) = \frac{1}{2\pi} \int_{-\pi}^{\pi} \frac{1 - r^2}{1 + r^2 - 2r\cos(\theta - t)}\, u(e^{it})\, dt, \qquad (re^{i\theta} \in \mathbb{D}).$$

Hence
$$U(z) > 0, \qquad (z \in \mathbb{D}).$$
Moreover, $F(z) = U(z) + iV(z)$ is analytic on $\mathbb{D}$ with $F(0) = U(0)$ and
$$|F(z)| > 0, \qquad (z \in \mathbb{D}).$$
Since U and $|F|$ are strictly positive on $\mathbb{D}$, they are infinitely differentiable at each point of $\mathbb{D}$. Moreover, by the Cauchy–Riemann equations,
$$\nabla^2 \, U^p \,(z) = p(p-1) \, |F'(z)|^2 \, |U(z)|^{p-2}$$
and
$$\nabla^2 \, |F|^p \,(z) = p^2 \, |F'(z)|^2 \, |F(z)|^{p-2},$$
and thus (note that $p - 2 < 0$)
$$\nabla^2 \, |F|^p \,(z) \le \frac{p}{p-1} \, \nabla^2 \, U^p \,(z) \tag{6.2}$$
for all $z \in \mathbb{D}$. Now, by Green's theorem,
$$\int_{-\pi}^{\pi} \frac{\partial}{\partial r} |F|^p \, (re^{i\theta}) \, rd\theta \;=\; \int_0^r \int_{-\pi}^{\pi} \nabla^2 \, |F|^p \, (\rho e^{i\theta}) \, \rho d\rho d\theta,$$
$$\int_{-\pi}^{\pi} \frac{\partial}{\partial r} U^p \, (re^{i\theta}) \, rd\theta \;=\; \int_0^r \int_{-\pi}^{\pi} \nabla^2 \, U^p \, (\rho e^{i\theta}) \, \rho d\rho d\theta.$$
Therefore, by (6.2), we have
$$\frac{d}{dr}\left(\int_{-\pi}^{\pi} |F(re^{i\theta})|^p \, d\theta \right) \le \frac{p}{p-1} \frac{d}{dr}\left(\int_{-\pi}^{\pi} U^p(re^{i\theta}) \, d\theta \right).$$
Integrating both sides over $[0, r]$ gives us
$$\int_{-\pi}^{\pi} |F(re^{i\theta})|^p \, d\theta - \int_{-\pi}^{\pi} |F(0)|^p \, d\theta \le \frac{p}{p-1} \left(\int_{-\pi}^{\pi} U^p(re^{i\theta}) \, d\theta - \int_{-\pi}^{\pi} U^p(0) \, d\theta \right).$$
Thus
$$\frac{1}{2\pi} \int_{-\pi}^{\pi} |V(re^{i\theta})|^p \, d\theta \;\le\; \frac{1}{2\pi} \int_{-\pi}^{\pi} |F(re^{i\theta})|^p \, d\theta$$
$$\le\; \frac{p}{p-1} \frac{1}{2\pi} \int_{-\pi}^{\pi} U^p(re^{i\theta}) \, d\theta - \frac{p}{p-1} U^p(0).$$
Therefore, by Corollary 2.15,
$$\| V_r \|_p \le \left(\frac{p}{p-1} \right)^{\frac{1}{p}} \| U_r \|_p \le \left(\frac{p}{p-1} \right)^{\frac{1}{p}} \| u \|_p. \tag{6.3}$$

Case 2: $1 < p < 2$.

For a general $u \in L^p(\mathbb{T})$, write $u = (u^{(1)} - u^{(2)}) + i(u^{(3)} - u^{(4)})$, where $u^{(k)} \geq 0$ and $u^{(1)} u^{(2)} \equiv u^{(3)} u^{(4)} \equiv 0$. Let $V^{(k)} = Q * u^{(k)}$. Hence $V_r = (V_r^{(1)} - V_r^{(2)}) + i(V_r^{(3)} - V_r^{(4)})$, and by (6.3),

$$
\begin{aligned}
\|V_r\|_p &\leq \|V_r^{(1)}\|_p + \|V_r^{(2)}\|_p + \|V_r^{(3)}\|_p + \|V_r^{(4)}\|_p \\
&\leq \left(\frac{p}{p-1}\right)^{\frac{1}{p}} \left(\|u^{(1)}\|_p + \|u^{(2)}\|_p + \|u^{(3)}\|_p + \|u^{(4)}\|_p\right).
\end{aligned}
$$

Therefore,

$$
\begin{aligned}
\|V_r\|_p^p &\leq 4^p \left(\frac{p}{p-1}\right) \left(\|u^{(1)}\|_p^p + \|u^{(2)}\|_p^p + \|u^{(3)}\|_p^p + \|u^{(4)}\|_p^p\right) \\
&= 4^p \left(\frac{p}{p-1}\right) \left(\|u^{(1)} - u^{(2)}\|_p^p + \|u^{(3)} - u^{(4)}\|_p^p\right) \\
&\leq 2 \times 4^p \left(\frac{p}{p-1}\right) \|u\|_p^p.
\end{aligned}
$$

This finishes the case $1 < p < 2$.

Case 3: $2 < p < \infty$.

Let q be its conjugate exponent of p (note that $1 < q < 2$). Let

$$
U^{(1)}(re^{i\theta}) = \sum_{n=-N}^{M} a_n r^{|n|} e^{in\theta}, \qquad (re^{i\theta} \in \mathbb{D}),
$$

be an arbitrary trigonometric polynomial. The harmonic conjugate of $U^{(1)}$ is

$$
V^{(1)}(re^{i\theta}) = \sum_{n=-N}^{M} -i\,\mathrm{sgn}(n)\, a_n r^{|n|} e^{in\theta}, \qquad (re^{i\theta} \in \mathbb{D}).
$$

By Lemma 6.5, by Hölder's inequality, and then by (6.3), we have

$$
\begin{aligned}
\left| \frac{1}{2\pi} \int_{-\pi}^{\pi} V(re^{i\theta}) U^{(1)}(re^{i\theta})\, d\theta \right| &\leq \left| \frac{1}{2\pi} \int_{-\pi}^{\pi} U(re^{i\theta}) V^{(1)}(re^{i\theta})\, d\theta \right| \\
&\leq \|U_r\|_p \|V_r^{(1)}\|_q \\
&\leq \|u\|_p \times c_q \|U_r^{(1)}\|_q.
\end{aligned}
$$

Since trigonometric polynomials are dense in $L^q(\mathbb{T})$, if we take the supremum over all nonzero polynomials $U^{(1)}$ in the inequality

$$
\frac{1}{\|U_r^{(1)}\|_q} \left| \frac{1}{2\pi} \int_{-\pi}^{\pi} V(re^{i\theta}) U^{(1)}(re^{i\theta})\, d\theta \right| \leq c_q \|u\|_p,
$$

we obtain

$$
\|V_r\|_p \leq c_q \|u\|_p.
$$

Therefore, we now know that $V \in h^p(\mathbb{D})$, $1 < p < \infty$, with $\|V\|_p \leq c_p \|u\|_p$. The rest of the proof is as given for Theorem 6.4. $\square$

If $u \in L^1(\mathbb{T})$ is real, then certainly

$$F(z) = \frac{1}{2\pi} \int_{-\pi}^{\pi} \frac{e^{it} + z}{e^{it} - z} \, u(e^{it}) \, dt, \qquad (z \in \mathbb{D}),$$

is an analytic function on the unit disc satisfying

$$\Re F(re^{i\theta}) = \frac{1}{2\pi} \int_{-\pi}^{\pi} \frac{1 - r^2}{1 + r^2 - 2r\cos(\theta - t)} \, u(e^{it}) \, dt, \qquad (re^{i\theta} \in \mathbb{D}).$$

If we know that $F \in H^p(\mathbb{D})$, $1 \leq p \leq \infty$, then we necessarily have $\Re F \in h^p(\mathbb{D})$, and thus, by Theorems 3.5, 3.6 and 3.7, $u \in L^p(\mathbb{T})$. M. Riesz's theorem enables us to obtain the inverse of this result whenever $1 < p < \infty$. It is not difficult to construct $u \in L^1(\mathbb{T})$ such that $F \notin H^1(\mathbb{D})$, or similarly a function $u \in L^\infty(\mathbb{T})$ which gives $F \notin H^\infty(\mathbb{D})$.

Corollary 6.7 *Let $u \in L^p(\mathbb{T})$, $1 < p < \infty$, be real, and let*

$$F(z) = \frac{1}{2\pi} \int_{-\pi}^{\pi} \frac{e^{it} + z}{e^{it} - z} \, u(e^{it}) \, dt, \qquad (z \in \mathbb{D}).$$

Then $F \in H^p(\mathbb{D})$ and

$$\|u\|_p \leq \|F\|_p \leq c_p \|u\|_p,$$

where c_p is a constant just depending on p. Moreover, the boundary values of F are given by

$$f = u + i\tilde{u} \in H^p(\mathbb{T}).$$

Proof. By Corollary 2.15, $\Re F \in h^p(\mathbb{D})$ and by M. Riesz's theorem, $\Im F \in h^p(\mathbb{D})$. Hence, $F \in H^p(\mathbb{D})$. Moreover, by Corollary 2.15 and Theorem 6.6, we have

$$\|u\|_p = \|\Re F\|_p \leq \|F\|_p \leq \|\Re F\|_p + \|\Im F\|_p \leq c_p \|u\|_p.$$

Then, by Corollary 5.18,

$$f(e^{i\theta}) = \lim_{r \to 1} F(re^{i\theta}) = u(e^{i\theta}) + i\tilde{u}(e^{i\theta})$$

for almost all $e^{i\theta} \in \mathbb{T}$. In other words, there is an $f \in H^p(\mathbb{T})$ such that $u = \Re f$ and $\tilde{u} = \Im f$. □

Exercises

Exercise 6.3.1   Find $u \in L^1(\mathbb{T})$ such that $F \notin H^1(\mathbb{D})$, where

$$F(z) = \frac{1}{2\pi} \int_{-\pi}^{\pi} \frac{e^{it} + z}{e^{it} - z} \, u(e^{it}) \, dt, \qquad (z \in \mathbb{D}).$$

Exercise 6.3.2   Find $u \in L^\infty(\mathbb{T})$ such that $F \notin H^\infty(\mathbb{D})$, where

$$F(z) = \frac{1}{2\pi} \int_{-\pi}^{\pi} \frac{e^{it} + z}{e^{it} - z} \, u(e^{it}) \, dt, \qquad (z \in \mathbb{D}).$$

Exercise 6.3.3   Let $u \in L^p(T)$, $1 < p < \infty$. Show that

$$\hat{\tilde{u}}(n) = -i \operatorname{sgn}(n) \, \hat{u}(n), \qquad (n \in \mathbb{Z}).$$

Hint: Use both representations given in Theorem 6.6.

Exercise 6.3.4   Let $u \in L^p(\mathbb{T})$, $1 < p < \infty$. Let

$$Pu(e^{it}) = \sum_{n=0}^{\infty} \hat{u}(n) \, e^{int}.$$

Show that $Pu \in H^p(\mathbb{T})$ and that

$$\|Pu\|_p \leq C_p \, \|u\|_p,$$

where C_p is an absolute constant.
Hint: Use Theorem 6.6 and Exercise 6.3.3. Note that

$$2Pu = \hat{u}(0) + u + i\tilde{u}.$$

Remark: The operator $P : L^p(\mathbb{T}) \longrightarrow H^p(\mathbb{T})$ is called the Riesz projection.

Exercise 6.3.5   Let $u \in L^p(\mathbb{T})$, $1 < p < \infty$, with

$$\int_{-\pi}^{\pi} u(e^{it}) \, dt = 0.$$

Show that

$$\tilde{\tilde{u}} = -u.$$

Hint: By Corollary 6.7, there is an $f \in H^p(\mathbb{T})$ such that $f = u + i\tilde{u}$. Replace f by if.
Remark: Compare with Exercise 5.7.4. A real $u \in L^1(\mathbb{T})$ is not necessarily the real part of an $f \in H^1(\mathbb{T})$.

Exercise 6.3.6   Let $u \in L^p(\mathbb{T})$, $1 < p < \infty$. Suppose that

$$\int_{-\pi}^{\pi} u(e^{it}) \, dt = 0.$$

Show that there are constants c_p and C_p such that

$$c_p \, \|\tilde{u}\|_p \leq \|u\|_p \leq C_p \, \|\tilde{u}\|_p.$$

Hint: Use Exercise 6.3.5 and Theorem 6.6.

6.4   The Hilbert transform of bounded functions

If $u \in L^{\infty}(\mathbb{T})$, then $u \in L^p(\mathbb{T})$ for all $1 < p < \infty$, and thus by Theorem 6.6,

$$\tilde{u} \in \bigcap_{1<p<\infty} L^p(\mathbb{T}). \tag{6.4}$$

However, by a simple example, we see that $\tilde{u}$ is not necessarily in $L^{\infty}(\mathbb{T})$ for an arbitrary $u \in L^{\infty}(\mathbb{T})$. Let us consider the analytic function

$$\begin{aligned}
F(z) &= i \log\left(\frac{1+z}{1-z}\right) \\
&= -\arctan\left(\frac{2r\sin\theta}{1-r^2}\right) + \frac{i}{2}\log\left(\frac{1+r^2+2r\cos\theta}{1+r^2-2r\cos\theta}\right),
\end{aligned}$$

where $z = re^{i\theta} \in \mathbb{D}$. Hence

$$u(e^{i\theta}) = \lim_{r\to 1} \Re F(re^{i\theta}) = \begin{cases} -\dfrac{\pi}{2} & \text{if} \quad 0 < \theta < \pi, \\[2mm] \dfrac{\pi}{2} & \text{if} \quad -\pi < \theta < 0, \end{cases}$$

and

$$v(e^{i\theta}) = \lim_{r\to 1} \Im F(re^{i\theta}) = -\log|\tan(\theta/2)|.$$

Since $F(z) = 2\sum_{n=0}^{\infty} z^{2n+1}/(2n+1)$, by Theorem 5.1, $F \in H^2(\mathbb{D})$. Thus, by Theorem 5.9,

$$\tilde{u}(e^{i\theta}) = v(e^{i\theta}) = -\log|\tan(\theta/2)| \tag{6.5}$$

for $e^{i\theta} \in \mathbb{T} \setminus \{1, -1\}$. It is clear that $\tilde{u}$ is unbounded. Nevertheless, $\tilde{u} \in L^p(\mathbb{T})$ for all $1 < p < \infty$. Now we show that the Hilbert transform of a bounded function satisfies a stronger condition than (6.4).

Theorem 6.8 (Zygmund) *Let $u \in L^{\infty}(\mathbb{T})$ be real with*

$$\|u\|_{\infty} \leq \frac{\pi}{2}.$$

Then

$$\frac{1}{2\pi} \int_{-\pi}^{\pi} e^{|\tilde{u}(e^{i\theta})|} \cos(u(e^{i\theta})) \, d\theta \leq 2.$$

Proof. Let

$$F(z) = \frac{1}{2\pi} \int_{-\pi}^{\pi} \frac{e^{it}+z}{e^{it}-z} u(e^{it}) \, dt = U(z) + iV(z), \qquad (z \in \mathbb{D}).$$

The condition $\|u\|_{\infty} \leq \pi/2$ implies

$$-\frac{\pi}{2} \leq U(re^{i\theta}) \leq \frac{\pi}{2},$$

and thus $\cos(U(re^{i\theta})) \geq 0$, for all $re^{i\theta} \in \mathbb{D}$. Fix $r < 1$. Then

$$\frac{1}{2\pi} \int_{-\pi}^{\pi} e^{-i F(re^{i\theta})} \, d\theta = e^{-i F(0)}.$$

Taking the real part of both sides gives

$$\frac{1}{2\pi} \int_{-\pi}^{\pi} e^{V(re^{i\theta})} \cos(U(re^{i\theta})) \, d\theta = \cos(U(0)) \leq 1.$$

Replace F by $-F$ and repeat the preceding argument to get

$$\frac{1}{2\pi} \int_{-\pi}^{\pi} e^{-V(re^{i\theta})} \cos(U(re^{i\theta})) \, d\theta \leq 1.$$

The last two inequalities together imply

$$\frac{1}{2\pi} \int_{-\pi}^{\pi} e^{|V(re^{i\theta})|} \cos(U(re^{i\theta})) \, d\theta \leq 2.$$

Now, by Fatou's lemma,

$$\frac{1}{2\pi} \int_{-\pi}^{\pi} e^{|\tilde{u}(e^{i\theta})|} \cos(u(e^{i\theta})) \, d\theta$$

$$= \frac{1}{2\pi} \int_{-\pi}^{\pi} \liminf_{r \to 1} \left\{ e^{|V(re^{i\theta})|} \cos(U(re^{i\theta})) \right\} d\theta$$

$$\leq \liminf_{r \to 1} \frac{1}{2\pi} \int_{-\pi}^{\pi} e^{|V(re^{i\theta})|} \cos(U(re^{i\theta})) \, d\theta \leq 2.$$

$\square$

Corollary 6.9 *Let $u \in L^{\infty}(\mathbb{T})$ be real. Then*

$$\frac{1}{2\pi} \int_{-\pi}^{\pi} e^{\lambda |\tilde{u}(e^{i\theta})|} \, d\theta \leq 2 \sec(\lambda \|u\|_{\infty})$$

for

$$0 \leq \lambda < \frac{\pi}{2 \|u\|_{\infty}}.$$

Proof. By Theorem 6.8,

$$\frac{1}{2\pi} \int_{-\pi}^{\pi} e^{\lambda |\tilde{u}(e^{i\theta})|} \cos(\lambda u(e^{i\theta})) \, d\theta \leq 2$$

provided that $0 \leq \lambda < \pi/(2 \|u\|_{\infty})$. Since

$$-\|u\|_{\infty} \leq U(re^{i\theta}) \leq \|u\|_{\infty},$$

for all $re^{i\theta} \in \mathbb{D}$, and

$$\lambda \|u\|_{\infty} < \frac{\pi}{2},$$

then
$$\cos(\lambda\, U(re^{i\theta})) \geq \cos(\lambda\, \|u\|_{\infty}) > 0.$$
Therefore,
$$\frac{1}{2\pi}\int_{-\pi}^{\pi} e^{\lambda\,|\tilde{u}(e^{i\theta})|}\, d\theta \leq 2/\cos(\lambda\, \|u\|_{\infty}).$$

$\square$

As (6.5) shows, the condition $\lambda < \pi/(2\, \|u\|_{\infty})$ is sharp. In that example, if $\lambda = \pi/(2\, \|u\|_{\infty}) = 1$ then
$$\exp\left(\lambda\,|\tilde{u}(e^{i\theta})|\right) = 1/|\tan\theta/2|$$
for $\theta \in (-\pi/2, \pi/2)$, and this function is not integrable.

Corollary 6.10 *Let $u \in L^{\infty}(\mathbb{T})$ be real. Then*
$$\limsup_{n\to\infty} \frac{\|\tilde{u}\|_n}{n} \leq \frac{2\, \|u\|_{\infty}}{\pi e}.$$

Proof. On the one hand, for $\lambda > 0$,
$$\frac{1}{2\pi}\int_{-\pi}^{\pi} e^{\lambda\,|\tilde{u}(e^{i\theta})|}\, d\theta = \sum_{n=0}^{\infty} \frac{\|\tilde{u}\|_n^n}{n!}\, \lambda^n$$
and the series is convergent if and only if
$$\limsup_{n\to\infty} \frac{\|\tilde{u}\|_n}{(n!)^{1/n}}\, \lambda < 1.$$
On the other hand, by Corollary 6.9,
$$\frac{1}{2\pi}\int_{-\pi}^{\pi} e^{\lambda\,|\tilde{u}(e^{i\theta})|}\, d\theta < \infty,$$
provided that $0 \leq \lambda < \pi/(2\, \|u\|_{\infty})$. Therefore,
$$\limsup_{n\to\infty} \frac{\|\tilde{u}\|_n}{(n!)^{1/n}} \leq \frac{2\, \|u\|_{\infty}}{\pi}.$$
Finally, by Stirling's formula $\lim_{n\to\infty}\frac{n}{(n!)^{1/n}} = e$ and hence the result follows.

$\square$

6.5   The Hilbert transform of Dini continuous functions

A continuous function on $\mathbb{T}$ is certainly bounded. Hence, at least its Hilbert transform is integrable as described in Corollary 6.9. But continuity on $\mathbb{T}$ and the fact that trigonometric polynomials are dense in $\mathcal{C}(\mathbb{T})$ enable us to get rid of the restriction $\lambda < \pi/(2\, \|u\|_{\infty})$ in this case.

Theorem 6.11 *Let $u \in C(\mathbb{T})$. Then*

$$\int_{-\pi}^{\pi} \exp\left(\lambda \, |\tilde{u}(e^{i\theta})| \right) d\theta < \infty$$

for all $\lambda \in \mathbb{R}$.

Proof. The assertion for $\lambda \leq 0$ is trivial. Hence suppose that $\lambda > 0$. Fix $\varepsilon > 0$. There is a trigonometric polynomial $\sum_{n=-M}^{N} a_n \, e^{in\theta}$ such that $\|\varphi\|_\infty < \varepsilon$, where

$$\varphi(e^{i\theta}) = u(e^{i\theta}) - \sum_{n=-M}^{N} a_n \, e^{in\theta}, \qquad (e^{i\theta} \in \mathbb{T}).$$

Hence

$$\tilde{u}(e^{i\theta}) = \tilde{\varphi}(e^{i\theta}) + \sum_{n=-M}^{N} -i \, \mathrm{sgn}\, n \, a_n \, e^{in\theta}, \qquad (e^{i\theta} \in \mathbb{T}),$$

and thus

$$|\tilde{u}(e^{i\theta})| \leq |\tilde{\varphi}(e^{i\theta})| + \sum_{n=-M}^{N} |a_n|, \qquad (e^{i\theta} \in \mathbb{T}).$$

By Corollary 6.9, we have

$$\int_{-\pi}^{\pi} e^{\lambda \, |\tilde{u}(e^{i\theta})|} \, d\theta \leq e^{\sum_{n=-M}^{N} |a_n|} \int_{-\pi}^{\pi} e^{\lambda \, |\tilde{\varphi}(e^{i\theta})|} \, d\theta < \infty,$$

whenever

$$0 \leq \lambda < \frac{\pi}{2\,\varepsilon} < \frac{\pi}{2\,\|\varphi\|_\infty}.$$

Since ε is arbitrary, $e^{\lambda \, |\tilde{u}|}$ is integrable for all λ. □

The *modulus of continuity* of $u \in C(\mathbb{T})$ is defined by

$$\omega(t) = \omega_u(t) = \sup_{|\theta - \theta'| < t} |u(e^{i\theta}) - u(e^{i\theta'})|. \qquad (6.6)$$

One can easily show that ω is a positive decreasing function with

$$\lim_{t \to 0} \omega(t) = 0$$

and

$$\omega(t_1 + t_2) \leq \omega(t_1) + \omega(t_2)$$

for all $t_1, t_2 > 0$. Moreover, if we define ω_f for an arbitrary function f as above, then f is continuous on $\mathbb{T}$ if and only if $\lim_{t \to 0} \omega_f(t) = 0$.

The function u is called *Dini continuous* on $\mathbb{T}$ if

$$\int_0^{\pi} \frac{\omega(t)}{t} \, dt < \infty.$$

Clearly, the upper bound π can be replaced by any other positive number. If u is Dini continuous on $\mathbb{T}$, then

$$\lim_{\delta \to 0} \int_0^\delta \frac{\omega(t)}{t} \, dt = 0 \tag{6.7}$$

and the inequality

$$\delta \int_\delta^\pi \frac{\omega(t)}{t^2} \, dt \le \int_\delta^{\delta'} \frac{\omega(t)}{t} \, dt + \frac{\delta}{\delta'} \int_{\delta'}^\pi \frac{\omega(t)}{t} \, dt$$

shows that we also have

$$\lim_{\delta \to 0} \delta \int_\delta^\pi \frac{\omega(t)}{t^2} \, dt = 0. \tag{6.8}$$

Theorem 6.12 *Let u be Dini continuous on $\mathbb{T}$. Then $\tilde{u}$ exists at all points of $\mathbb{T}$ and moreover*

$$\omega_{\tilde{u}}(\delta) \le C \left(\int_0^\delta \frac{\omega_u(t)}{t} \, dt + \delta \int_\delta^\pi \frac{\omega_u(t)}{t^2} \, dt \right),$$

where C is an absolute constant.

Proof. Since

$$\left| \frac{u(e^{i(\theta-t)}) - u(e^{i(\theta+t)})}{2\tan(t/2)} \right| \le C \, \frac{\omega_u(t)}{t}$$

then, by (5.10), Dini continuity ensures that $\tilde{u}(e^{i\theta})$ exists at all $e^{i\theta} \in \mathbb{T}$ and it is given by

$$\tilde{u}(e^{i\theta}) = \frac{1}{\pi} \int_0^\pi \frac{u(e^{i(\theta-t)}) - u(e^{i(\theta+t)})}{2\tan(t/2)} \, dt. \tag{6.9}$$

For a similar reason, $Hu(e^{i\theta})$ also exists for all $e^{i\theta} \in \mathbb{T}$ and

$$Hu(e^{i\theta}) = \frac{1}{\pi} \int_0^\pi \frac{u(e^{i(\theta-t)}) - u(e^{i(\theta+t)})}{t} \, dt. \tag{6.10}$$

Representations (6.9) and (6.10), along with the property (5.8), imply that

$$\omega_{\tilde{u}-Hu}(\delta) \le C \, \omega_u(\delta) \le C \delta \int_\delta^\pi \frac{\omega_u(t)}{t^2} \, dt \tag{6.11}$$

at least for $\delta < \pi/2$. Since

$$\omega_{\tilde{u}} \le \omega_{\tilde{u}-Hu} + \omega_{Hu}$$

it is enough to prove the theorem for Hu.

Fix $\delta > 0$. Consider two arbitrary points θ_1 and θ_2 with $|\theta_2 - \theta_1| < \delta$ and let $\theta_3 = (\theta_1 + \theta_2)/2$. Without loss of generality, assume that $\theta_1 < \theta_2$. Since the

H transform of the constant function is zero, replacing u by $u - u(\theta_3)$, we also assume that $u(\theta_3) = 0$. According to the main definition (5.7),

$$Hu(e^{i\theta}) = \lim_{\varepsilon \to 0} \frac{1}{\pi} \int_{\varepsilon < |\theta - t| < \pi} \frac{u(e^{it})}{\theta - t} \, dt,$$

and thus, for any $\delta > 0$ and any θ,

$$
\begin{aligned}
Hu(e^{i\theta}) &= \lim_{\varepsilon \to 0} \frac{1}{\pi} \left(\int_{\varepsilon < |t - \theta| < \delta} + \int_{\delta < |t - \theta| < \pi} \right) \frac{u(e^{it})}{\theta - t} \, dt \\
&= \frac{1}{\pi} \int_{\delta < |t - \theta| < \pi} \frac{u(e^{it})}{\theta - t} \, dt + \lim_{\varepsilon \to 0} \frac{1}{\pi} \int_{\varepsilon < |t - \theta| < \delta} \frac{u(e^{it}) - u(e^{i\theta})}{\theta - t} \, dt \\
&= \frac{1}{\pi} \int_{\delta < |t - \theta| < \pi} \frac{u(e^{it})}{\theta - t} \, dt + \frac{1}{\pi} \int_{|t - \theta| < \delta} \frac{u(e^{it}) - u(e^{i\theta})}{\theta - t} \, dt.
\end{aligned}
$$

For the last integral we have

$$\left| \int_{|t - \theta| < \delta} \frac{u(e^{it}) - u(e^{i\theta})}{\theta - t} \, dt \right| \leq 2 \int_0^\delta \frac{\omega_u(\tau)}{\tau} \, d\tau.$$

Hence

$$|Hu(e^{i\theta_1}) - Hu(e^{i\theta_2})| \leq \frac{4}{\pi} \int_0^\delta \frac{\omega_u(t)}{t} \, dt + |\mathcal{I}_1| + |\mathcal{I}_2| + |\mathcal{I}_3|,$$

where

$$
\begin{aligned}
\mathcal{I}_1 &= \frac{1}{\pi} \int_{\theta_1 + \delta}^{\theta_2 + \delta} \frac{u(e^{it})}{\theta_1 - t} \, dt, \\
\mathcal{I}_2 &= \frac{1}{\pi} \int_{\theta_1 - \delta}^{\theta_2 - \delta} \frac{u(e^{it})}{\theta_2 - t} \, dt, \\
\mathcal{I}_3 &= \frac{1}{\pi} \int_{\delta + \frac{(\theta_2 - \theta_1)}{2} < |t - \theta_3| < \pi} u(e^{it}) \left(\frac{1}{\theta_1 - t} - \frac{1}{\theta_2 - t} \right) dt.
\end{aligned}
$$

We now estimate each of these integrals:

$$
\begin{aligned}
|\mathcal{I}_1| &\leq \frac{1}{\pi} \int_{\theta_1 + \delta}^{\theta_2 + \delta} \left| \frac{u(e^{it}) - u(e^{i\theta_3})}{\theta_1 - t} \right| dt \\
&\leq \frac{1}{\pi} \int_{\theta_1 + \delta}^{\theta_2 + \delta} \left| \frac{u(e^{it}) - u(e^{i\theta_3})}{t - \theta_3} \right| dt \\
&\leq \frac{\omega_u(3\delta/2)}{\pi} \int_{\delta/2}^{3\delta/2} \frac{dt}{t} \\
&\leq \left(\frac{3}{2\pi} \log 3 \right) \omega_u(\delta).
\end{aligned}
$$

A similar estimate holds for $\mathcal{I}_2$ and thus, by (6.11),

$$|\mathcal{I}_k| \leq C \delta \int_\delta^\pi \frac{\omega(t)}{t^2}\, dt, \qquad (k = 1, 2).$$

Finally, for the third integral we have

$$
\begin{aligned}
|\mathcal{I}_3| &\leq \frac{(\theta_2 - \theta_1)}{\pi} \int_{\delta + \frac{(\theta_2 - \theta_1)}{2} < |t - \theta_3| < \pi} \left| \frac{u(e^{it}) - u(e^{i\theta_3})}{(\theta_1 - t)(\theta_2 - t)} \right| dt \\
&\leq \frac{9\delta}{4\pi} \int_{\delta + \frac{(\theta_2 - \theta_1)}{2} < |t - \theta_3| < \pi} \left| \frac{u(e^{it}) - u(e^{i\theta_3})}{(\theta_3 - t)^2} \right| dt \\
&\leq \frac{9\delta}{4\pi} \int_\delta^\pi \frac{\omega(t)}{t^2}\, dt.
\end{aligned}
$$

$\square$

Corollary 6.13 *Let u be Dini continuous on $\mathbb{T}$. Then $\tilde{u}$ exists at all points of $\mathbb{T}$ and moreover $\tilde{u} \in C(\mathbb{T})$.*

Proof. By (6.7) and (6.8) and by Theorem 6.12, we have

$$\lim_{\delta \to 0} \omega_{\tilde{u}}(\delta) = 0.$$

Hence, $\tilde{u} \in C(\mathbb{T})$. $\square$

A function $u \in C(\mathbb{T})$ is said to be Lip_α, $0 < \alpha \leq 1$, if

$$\omega_u(t) = O(t^\alpha)$$

as $t \to 0$. Clearly every Lip_α function is Dini continuous on $\mathbb{T}$. Hence $\tilde{u}$ exists at all points of $\mathbb{T}$. Moreover, $\tilde{u}$ is continuous on $\mathbb{T}$. The following results provide more information about the modulus of continuity of $\tilde{u}$.

Corollary 6.14 (Privalov) *Let u be Lip_α on $\mathbb{T}$ with $0 < \alpha < 1$. Then $\tilde{u}$ is also Lip_α on $\mathbb{T}$.*

Proof. By Theorem 6.12, we have

$$
\begin{aligned}
\omega_{\tilde{u}}(\delta) &\leq C \left(\int_0^\delta \frac{\omega(t)}{t}\, dt + \delta \int_\delta^\pi \frac{\omega(t)}{t^2}\, dt \right) \\
&\leq C \left(\int_0^\delta \frac{c t^\alpha}{t}\, dt + \delta \int_\delta^\infty \frac{c t^\alpha}{t^2}\, dt \right) \leq C' \delta^\alpha.
\end{aligned}
$$

Hence, $\tilde{u}$ is also Lip_α on $\mathbb{T}$. $\square$

Corollary 6.15 *Let u be Lip_1 on $\mathbb{T}$. Then*

$$\omega_{\tilde{u}}(\delta) = O(\delta \log 1/\delta)$$

as $\delta \to 0$.

Proof. By Theorem 6.12, we have

$$
\begin{aligned}
\omega_{\tilde{u}}(\delta) &\leq C \left(\int_0^\delta \frac{\omega(t)}{t}\, dt + \delta \int_\delta^\pi \frac{\omega(t)}{t^2}\, dt \right) \\
&\leq C \left(\int_0^\delta \frac{ct}{t}\, dt + \delta \int_\delta^\pi \frac{ct}{t^2}\, dt \right) \\
&\leq C' \left(\delta + \delta \log \pi/\delta \right).
\end{aligned}
$$

$\square$

Exercises

Exercise 6.5.1 Let $0 < \alpha < 1$, and let $\beta_1, \ldots, \beta_n \in \mathbb{R}$. Let $u \in \mathcal{C}(\mathbb{T})$ and suppose that

$$
\omega_u(t) = O(\, t^\alpha (\, \log 1/t \,)^{\beta_1}\, (\, \log \log 1/t \,)^{\beta_2} \cdots (\, \log \log \cdots \log 1/t \,)^{\beta_n}\,)
$$

as $t \to 0$. Show that

$$
\omega_{\tilde{u}}(t) = O(\, t^\alpha (\, \log 1/t \,)^{\beta_1}\, (\, \log \log 1/t \,)^{\beta_2} \cdots (\, \log \log \cdots \log 1/t \,)^{\beta_n}\,)
$$

as $t \to 0$.

Exercise 6.5.2 Let $\beta_1 > 0$, and let $\beta_2, \ldots, \beta_n \in \mathbb{R}$. Let $u \in \mathcal{C}(\mathbb{T})$ and suppose that

$$
\omega_u(t) = O(\, t(\, \log 1/t \,)^{\beta_1}\, (\, \log \log 1/t \,)^{\beta_2} \cdots (\, \log \log \cdots \log 1/t \,)^{\beta_n}\,)
$$

as $t \to 0$. Show that

$$
\omega_{\tilde{u}}(t) = O(\, t(\, \log 1/t \,)^{1+\beta_1}\, (\, \log \log 1/t \,)^{\beta_2} \cdots (\, \log \log \cdots \log 1/t \,)^{\beta_n}\,)
$$

as $t \to 0$.

6.6 Zygmund's $L \log L$ theorem

As we mentioned before, the assumption $u \in L^1(\mathbb{T})$ is not enough to ensure that $\tilde{u}$ is also in $L^1(\mathbb{T})$. A. Zygmund's $L \log L$ theorem provides a sufficient condition which ensures $\tilde{u} \in L^1(\mathbb{T})$.

Theorem 6.16 (Zygmund) *Let u be real, and suppose that*

$$
\| u \log^+ |u| \,\|_1 = \frac{1}{2\pi} \int_{-\pi}^\pi |u(e^{it})| \, \log^+ |u(e^{it})| \, dt < \infty.
$$

Let

$$V(re^{i\theta}) = \frac{1}{2\pi} \int_{-\pi}^{\pi} \frac{2r \sin(\theta - t)}{1 + r^2 - 2r \cos(\theta - t)} u(e^{it}) \, dt, \qquad (re^{i\theta} \in \mathbb{D}).$$

Then $V \in h^1(\mathbb{D})$, $\tilde{u} \in L^1(\mathbb{T})$ *and*

$$\|V\|_1 = \|\tilde{u}\|_1 \le \| u \log^+ |u| \|_1 + 3e.$$

Moreover,

$$V(re^{i\theta}) = \frac{1}{2\pi} \int_{-\pi}^{\pi} \frac{1 - r^2}{1 + r^2 - 2r \cos(\theta - t)} \tilde{u}(e^{it}) \, dt, \qquad (re^{i\theta} \in \mathbb{D}).$$

Proof. First suppose that $u \ge e$, and let

$$U(re^{i\theta}) = \frac{1}{2\pi} \int_{-\pi}^{\pi} \frac{1 - r^2}{1 + r^2 - 2r \cos(\theta - t)} u(e^{it}) \, dt, \qquad (re^{i\theta} \in \mathbb{D}). \quad (6.12)$$

Thus $U(z) \ge e$, and $F(z) = U(z) + iV(z)$ is analytic on $\mathbb{D}$ with $F(0) = U(0)$.
    The function

$$\varphi(t) = \begin{cases} t \, \log t & \text{if} \quad t \ge 1, \\[2mm] 0 & \text{if} \quad t \le 1 \end{cases}$$

is nondecreasing and convex on $\mathbb{R}$. Thus, applying Jensen's inequality to (6.12) gives us

$$U(re^{i\theta}) \log U(re^{i\theta}) \le \frac{1}{2\pi} \int_{-\pi}^{\pi} \frac{1 - r^2}{1 + r^2 - 2r \cos(\theta - t)} u(e^{it}) \log u(e^{it}) \, dt$$

for all $re^{i\theta} \in \mathbb{D}$. Hence, by Fubini's theorem,

$$\int_{-\pi}^{\pi} U(re^{i\theta}) \log U(re^{i\theta}) \, d\theta \le \int_{-\pi}^{\pi} u(e^{it}) \log u(e^{it}) \, dt \quad (6.13)$$

for each $0 \le r < 1$.
    Since $|F| \ge U > e$, both $U \log U$ and $|F|$ are infinitely differentiable on $\mathbb{D}$. Moreover, by the Cauchy–Riemann equations,

$$\nabla^2 (U \log U)(z) = \frac{|F'(z)|^2}{U(z)}$$

and

$$\nabla^2 (|F|)(z) = \frac{|F'(z)|^2}{|F(z)|},$$

and thus we have

$$\nabla^2 (|F|)(z) \le \nabla^2 (U \log U)(z)$$

for all $z \in \mathbb{D}$. Now, Green's theorem says that

$$\int_{-\pi}^{\pi} \frac{\partial}{\partial r} |F| \, (re^{i\theta}) \, rd\theta \;=\; \int_0^r \int_{-\pi}^{\pi} \nabla^2 |F| \, (\rho e^{i\theta}) \, \rho d\rho d\theta,$$

$$\int_{-\pi}^{\pi} \frac{\partial}{\partial r} \Big(U \log U \Big) (re^{i\theta}) \, rd\theta \;=\; \int_0^r \int_{-\pi}^{\pi} \nabla^2 (U \log U) \, (\rho e^{i\theta}) \, \rho d\rho d\theta.$$

Therefore,

$$\frac{d}{dr} \bigg(\int_{-\pi}^{\pi} |F(re^{i\theta})| \, d\theta \bigg) \le \frac{d}{dr} \bigg(\int_{-\pi}^{\pi} U(re^{i\theta}) \log U(re^{i\theta}) \, d\theta \bigg).$$

Integrating both sides over $[0, r]$ gives us

$$\int_{-\pi}^{\pi} |F(re^{i\theta})| \, d\theta - \int_{-\pi}^{\pi} |F(0)| \, d\theta \le \int_{-\pi}^{\pi} U(re^{i\theta}) \log U(re^{i\theta}) \, d\theta - \int_{-\pi}^{\pi} U(0) \log U(0) \, d\theta.$$

Hence,

$$\begin{aligned}
\frac{1}{2\pi} \int_{-\pi}^{\pi} |V(re^{i\theta})| \, d\theta \;&\le\; \frac{1}{2\pi} \int_{-\pi}^{\pi} |F(re^{i\theta})| \, d\theta \\[4pt]
&\le\; \frac{1}{2\pi} \int_{-\pi}^{\pi} U(re^{i\theta}) \log U(re^{i\theta}) \, d\theta + U(0)\,(1 - \log U(0)) \\[4pt]
&\le\; \frac{1}{2\pi} \int_{-\pi}^{\pi} U(re^{i\theta}) \log U(re^{i\theta}) \, d\theta.
\end{aligned}$$

(Here we used the fact that $U \ge e$.) Finally, by (6.13),

$$\frac{1}{2\pi} \int_{-\pi}^{\pi} |V(re^{i\theta})| \, d\theta \le \frac{1}{2\pi} \int_{-\pi}^{\pi} u(e^{it}) \log u(e^{it}) \, dt \tag{6.14}$$

for all $0 \le r < 1$. This settles the case whenever $u \ge e$.

For an arbitrary u, write

$$u = u^{(1)} + u^{(2)} - u^{(3)},$$

where

$$u^{(1)}(e^{i\theta}) = \begin{cases} u(e^{i\theta}) & \text{if} \quad u(e^{i\theta}) > e, \\[4pt] e & \text{if} \quad -e \le u(e^{i\theta}) \le e, \\[4pt] e & \text{if} \quad u(e^{i\theta}) < -e \end{cases}$$

and

$$u^{(2)}(e^{i\theta}) = \begin{cases} e & \text{if} \quad u(e^{i\theta}) > e, \\[4pt] u(e^{i\theta}) & \text{if} \quad -e \le u(e^{i\theta}) \le e, \\[4pt] -e & \text{if} \quad u(e^{i\theta}) < -e \end{cases}$$

and

$$
u^{(3)}(e^{i\theta}) = \begin{cases} e & \text{if} & u(e^{i\theta}) > e, \\[2mm] e & \text{if} & -e \le u(e^{i\theta}) \le e, \\[2mm] -u(e^{i\theta}) & \text{if} & u(e^{i\theta}) < -e. \end{cases}
$$

The functions $u^{(1)}$, $u^{(2)}$ and $u^{(3)}$ are defined such that $u^{(1)} \ge e$, $u^{(3)} \ge e$ and $-e \le u^{(2)} \le e$. Let $V^{(k)} = Q * u^{(k)}$. Hence $V_r = V_r^{(1)} + V_r^{(2)} - V_r^{(3)}$. Thus, by (6.14), by Corollary 2.19 and Theorem 6.4, we have

$$
\begin{aligned}
\|V_r\|_1 &\le \|V_r^{(1)}\|_1 + \|V_r^{(2)}\|_1 + \|V_r^{(3)}\|_1 \\
&\le \|u^{(1)} \log u^{(1)}\|_1 + \|V_r^{(2)}\|_2 + \|u^{(3)} \log u^{(3)}\|_1 \\
&\le \| u \log^+ |u| \|_1 + 2e + \|U_r^{(2)}\|_2 \\
&\le \| u \log^+ |u| \|_1 + 2e + \|U_r^{(2)}\|_\infty \le \| u \log^+ |u| \|_1 + 3e.
\end{aligned}
$$

Therefore, $V \in h^1(\mathbb{D})$ with $\|V\|_1 \le \| u \log^+ |u| \|_1 + 3e$. Hence

$$
F = U + iV \in H^1(\mathbb{D})
$$

and, by Lemma 3.10 and Theorem 5.17,

$$
f(e^{i\theta}) = \lim_{r \to 1} F(re^{i\theta}) = u(e^{i\theta}) + i\tilde{u}(e^{i\theta})
$$

for almost all $e^{i\theta} \in \mathbb{T}$. Thus, by Theorem 5.11,

$$
F(re^{i\theta}) = \frac{1}{2\pi} \int_{-\pi}^{\pi} \frac{1 - r^2}{1 + r^2 - 2r\cos(\theta - t)} f(e^{it}) \, dt, \qquad (re^{i\theta} \in \mathbb{D}).
$$

Taking the imaginary part of both sides gives

$$
V(re^{i\theta}) = \frac{1}{2\pi} \int_{-\pi}^{\pi} \frac{1 - r^2}{1 + r^2 - 2r\cos(\theta - t)} \tilde{u}(e^{it}) \, dt, \qquad (re^{i\theta} \in \mathbb{D}).
$$

Finally, by Corollary 2.15, $\|V\|_1 = \|\tilde{u}\|_1$. $\qquad\square$

Zygmund's theorem can easily be generalized for complex functions u satisfying $u \log^+ |u| \in L^1(\mathbb{T})$. It also enables us to construct functions in $H^1(\mathbb{D})$.

Corollary 6.17 *Let*

$$
\int_{-\pi}^{\pi} |u(e^{it})| \log^+ |u(e^{it})| \, dt < \infty,
$$

and let

$$
F(z) = \frac{1}{2\pi} \int_{-\pi}^{\pi} \frac{e^{it} + z}{e^{it} - z} u(e^{it}) \, dt, \qquad (z \in \mathbb{D}).
$$

Then $F \in H^1(\mathbb{D})$.

Proof. Without loss of generality, suppose that u is real. Since $u \in L^1(\mathbb{T})$, by Corollary 2.15, $\Re F \in h^1(\mathbb{D})$, and since $u \log^+ |u| \in L^1(\mathbb{T})$, by Zygmund's theorem, we also have $\Im F \in h^1(\mathbb{D})$. Hence $F \in H^1(\mathbb{D})$. $\qquad\square$

6.7   M. Riesz's $L \log L$ theorem

M. Riesz showed that Zygmund's sufficient condition given in Theorem 6.16 is also necessary if u is lower bounded. However, this condition cannot be relaxed and the inverse of Zygmund's theorem, in the general case, is wrong.

Theorem 6.18 (M. Riesz) *Let u be real, $u \geq C > -\infty$, and let $u \in L^1(\mathbb{T})$. Let*

$$V(re^{i\theta}) = \frac{1}{2\pi} \int_{-\pi}^{\pi} \frac{2r \sin(\theta - t)}{1 + r^2 - 2r \cos(\theta - t)} u(e^{it}) \, dt, \qquad (re^{i\theta} \in \mathbb{D}).$$

Suppose that $V \in h^1(\mathbb{D})$. Then $u \log^+ |u| \in L^1(\mathbb{T})$.

Proof. Write $u = u^{(1)} + u^{(2)}$, where

$$u^{(1)}(e^{i\theta}) = \begin{cases} u(e^{i\theta}) - 1 & \text{if} \quad u(e^{i\theta}) < 1, \\ \\ 0 & \text{if} \quad u(e^{i\theta}) \geq 1 \end{cases}$$

and

$$u^{(2)}(e^{i\theta}) = \begin{cases} 1 & \text{if} \quad u(e^{i\theta}) < 1, \\ \\ u(e^{i\theta}) & \text{if} \quad u(e^{i\theta}) \geq 1. \end{cases}$$

Since

$$|u| \log^+ |u| \leq |C| \log^+ |C| + |u^{(2)}| \log^+ |u^{(2)}|,$$

it is enough to show that $|u^{(2)}| \log^+ |u^{(2)}| \in L^1(\mathbb{T})$. Therefore, without loss of generality, we assume that $u \geq 1$.

Let

$$U(re^{i\theta}) = \frac{1}{2\pi} \int_{-\pi}^{\pi} \frac{1 - r^2}{1 + r^2 - 2r \cos(\theta - t)} u(e^{it}) \, dt, \qquad (re^{i\theta} \in \mathbb{D}),$$

and let $F = U + iV$. Then $F(0) = U(0)$ and F maps the unit disc into the half plane $\Re z \geq 1$. Using the main branch of the logarithm on $\mathbb{C} \setminus (-\infty, 0]$, i.e. $\log z = \log |z| + i \arg z$, where $-\pi < \arg z < \pi$, we have

$$\frac{1}{2\pi} \int_{-\pi}^{\pi} F(re^{i\theta}) \log F(re^{i\theta}) \, d\theta = F(0) \log F(0) = U(0) \log U(0).$$

Hence,

$$\frac{1}{2\pi} \int_{-\pi}^{\pi} \left(U(re^{i\theta}) + i V(re^{i\theta}) \right) \left(\log |F(re^{i\theta})| + i \arg F(re^{i\theta}) \right) d\theta = U(0) \log U(0).$$

Taking the real part of both sides gives

$$\frac{1}{2\pi} \int_{-\pi}^{\pi} \left(U(re^{i\theta}) \log |F(re^{i\theta})| - V(re^{i\theta}) \arg F(re^{i\theta}) \right) d\theta = U(0) \log U(0).$$

Therefore,

$$\frac{1}{2\pi} \int_{-\pi}^{\pi} U(re^{i\theta}) \log U(re^{i\theta}) \, d\theta$$

$$\leq \frac{1}{2\pi} \int_{-\pi}^{\pi} U(re^{i\theta}) \log |F(re^{i\theta})| \, d\theta$$

$$\leq \frac{1}{2\pi} \int_{-\pi}^{\pi} |V(re^{i\theta})| \, |\arg F(re^{i\theta})| \, d\theta + U(0) \log U(0)$$

$$\leq \frac{\pi}{2} \frac{1}{2\pi} \int_{-\pi}^{\pi} |V(re^{i\theta})| \, d\theta + U(0) \log U(0)$$

$$\leq \frac{\pi}{2} \|V\|_1 + U(0) \log U(0).$$

Hence, by Fatou's lemma,

$$\frac{1}{2\pi} \int_{-\pi}^{\pi} u(e^{i\theta}) \log u(e^{i\theta}) \, d\theta \leq \liminf_{r \to 1} \frac{1}{2\pi} \int_{-\pi}^{\pi} U(re^{i\theta}) \log U(re^{i\theta}) \, d\theta$$

$$\leq \frac{\pi}{2} \|V\|_1 + U(0) \log U(0) < \infty.$$

$$\square$$

Exercises

Exercise 6.7.1   Construct $u \geq 1$, $u \in L^1(\mathbb{T})$, such that $u \log u \notin L^1(I)$ for any open arc $I \subset \mathbb{T}$.

Exercise 6.7.2   Construct $u \in L^1(\mathbb{T})$ such that $\tilde{u} \notin L^1(I)$ for any open arc $I \subset \mathbb{T}$.

Hint: Use Exercise 6.7.1 and Theorems 6.16 and 6.18.

Chapter 7

Blaschke products and their applications

7.1 Automorphisms of the open unit disc

Let $(z_n)_{n \geq 1}$ be a sequence in the open unit disc $\mathbb{D}$ with

$$\lim_{n \to \infty} |z_n| = 1.$$

Such a sequence has no accumulation point inside $\mathbb{D}$. Therefore, according to a classical theorem of Weierstrass, there is a function G, analytic on the open unit disc, such that $G(z_n) = 0$ for all $n \geq 1$. We are even able to give an explicit formula for G. If F is another analytic function with the same set of zeros, then $K = F/G$ is a well-defined *zero-free* analytic function on $\mathbb{D}$. In other words, we can write

$$F = GK,$$

where G is usually an infinite product constructed using the zeros of F, and K is zero-free. This decomposition is fine and suitable for elementary studies of analytic functions. However, if we assume that $F \in H^p(\mathbb{D})$, then we are not able to conclude that G or K are in any Hardy spaces. Of course, the choice of G is not unique. F. Riesz observed that the zeros of a function $F \in H^p(\mathbb{D})$ satisfy the Blaschke condition

$$\sum_n (1 - |z_n|) < \infty.$$

Hence he was able to take a Blaschke product in order to extract the zeros of F. More importantly, he showed that the zero-free function that we obtain stays in the same Hardy space and has the same norm as F.

A bijective analytic function $f : \mathbb{D} \longrightarrow \mathbb{D}$ is called an automorphism of the open unit disc. Clearly, $f^{-1} : \mathbb{D} \longrightarrow \mathbb{D}$, as a function, is well-defined and bijective. Moreover, based on elementary facts from complex analysis, we know

155

that f^{-1} is analytic too. In other words, f^{-1} is also an automorphism of the open unit disc. It is easy to see that the family of all automorphisms of the open unit disc equipped with the law of composition of functions is a group. As a matter of fact, the preceding assertion is true for the class of automorphisms of any domain in the complex plane $\mathbb{C}$. What makes this specific case more interesting is that we are able to give an explicit formula for all its elements.

Let $w \in \mathbb{D}$ and let

$$b_w(z) = \frac{w - z}{1 - \bar{w}\, z}.$$

The function b_w is called a *Blaschke factor* or a *Möbius transformation* for the open unit disc. The first important observation is that

$$b_w \circ b_w = id \tag{7.1}$$

for any $w \in \mathbb{D}$, where id is the identity element defined by $id(z) = z$.

For each $e^{i\theta} \in \mathbb{T}$ and $w = \rho e^{i\vartheta} \in \mathbb{D}$, we have

$$b_w(e^{i\theta}) = \frac{\rho e^{i\vartheta} - e^{i\theta}}{1 - \rho e^{-i\vartheta} e^{i\theta}} = -e^{i\theta} \frac{1 - \rho e^{-i(\theta - \vartheta)}}{1 - \rho e^{i(\theta - \vartheta)}},$$

and thus

$$|b_w(e^{i\theta})| = 1. \tag{7.2}$$

Hence, by the maximum principle,

$$|b_w(z)| < 1 \tag{7.3}$$

for all $z \in \mathbb{D}$. The properties (7.1) and (7.3) show that each b_w is an automorphism of the open unit disc. Similarly, (7.1) and (7.2) show that the restriction

$$b_w : \mathbb{T} \longrightarrow \mathbb{T}$$

is a bijective map.

Let f be any automorphism of the open unit disc. Then, by definition, there is a unique $w \in \mathbb{D}$ such that $f(w) = 0$. Define $g = f \circ b_w$. Then g is also an automorphism of the open unit disc and $g(0) = 0$. Hence, by Schwarz's lemma, we have

$$|g(z)| \leq |z|, \qquad (z \in \mathbb{D}).$$

The essential observation here is that exactly the same argument applies to g^{-1}. Since g^{-1} is an automorphism of the open unit disc and $g^{-1}(0) = 0$, then

$$|g^{-1}(z)| \leq |z|, \qquad (z \in \mathbb{D}).$$

Replace z by $g(z)$ in the last inequality to get $|z| \leq |g(z)|$, and thus

$$|g(z)| = |z|, \qquad (z \in \mathbb{D}).$$

Therefore, there is a constant $\gamma \in \mathbb{T}$ such that $g(z) = \gamma z$. Hence

$$(f \circ b_w)(z) = \gamma z, \qquad (z \in \mathbb{D}).$$

Finally, replace z by $b_w(z)$ in the last identity to obtain

$$f = \gamma \, b_w.$$

Clearly, any such function is also an automorphism of the open unit disc.

Exercises

Exercise 7.1.1 Let $U \in h(\mathbb{D})$. Show that

(a)
$$U(z^n) \in h(\mathbb{D}), \qquad (n \geq 1).$$

(b)
$$U\left(e^{i\beta} \frac{z - \alpha}{1 - \bar{\alpha}z} \right) \in h(\mathbb{D}), \qquad (|\alpha| < 1, \ \beta \in \mathbb{R}).$$

Hint: Use Exercise 1.1.4.

Exercise 7.1.2 [Schwarz's lemma] Let $F \in H^\infty(\mathbb{D})$, $\|F\|_\infty \leq 1$ and $F(0) = 0$. Show that
$$|F(z)| \leq |z|, \qquad (z \in \mathbb{D}),$$
and that
$$|F'(0)| \leq 1.$$
Moreover, if $|F(z)| = |z|$ for one $z \in \mathbb{D} \setminus \{0\}$, or if $|F'(0)| = 1$, then $F(z) = \gamma z$, where γ is a constant of modulus one.
Hint: Consider $G(z) = F(z)/z$ and apply the maximum modulus principle.

Exercise 7.1.3 Verify that

$$b'_w(0) = -(1 - |w|^2)$$

and that

$$b'_w(w) = \frac{-1}{1 - |w|^2}.$$

Exercise 7.1.4 Let $\gamma \in \mathbb{T}$, and let $w \in \mathbb{D}$. Define

$$f(z) = b_w(\gamma z), \qquad (z \in \mathbb{D}).$$

Show that f is an automorphism of the open unit disc. Find $\gamma' \in \mathbb{T}$ and $w' \in \mathbb{D}$ such that

$$f = \gamma' \, b_{w'}.$$

Exercise 7.1.5   Let $w, w' \in \mathbb{D}$. Compute $b_w \circ b_{w'}$.

Exercise 7.1.6   Let $\alpha, \beta \in \mathbb{D}$, and let

$$\mathcal{A}_{\alpha,\beta} = \{\, F \in H^\infty(\mathbb{D}) \ : \ \|F\|_\infty \leq 1, \ F(\alpha) = \beta \,\}.$$

Find

$$\sup_{F \in \mathcal{A}_{\alpha,\beta}} |F'(\alpha)|.$$

Hint: Consider $G = b_\beta \circ F \circ b_\alpha$. Then $G(0) = 0$.

7.2   Blaschke products for the open unit disc

Let $\{z_n\}_{1 \leq n \leq N}$ be a finite sequence of complex numbers inside the unit disc $\mathbb{D}$ and let $\alpha \in \mathbb{R}$. Then

$$B(z) = e^{i\alpha} \prod_{n=1}^{N} \frac{z_n - z}{1 - \bar{z}_n z}$$

is called a *finite Blaschke product* for the unit disc. In the sequence $\{z_n\}_{1 \leq n \leq N}$, repetition is allowed. Based on the observation for Blaschke factors, we immediately see that

$$|B(e^{i\theta})| = 1 \tag{7.4}$$

for all $e^{i\theta} \in \mathbb{T}$, and

$$|B(z)| < 1 \tag{7.5}$$

for all $z \in \mathbb{D}$. But, if $N \geq 2$, B is not one-to-one.

More generally, consider an infinite sequence of complex numbers in the unit disc $\{z_n\}_{n \geq 1}$, which is indexed such that

$$0 \leq |z_1| \leq |z_2| \leq \cdots$$

and $\lim_{n \to \infty} |z_n| = 1$. Let $\sigma = \sigma_{\{z_n\}_{n \geq 1}}$ denote the set of all accumulation points of $\{z_n\}_{n \geq 1}$. Since $\lim_{n \to \infty} |z_n| = 1$, σ is a closed nonempty subset of $\mathbb{T}$. Moreover, σ also coincides with the accumulation points of the set $\{1/\bar{z}_n : n \geq 1\}$. Let

$$\Omega = \Omega_{\{z_n\}_{n \geq 1}} = \mathbb{C} \setminus \left(\sigma \cup \{1/\bar{z}_n : n \geq 1\} \right).$$

Note that we always have $\mathbb{D} \subset \Omega$ for any possible choice of $\{z_n\}_{n \geq 1}$. Besides, $\mathbb{T} \setminus \sigma$ is also a subset of Ω and if it is nonempty, being a countable union of disjoint open arcs of $\mathbb{T}$, it provides the connection between the disjoint open sets $\mathbb{D}$ and $\Omega \setminus \bar{\mathbb{D}}$.

Let

$$B_N(z) = \prod_{n=1}^{N} \frac{|z_n|}{z_n} \frac{z_n - z}{1 - \bar{z}_n z}. \tag{7.6}$$

By convention, we put $|z_n|/z_n = 1$ whenever $z_n = 0$. Under a certain condition on the rate of growth of $\{|z_n|\}_{n\geq 1}$, the partial products B_N converge uniformly on compact subsets of Ω to a nonzero analytic function, which is denoted by

$$B(z) = \prod_{n=1}^{\infty} \frac{|z_n|}{z_n} \frac{z_n - z}{1 - \bar{z}_n z}. \tag{7.7}$$

The function B is called an *infinite Blaschke product* for the open unit disc.

Theorem 7.1 (Blaschke [3]) *Let $\{z_n\}_{n\geq 1}$ be a sequence in the open unit disc such that $\lim_{n\to\infty} |z_n| = 1$. Let $\sigma = \sigma_{\{z_n\}_{n\geq 1}}$ denote the set of accumulation points of $\{z_n\}_{n\geq 1}$ and let $\Omega = \mathbb{C} \setminus (\sigma \cup \{1/\bar{z}_n : n \geq 1\})$. Then the partial products*

$$B_N(z) = \prod_{n=1}^{N} \frac{|z_n|}{z_n} \frac{z_n - z}{1 - \bar{z}_n z}$$

are uniformly convergent on compact subsets of Ω if and only if

$$\sum_{n=1}^{\infty} (1 - |z_n|) < \infty.$$

Remark: There is no restriction on $\arg z_n$.

Proof. Suppose that the sequence $\{B_N\}_{N\geq 1}$ is uniformly convergent to B on compact subsets of Ω. Suppose that $z_1 = z_2 = \cdots = z_k = 0$ and $z_{k+1} \neq 0$. By the maximum principle, $\{B_N\}_{N\geq 1}$ is uniformly convergent on compact subsets of Ω if and only if $\{B_N(z)/z^k\}_{N\geq 1}$ does the same. Hence, without loss of generality, we assume that $|z_1| > 0$. According to this assumption, we have

$$|B(0)| = \prod_{n=1}^{\infty} |z_n| > 0.$$

On the other hand, the elementary inequality

$$t \leq e^{t-1}, \quad 0 \leq t \leq 1,$$

implies

$$|B(0)| = \prod_{n=1}^{\infty} |z_n| \leq \exp\left(-\sum_{n=1}^{\infty} (1 - |z_n|)\right),$$

and thus

$$\sum_{n=1}^{\infty} (1 - |z_n|) < \infty.$$

Now, suppose that the last inequality holds. First of all, the identity

$$\frac{|z_n|}{z_n} \frac{z_n - z}{1 - \bar{z}_n z} = 1 - (1 - |z_n|)\left(1 + \frac{|z_n|(1 + |z_n|)z}{z_n(1 - \bar{z}_n z)}\right) \tag{7.8}$$

implies

$$\left| 1 - \frac{|z_n|}{z_n} \frac{z_n - z}{1 - \bar{z}_n z} \right| \leqslant (1 - |z_n|) \left(1 + \frac{|z|}{|z - 1/\bar{z}_n|} (1 + 1/|z_1|) \right).$$

But, on a compact set $K \subset \Omega$, we have

$$1 + \frac{|z|}{|z - 1/\bar{z}_n|} (1 + 1/|z_1|) \leqslant C_K,$$

where C_K is a constant independent of n. Therefore, for all $z \in K$,

$$\left| 1 - \frac{|z_n|}{z_n} \frac{z_n - z}{1 - \bar{z}_n z} \right| \leqslant C_K (1 - |z_n|), \qquad (n \geqslant 1).$$

This inequality establishes the uniform convergence of B_N on K. □

A sequence $\{z_n\}$ of complex numbers in the open unit disc satisfying

$$\sum_{n=1}^{\infty} \left(1 - |z_n| \right) < \infty \tag{7.9}$$

is called a Blaschke sequence. We already saw that the set of accumulation points of any Blaschke sequence is a subset of the unit circle $\mathbb{T}$. Hence, under the conditions of Theorem 7.1, B_N converges uniformly on each compact subset of $\mathbb{D}$ to the Blaschke product

$$B(z) = \prod_{n=1}^{\infty} \frac{|z_n|}{z_n} \frac{z_n - z}{1 - \bar{z}_n z}.$$

Since $|B_N(z)| < 1$ on the open unit disc, we also have

$$|B(z)| < 1, \qquad (z \in \mathbb{D}). \tag{7.10}$$

According to (7.10), a Blaschke product B, restricted to $\mathbb{D}$, is in $H^\infty(\mathbb{D})$. Hence, by Theorem 5.2, there is a unique function $b \in L^\infty(\mathbb{T})$ such that

$$B(re^{i\theta}) = \frac{1}{2\pi} \int_{-\pi}^{\pi} \frac{1 - r^2}{1 + r^2 - 2r\cos(\theta - t)} b(e^{it}) \, dt, \qquad (re^{i\theta} \in \mathbb{D}).$$

Moreover,

$$\| B \|_{H^\infty(\mathbb{D})} = \| b \|_{L^\infty(\mathbb{T})}$$

and, by Lemma 3.10,

$$\lim_{r \to 1^-} B(re^{i\theta}) = b(e^{i\theta})$$

for almost all $e^{i\theta} \in \mathbb{T}$.

If B is a finite Blaschke product, then b is well-defined and analytic at all points of $\mathbb{T}$. But, if B is an infinite Blaschke product, then b, as a bounded measurable function, is defined almost everywhere on $\mathbb{T}$. Nevertheless, if an arc of $\mathbb{T}$ is free of the accumulation points of the zeros of B, then Theorem 7.1 ensures that b is a well-defined analytic function on this arc and $|b| \equiv 1$ there.

Exercises

Exercise 7.2.1   Let E be a closed nonempty subset of $\mathbb{T}$. Construct a Blaschke sequence whose accumulation set is exactly E. In particular, construct a Blaschke sequence that accumulates at all points of $\mathbb{T}$.

Exercise 7.2.2   Let $\{z_n\}_{1 \leqslant n \leqslant N} \subset \mathbb{D}$, let $\gamma \in \mathbb{T}$, and let

$$B(z) = \gamma \prod_{n=1}^{N} \frac{z_n - z}{1 - \bar{z}_n z}.$$

Show that, for any $w \in \mathbb{D}$, the equation

$$B(z) = w$$

has exactly N solutions in $\mathbb{D}$.
Hint: Do it by induction. The function $B \circ b_{z_1}$ might be useful too.
Remark: Repetition is allowed among z_n and also among the solutions of the preceding equation.

Exercise 7.2.3   Construct an infinite Blaschke product B with zeros on the interval $[0, 1)$ such that
$$\limsup_{r \to 1} |B(r)| = 1.$$
Remark: Note that for any such product, $\liminf_{r \to 1} |B(r)| = 0$.

Exercise 7.2.4   Construct an infinite Blaschke product B with zeros on the interval $[0, 1)$ such that
$$\lim_{r \to 1} B(r) = 0.$$
Hint: Take $z_n = 1 - n^{-2}$, $n \geq 1$.

Exercise 7.2.5   Let B be a Blaschke product with zeros on the interval $[0, 1)$, and let
$$F(z) = (z - 1)^2 B(z), \qquad (z \in \mathbb{D}).$$
Show that $F' \in H^\infty(\mathbb{D})$.

7.3   Jensen's formula

Let $F \in H^\infty(\mathbb{D})$. Then, by Lemma 3.10, we know that $f(e^{i\theta}) = \lim_{r \to 1} F(re^{i\theta})$ exists for almost all $e^{i\theta} \in \mathbb{T}$. If

$$|f(e^{i\theta})| = 1,$$

for almost all $e^{i\theta} \in \mathbb{T}$, we say that F is an *inner function* for the open unit disc. A Blaschke product is an inner function. To establish this result we need Jensen's formula.

Jensen's formula provides a connection between the moduli of the zeros of a function F inside a disc and the values of $|F|$ on the boundaries of the disc. This result is indeed a generalization of the mean value property of harmonic functions (Corollary 3.2). Jensen's formula is one of the most useful results of function theory.

Theorem 7.2 (Jensen's formula) *Let F be analytic on a domain containing the closed disc $\overline{D}_r$ and let $z_1, z_2, \ldots, z_n$ be the zeros of F inside D_r, repeated according to their multiplicities. Suppose that $F(0) \neq 0$. Then*

$$\log |F(0)| = -\sum_{k=1}^{n} \log \left(\frac{r}{|z_k|} \right) + \frac{1}{2\pi} \int_{-\pi}^{\pi} \log |F(re^{it})| \, dt.$$

Proof. First, suppose that F has no zeros on the circle ∂D_r. Let

$$B(z) = \prod_{k=1}^{n} \frac{r(z_k - z)}{r^2 - \bar{z}_k z},$$

and let $G = F/B$. Then G is a well-defined analytic function with no zeros on the closed disc $\overline{D}_r$. Hence, $\log |G|$ is harmonic there and, by Corollary 3.2, we have

$$\log |G(0)| = \frac{1}{2\pi} \int_{-\pi}^{\pi} \log |G(re^{it})| \, dt. \tag{7.11}$$

But, B is defined such that $|B(\zeta)| = 1$ for all $\zeta \in \partial D_r$. Therefore,

$$|G(re^{it})| = |F(re^{it})|$$

for all t. We also have

$$G(0) = F(0) \prod_{k=1}^{n} \frac{r}{z_k}.$$

Plugging the last two identities into (7.11) gives Jensen's formula whenever F has no zeros on the circle ∂D_r. But both sides of the formula are continuous with respect to r, and the sequence $\{|z_n|\}_{n \geq 1}$ has no accumulation point inside the interval $[0, 1)$. Hence, the equality holds for all $r \in [0, 1)$. $\qquad \square$

In the proof of Jensen's formula, we used Corollary 3.2 to obtain (7.11). However, based on Lemma 3.4 and a simple change of variable, we obtain

$$\log|G(r_0 e^{i\theta_0})| = \frac{1}{2\pi} \int_{-\pi}^{\pi} \frac{r^2 - r_0^2}{r^2 + r_0^2 - 2rr_0 \cos(\theta_0 - t)} \log|G(re^{it})| \, dt$$

for all $r_0 e^{i\theta_0} \in D_r$. This identity implies

$$
\begin{aligned}
\log|F(z_0)| &= -\sum_{k=1}^{n} \log \left| \frac{r^2 - \bar{z}_k z_0}{r(z_0 - z_k)} \right| \\
&+ \frac{1}{2\pi} \int_{-\pi}^{\pi} \frac{r^2 - r_0^2}{r^2 + r_0^2 - 2r_0 r \cos(\theta_0 - t)} \log|F(re^{it})| \, dt,
\end{aligned}
$$

which is called the *Poisson–Jensen formula*. We state this formula slightly differently in the following corollary. As a matter of fact, this result is a special case of the canonical factorization that will be discussed in Section 7.6.

Corollary 7.3 *Let F be analytic on a domain containing the closed disc $\overline{D}_r$. Let $z_1, \ldots, z_n$ denote the zeros of F in D_r repeated according to their multiplicities. Then, for each $z_0 = r_0 e^{i\theta_0} \in D_r$, we have*

$$F(z_0) = \gamma \prod_{k=1}^{n} \frac{r(z_0 - z_k)}{r^2 - \bar{z}_k z_0} \exp\left\{ \frac{1}{2\pi} \int_{-\pi}^{\pi} \frac{re^{it} + z_0}{re^{it} - z_0} \log|F(re^{it})| \, dt \right\},$$

where γ is a constant of modulus one.

Now we are ready to show that each Blaschke product is an inner function. Note that this fact is trivial for finite Blaschke products.

Theorem 7.4 (F. Riesz [18]) *Let B be a Blaschke product for the open unit disc. Let*

$$b(e^{i\theta}) = \lim_{r \to 1} B(re^{i\theta})$$

wherever the limit exists. Then

$$|b(e^{i\theta})| = 1$$

for almost all $e^{i\theta} \in \mathbb{T}$.

Proof. First of all, by (7.10), we have

$$\frac{1}{2\pi} \int_{-\pi}^{\pi} \log|B(re^{i\theta})| \, d\theta \leq 0 \tag{7.12}$$

for all r, $0 \leq r < 1$. Without loss of generality, we assume that $B(0) \neq 0$. Since otherwise, we can divide B by the factor z^m and this modification does not change $|b|$. Then, by Jensen's formula,

$$\log|B(0)| = -\sum_{|z_n| < r} \log\left(\frac{r}{|z_n|}\right) + \frac{1}{2\pi} \int_{-\pi}^{\pi} \log|B(re^{i\theta})| \, d\theta$$

for all r, $0 < r < 1$. Since $B(0) = \prod_{n=1}^{\infty} |z_n|$, we thus have

$$\frac{1}{2\pi} \int_{-\pi}^{\pi} \log |B(re^{i\theta})| \, d\theta = \sum_{|z_n| < r} \log \left(\frac{r}{|z_n|} \right) - \sum_{n=1}^{\infty} \log \left(\frac{1}{|z_n|} \right).$$

Given $\varepsilon > 0$, choose N so large that $\sum_{n=N+1}^{\infty} \log 1/|z_n| < \varepsilon$, and let

$$R = \max\{ |z_1|, \ldots, |z_N| \}.$$

Therefore, for all $r > R$, we have

$$\frac{1}{2\pi} \int_{-\pi}^{\pi} \log |B(re^{i\theta})| \, d\theta \geqslant \sum_{n=1}^{N} \log \left(\frac{r}{|z_n|} \right) - \sum_{n=1}^{N} \log \left(\frac{1}{|z_n|} \right) - \varepsilon,$$

which immediately implies

$$\liminf_{r \to 1} \frac{1}{2\pi} \int_{-\pi}^{\pi} \log |B(re^{i\theta})| \, d\theta \geqslant -\varepsilon.$$

Since ε is an arbitrary positive number and in the light of (7.12), the limit of integral means of $|B|$ exists and

$$\lim_{r \to 1} \frac{1}{2\pi} \int_{-\pi}^{\pi} \log |B(re^{i\theta})| \, d\theta = 0.$$

By (7.10), we have $|b(e^{i\theta})| \leqslant 1$ for almost all $e^{i\theta} \in \mathbb{T}$. Now, an application of Fatou's lemma gives

$$\frac{1}{2\pi} \int_{-\pi}^{\pi} -\log |b(e^{i\theta})| \, d\theta \leq \lim_{r \to 1} \frac{1}{2\pi} \int_{-\pi}^{\pi} -\log |B(re^{i\theta})| \, d\theta = 0.$$

Therefore, we necessarily have $|b(e^{i\theta})| = 1$ for almost all $e^{i\theta} \in \mathbb{T}$. □

In proving Riesz's theorem, we established the following result which by itself is interesting and will be needed later on in certain applications.

Corollary 7.5 *Let B be a Blaschke product for the open unit disc. Then*

$$\lim_{r \to 1} \int_{-\pi}^{\pi} \log |B(re^{i\theta})| \, d\theta = 0.$$

Exercises

Exercise 7.3.1 Show that

$$|b_{z_0}(z)|^2 = 1 - \frac{(1 - |z|^2)(1 - |z_0|^2)}{|1 - \bar{z}_0 z|^2}, \qquad (z \in \mathbb{D}).$$

Exercise 7.3.2   Let $z_0 \in \mathbb{D}$, and let $0 \leq r < 1$. Show that
$$\max_{|z|=r} \left| \frac{z_0 - z}{1 - \bar{z}_0 z} \right| \leq \frac{|z_0| + r}{1 + r|z_0|}.$$

Exercise 7.3.3   Let $z_0 \in \mathbb{D}$, $z_0 \neq 0$. Show that
$$\arg b_{z_0}(z) = \arcsin \frac{\Im(z_0 \, \bar{z}) \, (1 - |z_0|^2)}{|z_0| \, |z_0 - z| \, |1 - \bar{z}_0 z|}.$$

Exercise 7.3.4   Let $z_n \in \mathbb{D} \setminus \{0\}$, $n \geq 1$. Show that the following conditions are equivalent:

(i) $\sum_{n=1}^{\infty} \left(1 - |z_n|\right) < \infty$;

(ii) $\sum_{n=1}^{\infty} \log |z_n| > -\infty$;

(iii) $\prod_{n=1}^{\infty} |z_n| > 0$.

Exercise 7.3.5   Let F be analytic on the open unit disc $\mathbb{D}$ and continuous on the closed unit disc $\overline{\mathbb{D}}$. Suppose that
$$|F(e^{i\theta})| = 1$$
for all $e^{i\theta} \in \mathbb{T}$. Show that F is a finite Blaschke product.
Remark: This exercise shows that a finite Blaschke product is the solution of an extremal problem.

Exercise 7.3.6   Let F be meromorphic in the open unit disc $\mathbb{D}$ and continuous on the closed unit disc $\overline{\mathbb{D}}$. Suppose that
$$|F(e^{i\theta})| = 1$$
for all $e^{i\theta} \in \mathbb{T}$. Show that F is the quotient of two finite Blaschke products.

Exercise 7.3.7   Show that
$$\int_{-\pi}^{\pi} \log |1 - e^{i\theta}| \, d\theta = 0.$$

Remark: This fact was used implicitly in the proof of Jensen's formula.

Exercise 7.3.8   [Generalized Poisson–Jensen formula] Let F be a meromorphic function on a domain containing the closed disc $\overline{D}_r$. Let $z_1, \ldots, z_n$ and $p_1, \ldots, p_m$ denote respectively the zeros and poles of F in D_r repeated according to their multiplicities. Show that, for each $z_0 = r_0 e^{i\theta_0} \in D_r$, we have

$$
\begin{aligned}
\log |F(z_0)| &= -\sum_{k=1}^{n} \log \left| \frac{r^2 - \bar{z}_k z_0}{r(z_0 - z_k)} \right| + \sum_{k=1}^{m} \log \left| \frac{r^2 - \bar{p}_k z_0}{r(z_0 - p_k)} \right| \\
&\quad + \frac{1}{2\pi} \int_{-\pi}^{\pi} \frac{r^2 - r_0^2}{r^2 + r_0^2 - 2r_0 r \cos(\theta_0 - t)} \log |F(re^{it})| \, dt.
\end{aligned}
$$

7.4   Riesz's decomposition theorem

Let F be analytic, with no zeros on the open unit disc. Let p be any positive number. Then we can properly define the analytic function F^p on $\mathbb{D}$. This simple-looking property, which by the way is not fulfilled by harmonic functions, has profound applications in the theory of Hardy spaces. An element of $H^p(\mathbb{D})$ might have several zeros in the unit disc. F. Riesz showed that any such element can be written as the product of a Blaschke product and a nonvanishing element of $H^p(\mathbb{D})$. To prove this useful technique, we first verify that the zeros of functions in any Hardy space form a Blaschke sequence.

Lemma 7.6 *Let $F \in H^p(\mathbb{D})$, $0 < p \le \infty$, $F \not\equiv 0$, and let $\{z_n\}$ denote the sequence of its zeros in $\mathbb{D}$. Then*

$$\sum_n (1 - |z_n|) < \infty.$$

Proof. If F has a finite number of zeros, then the result is obvious. If it has infinitely many zeros, since $F \not\equiv 0$, they necessarily converge toward some points in the unit circle. In other words, $\lim_{r \to 1} |z_n| = 1$. Without loss of generality assume that $F(0) \neq 0$. By Jensen's formula, for each $0 < r < 1$, we have

$$\log |F(0)| = - \sum_{|z_n| < r} \log \left| \frac{r}{z_n} \right| + \frac{1}{2\pi} \int_{-\pi}^{\pi} \log |F(re^{i\theta})| \, d\theta.$$

If $0 < p < \infty$, then, by Jensen's inequality,

$$\frac{1}{2\pi} \int_{-\pi}^{\pi} \log |F(re^{i\theta})|^p \, d\theta \le \log \left(\frac{1}{2\pi} \int_{-\pi}^{\pi} |F(re^{i\theta})|^p \, d\theta \right) \le \log \|F\|_p^p$$

and thus

$$\frac{1}{2\pi} \int_{-\pi}^{\pi} \log |F(re^{i\theta})| \, d\theta \le \log \|F\|_p.$$

The last inequality clearly holds even if $p = \infty$. Hence,

$$\sum_{|z_n| < r} \log \left| \frac{r}{z_n} \right| \le \log \|F\|_p - \log |F(0)|.$$

Fix N and take R so large that the first N zeros are in the disc D_R. Therefore, if $R < r < 1$, we have

$$\sum_{n=1}^{N} \log \left| \frac{r}{z_n} \right| \le \log \|F\|_p - \log |F(0)|.$$

First let $r \to 1$ to obtain

$$\sum_{n=1}^{N} \log 1/|z_n| \le \log \|F\|_p - \log |F(0)|,$$

and then let $N \to \infty$ to get

$$\sum_{n=1}^{\infty} \log 1/|z_n| \leq \log \|F\|_p - \log |F(0)| < \infty.$$

But, the last inequality is equivalent to $\sum_n (1 - |z_n|) < \infty$. $\square$

Lemma 7.6 shows that we can form a Blaschke product with zeros of any nonzero element of a Hardy space. F. Riesz discovered that we can extract these zeros such that the remaining factor is still in the same space with the same norm.

Theorem 7.7 (F. Riesz [18]) *Let $F \in H^p(\mathbb{D})$, $0 < p \leq \infty$, $F \not\equiv 0$, and let B be the Blaschke product formed with the zeros of F in $\mathbb{D}$. Let*

$$G = F/B.$$

Then $G \in H^p(\mathbb{D})$, G is free of zeros in $\mathbb{D}$, and

$$\|G\|_p = \|F\|_p.$$

Proof. Clearly G is analytic and free of zeros on $\mathbb{D}$. Moreover, since $|B| \leq 1$, we necessarily have $\|F\|_p \leq \|G\|_p$.

Suppose that $0 < p < \infty$. Fix $N \geq 1$ and $0 < r < 1$. Let B_N be the finite Blaschke product formed with the first N zeros of F. Then $G_N = F/B_N$ is an analytic function on $\mathbb{D}$, and, by Corollary 4.14,

$$
\begin{aligned}
\frac{1}{2\pi} \int_{-\pi}^{\pi} |G_N(re^{i\theta})|^p \, d\theta \;&\leq\; \frac{1}{2\pi} \int_{-\pi}^{\pi} |G_N(\rho e^{i\theta})|^p \, d\theta \\
&=\; \frac{1}{2\pi} \int_{-\pi}^{\pi} \frac{|F_N(\rho e^{i\theta})|^p}{|B_N(\rho e^{i\theta})|^p} \, d\theta \\
&\leq\; \frac{1}{\inf_\theta |B_N(\rho e^{i\theta})|} \frac{1}{2\pi} \int_{-\pi}^{\pi} |F_N(\rho e^{i\theta})|^p \, d\theta \\
&\leq\; \frac{\|F\|_p^p}{\inf_\theta |B_N(\rho e^{i\theta})|}
\end{aligned}
$$

for all ρ, $r < \rho < 1$. Let $\rho \to 1$. Since $|B_N|$ *uniformly* tends to 1, then

$$\left(\frac{1}{2\pi} \int_{-\pi}^{\pi} |G_N(re^{i\theta})|^p \, d\theta \right)^{\frac{1}{p}} \leq \|F\|_p.$$

Now, let $N \to \infty$. On the circle $\{|z| = r\}$, B_N tends uniformly to B. Hence, we obtain

$$\|G_r\|_p \leq \|F\|_p.$$

Finally, take the supremum with respect to r to get $\|G\|_p \leq \|F\|_p$.

A small modification of this argument yields a proof for the case $p = \infty$. As a matter of fact, the proof is even simpler in this case. First, by the maximum principle,

$$|G_N(re^{i\theta})| \leq \sup_\theta |G_N(\rho e^{i\theta})| \leq \frac{\|F\|_\infty}{\inf_\theta |B_N(\rho e^{i\theta})|},$$

and thus, by letting $\rho \to 1$, we obtain

$$|G_N(re^{i\theta})| \leq \|F\|_\infty.$$

Then, if $N \to \infty$, we get

$$|G(re^{i\theta})| \leq \|F\|_\infty.$$

This inequality is equivalent to $\|G\|_\infty \leq \|F\|_\infty$. $\square$

In the succeeding sections, we provide several applications of Riesz's decomposition theorem.

Exercises

Exercise 7.4.1   (G. Julia [11]) Let $F \in H^p(\mathbb{D})$ and $F(0) = 0$. Show that

$$\|F_r\|_p \leq r \|F\|_p$$

for all $0 \leq r < 1$.
Hint: Apply Theorem 7.7.

Exercise 7.4.2   Let F, $F \not\equiv 0$, be analytic on the open unit disc and let $\{z_n\}$ denote the sequence of its zeros in $\mathbb{D}$. Show that

$$\sum_n (1 - |z_n|) < \infty$$

if and only if

$$\sup_{0 \leq r < 1} \int_{-\pi}^{\pi} \log |F(re^{i\theta})| \, d\theta < \infty.$$

7.5 Representation of $H^p(\mathbb{D})$ functions $(0 < p < 1)$

In this section, we show that certain results discussed in Sections 5.1 and 5.6 about $H^p(\mathbb{D})$ functions with $1 \leq p \leq \infty$ are valid even if $0 < p < 1$.

Theorem 7.8 *Let $F \in H^p(\mathbb{D})$, $0 < p < 1$. Then*

$$f(e^{i\theta}) = \lim_{r \to 1} F(re^{i\theta})$$

exists almost everywhere on $\mathbb{T}$. Moreover $f \in L^p(\mathbb{T})$,

$$\lim_{r \to 1} \|F_r - f\|_p = 0$$

and

$$\|F\|_p = \|f\|_p.$$

Proof. By Theorem 7.7, $F = BG$, where G is free of zeros on $\mathbb{D}$, $G \in H^p(\mathbb{D})$ and $\|F\|_p = \|G\|_p$. Pick a positive integer n such that $np \geq 1$ and let $K = G^{1/n}$. Then $K \in H^{np}(\mathbb{D}) \subset H^1(\mathbb{D})$ and

$$F = B K^n.$$

By Lemma 3.10, $b(e^{i\theta}) = \lim_{r \to 1} B(re^{i\theta})$ and $k(e^{i\theta}) = \lim_{r \to 1} K(re^{i\theta})$ exist almost everywhere on $\mathbb{T}$. Since n is an integer, $f(e^{i\theta}) = \lim_{r \to 1} F(re^{i\theta})$ also exist almost everywhere on $\mathbb{T}$ and

$$f = b k^n$$

with $b \in L^\infty(\mathbb{T})$ and $k \in L^{np}(\mathbb{T})$. Hence, in the first place, by Theorem 7.4, $|f|^p = |k|^{np}$ and thus $f \in L^p(\mathbb{T})$. Secondly, to show that $\lim_{r \to 1} \|F_r - f\|_p = 0$ we use the known fact that $\lim_{r \to 1} \|K_r - k\|_{np} = 0$. Indeed, using the binomial theorem, we have

$$
\begin{aligned}
|F_r - f|^p &= |B_r K_r^n - b k^n|^p \\
&= \left| B_r \left[(K_r - k) + k \right]^n - B_r k^n + (B_r - b) k^n \right|^p \\
&\leq \sum_{j=1}^{n} \binom{n}{j} |K_r - k|^{jp} |k|^{(n-j)p} + |B_r - b|^p |k|^{np}
\end{aligned}
$$

and thus, by Hölder's inequality,

$$\|F_r - f\|_p^p \leq \sum_{j=1}^{n} \binom{n}{j} \|K_r - k\|_{np}^{j/n} \|k\|_{np}^{(n-j)/n} + \frac{1}{2\pi} \int_{-\pi}^{\pi} |B_r(e^{i\theta}) - b(e^{i\theta})|^p |k(e^{i\theta})|^{np} \, d\theta.$$

By Theorem 5.1, $\|K_r - k\|_{np} \longrightarrow 0$ and, since $|B_r - b|^p |k|^{np} \leq 2|k|^{np}$, the dominated convergence theorem ensures that

$$\lim_{r \to 1} \int_{-\pi}^{\pi} |B_r(e^{i\theta}) - b(e^{i\theta})|^p |k(e^{i\theta})|^{np} \, d\theta = 0.$$

Therefore, $\|F_r - f\|_p \longrightarrow 0$ as $r \to 1$. Finally, by Corollary 4.14,

$$\|F\|_p = \lim_{r \to 1} \|F_r\|_p = \|f\|_p.$$

$\square$

At this point we are able to complete the proof of Kolmogorov's theorem (Corollary 6.2). In the proof of the corollary, we showed that $U, V \in h^p(\mathbb{D})$ and thus $F = U + iV \in H^p(\mathbb{D})$ with boundary values $f = u + i\tilde{u}$. By Theorem 7.8, as $r \longrightarrow 1$, $F_r \longrightarrow f$ in $L^p(\mathbb{T})$. Hence, $V_r \longrightarrow \tilde{u}$ in $L^p(\mathbb{T})$, which also implies $\|V\|_p = \|\tilde{u}\|_p$.

The correspondence established between $H^p(\mathbb{D})$ and $H^p(\mathbb{T})$, $1 \leq p \leq \infty$, in Theorems 5.1, 5.2 and 5.11, along with Fatou's theorem on radial limits (Lemma 3.10), enables us to give another characterization of $H^p(\mathbb{T})$, $1 \leq p \leq \infty$, classes. Functions in $H^p(\mathbb{T})$, $1 \leq p \leq \infty$, can be viewed as the boundary values of elements of $H^p(\mathbb{D})$. This interpretation is consistent with the original definition given in (5.2). However, the latter has one advantage. Based on Theorem 7.8, this point of view can equally be exploited as the definition of $H^p(\mathbb{T})$, $0 < p < 1$, classes. The following result gives more information about the correspondence between $H^p(\mathbb{D})$ and $H^p(\mathbb{T})$.

Theorem 7.9 *Let $F \in H^p(\mathbb{D})$, $0 < p \leq \infty$, $F \not\equiv 0$, and let*

$$f(e^{it}) = \lim_{r \to 1} F(re^{it})$$

wherever the limit exists. Then

$$\int_{-\pi}^{\pi} \left| \log |f(e^{it})| \right| \, dt < \infty,$$

$$\lim_{r \to 1} \int_{-\pi}^{\pi} \left| \log^+ |F(re^{it})| - \log^+ |f(e^{it})| \right| \, dt = 0$$

and

$$\log |F(re^{i\theta})| \leq \frac{1}{2\pi} \int_{-\pi}^{\pi} \frac{1 - r^2}{1 + r^2 - 2r\cos(\theta - t)} \log |f(e^{it})| \, dt, \qquad (re^{i\theta} \in \mathbb{D}).$$

Proof. Without loss of generality, assume that $p < \infty$ and that $F(0) \neq 0$. Since $\log^+ x \leq x$, $x > 0$, we have

$$p \log^+ |F(re^{i\theta})| \leq |F(re^{i\theta})|^p, \qquad (re^{i\theta} \in \mathbb{D}),$$

and thus

$$p \int_{-\pi}^{\pi} \log^+ |F(re^{i\theta})| \, d\theta \leq \int_{-\pi}^{\pi} |F(re^{i\theta})|^p \, d\theta \leq 2\pi \|F\|_p^p.$$

Hence, by Fatou's lemma,

$$\int_{-\pi}^{\pi} \log^+ |f(e^{i\theta})| \, d\theta \leq 2\pi \|F\|_p^p / p < \infty.$$

On the other hand, by Jensen's formula (Theorem 7.2),

$$\begin{aligned} \log |F(0)| &= -\sum_{|z_n| \leq r} \log \left(\frac{r}{|z_k|} \right) + \frac{1}{2\pi} \int_{-\pi}^{\pi} \log |F(re^{i\theta})| \, d\theta \\ &\leq \frac{1}{2\pi} \int_{-\pi}^{\pi} \log |F(re^{i\theta})| \, d\theta \end{aligned}$$

and thus

$$\frac{1}{2\pi} \int_{-\pi}^{\pi} \log^- |F(re^{i\theta})|\, d\theta \;\leq\; -\log|F(0)| + \frac{1}{2\pi} \int_{-\pi}^{\pi} \log^+ |F(re^{i\theta})|\, d\theta$$

$$\leq\; \frac{\|F\|_p^p}{p} - \log|F(0)|.$$

Another application of Fatou's lemma implies

$$\frac{1}{2\pi} \int_{-\pi}^{\pi} \log^- |f(e^{i\theta})|\, d\theta \leq \frac{\|F\|_p^p}{p} - \log|F(0)| < \infty.$$

Since $\big|\log|f|\big| = \log^+ |f| + \log^- |f|$, the first assertion is proved.

It is easy to verify that $\log x \leq (x - 1)^p/p$, for $x \geq 1$ and $p > 0$. Hence, we immediately obtain

$$|\log^+ a - \log^+ b| \leq |a - b|^p/p, \qquad (a, b > 0).$$

The second assertion is now a direct consequence of this inequality and Theorem 7.8.

Let $\rho < 1$ and consider $G(z) = F(\rho z)$. The function $\log|G|$ is subharmonic on a domain containing the closed unit disc. Hence, by Theorem 4.6,

$$\log|G(re^{i\theta})| \leq \frac{1}{2\pi} \int_{-\pi}^{\pi} \frac{1 - r^2}{1 + r^2 - 2r\cos(\theta - t)}\, \log|G(e^{it})|\, dt.$$

Thus, for $re^{i\theta} \in \mathbb{D}$,

$$\log|F(\rho re^{i\theta})| \;\leq\; \frac{1}{2\pi} \int_{-\pi}^{\pi} \frac{1 - r^2}{1 + r^2 - 2r\cos(\theta - t)}\, \log^+ |F(\rho e^{it})|\, dt$$

$$- \frac{1}{2\pi} \int_{-\pi}^{\pi} \frac{1 - r^2}{1 + r^2 - 2r\cos(\theta - t)}\, \log^- |F(\rho e^{it})|\, dt$$

$$=\; I_+(\rho) - I_-(\rho).$$

Let $\rho \to 1$. By the second assertion of the theorem, which was proved earlier, we have

$$I_+(\rho) \longrightarrow \int_{-\pi}^{\pi} \frac{1 - r^2}{1 + r^2 - 2r\cos(\theta - t)}\, \log^+ |f(e^{it})|\, dt$$

and, by Fatou's lemma,

$$\int_{-\pi}^{\pi} \frac{1 - r^2}{1 + r^2 - 2r\cos(\theta - t)}\, \log^- |f(e^{it})|\, dt \leq \liminf_{\rho \to 1} I_-(\rho).$$

Taking the difference of the last two relations gives the required inequality. □

Corollary 7.10 *Let $F \in H^p(\mathbb{D})$, $0 < p < \infty$, and let $f(e^{it}) = \lim_{r \to 1} F(re^{it})$ wherever the limit exists. Then*

$$|F(re^{i\theta})|^p \leq \frac{1}{2\pi} \int_{-\pi}^{\pi} \frac{1 - r^2}{1 + r^2 - 2r\cos(\theta - t)}\, |f(e^{it})|^p\, dt, \qquad (re^{i\theta} \in \mathbb{D}).$$

Proof. By Theorem 7.9,

$$\log |F(re^{i\theta})| \leq \frac{1}{2\pi} \int_{-\pi}^{\pi} \frac{1 - r^2}{1 + r^2 - 2r\cos(\theta - t)} \log |f(e^{it})| \, dt.$$

Hence

$$|F(re^{i\theta})|^p \leq \exp \left\{ \frac{1}{2\pi} \int_{-\pi}^{\pi} \frac{1 - r^2}{1 + r^2 - 2r\cos(\theta - t)} \log |f(e^{it})|^p \, dt \right\}.$$

Now, apply Jensen's inequality. □

In the second assertion of Theorem 7.9 we cannot replace $\log^+$ by $\log$. For example, consider

$$F(z) = \exp \left\{ -\frac{1 + z}{1 - z} \right\}, \qquad (z \in \mathbb{D}).$$

Then $F \in H^\infty(\mathbb{D})$ with boundary values

$$f(e^{i\theta}) = e^{-i\cot(\theta/2)}, \qquad (e^{i\theta} \in \mathbb{T}).$$

Hence

$$\left| \log |F(re^{i\theta})| - \log |f(e^{i\theta})| \right| = \frac{1 - r^2}{1 + r^2 - 2r\cos\theta}$$

and thus, for each $0 \leq r < 1$,

$$\int_{-\pi}^{\pi} \left| \log |F(re^{i\theta})| - \log |f(e^{i\theta})| \right| \, d\theta = 2\pi.$$

7.6 The canonical factorization in $H^p(\mathbb{D})$ $(0 < p \leq \infty)$

Let $F \in H^p(\mathbb{D})$, $0 < p \leq \infty$, $F \not\equiv 0$, and let f denote the boundary values of F on $\mathbb{T}$. By Theorem 7.9, $\log |f| \in L^1(\mathbb{T})$. Hence, we are enable to define

$$O_F(z) = \exp \left\{ \frac{1}{2\pi} \int_{-\pi}^{\pi} \frac{e^{it} + z}{e^{it} - z} \log |f(e^{it})| \, dt \right\}, \qquad (z \in \mathbb{D}), \qquad (7.13)$$

which is called the *outer part* of F. As a matter of fact, for any positive function h with $\log h \in L^1(\mathbb{T})$, the *outer function*

$$O_h(z) = \exp \left\{ \frac{1}{2\pi} \int_{-\pi}^{\pi} \frac{e^{it} + z}{e^{it} - z} \log |h(e^{it})| \, dt \right\}, \qquad (z \in \mathbb{D}), \qquad (7.14)$$

as an analytic function on the open unit disc, is well-defined. However, O_F has more interesting properties. First of all,

$$|O_F(re^{i\theta})| = \exp \left\{ \frac{1}{2\pi} \int_{-\pi}^{\pi} \frac{1 - r^2}{1 + r^2 - 2r\cos(\theta - t)} \log |f(e^{it})| \, dt \right\}$$

and thus, by Fatou's theorem (Lemma 3.10),

$$\lim_{r \to 1} |O_F(re^{i\theta})| = |f(e^{i\theta})| \tag{7.15}$$

for almost all $e^{i\theta} \in \mathbb{T}$. Secondly, assuming $p < \infty$,

$$|O_F(re^{i\theta})|^p = \exp\left\{ \frac{1}{2\pi} \int_{-\pi}^{\pi} \frac{1 - r^2}{1 + r^2 - 2r\cos(\theta - t)} \log |f(e^{it})|^p \, dt \right\}$$

and thus, by Jensen's inequality,

$$|O_F(re^{i\theta})|^p \leq \frac{1}{2\pi} \int_{-\pi}^{\pi} \frac{1 - r^2}{1 + r^2 - 2r\cos(\theta - t)} |f(e^{it})|^p \, dt.$$

Hence, by Fubini's theorem,

$$\frac{1}{2\pi} \int_{-\pi}^{\pi} |O_F(re^{i\theta})|^p \, d\theta \leq \frac{1}{2\pi} \int_{-\pi}^{\pi} |f(e^{it})|^p \, dt = \|F\|_p^p.$$

In other words, $O_F \in H^p(\mathbb{D})$ and $\|O_F\|_p \leq \|F\|_p$. This inequality clearly holds even if $p = \infty$. On the other hand, the inequality in Theorem 7.9 can be rewritten as

$$|F(z)| \leq |O_F(z)|, \qquad (z \in \mathbb{D}), \tag{7.16}$$

and thus we necessarily have

$$\|O_F\|_p = \|F\|_p. \tag{7.17}$$

The *inner part* of F is defined by

$$I_F(z) = \frac{F(z)}{O_F(z)}, \qquad (z \in \mathbb{D}).$$

By (7.16), $|I_F(z)| \leq 1$, $z \in \mathbb{D}$, and by (7.15), $|I_F(e^{i\theta})| = 1$ for almost all $e^{i\theta} \in \mathbb{T}$. Hence I_F is indeed an inner function. The factorization

$$F = I_F \, O_F$$

is called the canonical, or inner–outer, factorization of F.

We can say more about the inner factor I_F. According to Riesz's theorem (Theorem 7.7), we can write

$$I_F = B \, S,$$

where B is a Blaschke factor formed with the zeros of F and S is a zero-free inner function on $\mathbb{D}$. Therefore, $U = -\log|S|$ is a positive harmonic function on $\mathbb{D}$ satisfying

$$\lim_{r \to 1} U(re^{i\theta}) = 0$$

for almost all $e^{i\theta} \in \mathbb{T}$. Therefore, by Theorem 3.12, there is a positive measure $\sigma \in \mathcal{M}(\mathbb{T})$, singular with respect to the Lebesgue measure, such that

$$-\log |S(re^{i\theta})| = \frac{1}{2\pi} \int_{\mathbb{T}} \frac{1 - r^2}{1 + r^2 - 2r\cos(\theta - t)} \, d\sigma(e^{it}), \qquad (re^{i\theta} \in \mathbb{D}).$$

Hence

$$S(z) = S_\sigma(z) = \exp\left\{ -\frac{1}{2\pi} \int_{\mathbb{T}} \frac{e^{it} + z}{e^{it} - z} \, d\sigma(e^{it}) \right\}, \qquad (z \in \mathbb{D}). \qquad (7.18)$$

As a matter of fact, the counterexample given after Theorem 7.9 is a prototype of these singular inner functions obtained by a Dirac measure at point 1. Note that for singular inner functions, we have

$$\int_{-\pi}^{\pi} \log |S_\sigma(re^{i\theta})| \, d\theta = - \int_{\mathbb{T}} d\sigma(e^{it}) = -\sigma(\mathbb{T}) \qquad (7.19)$$

for each $0 < r \leq 1$. Compare with Corollary 7.5.

The complete canonical factorization of F now becomes

$$F = B S O,$$

where B is the Blaschke product formed with the zeros of F, the singular inner function S is given by (7.18), and the outer function O by (7.13). There is also a constant of modulus one which was absorbed in B.

Since σ is a positive Borel measure and singular with respect to the Lebesgue measure, we have $\sigma'(e^{i\theta}) = 0$ for almost all $e^{i\theta} \in \mathbb{T}$, which in turn implies

$$\lim_{r \to 1} |S(re^{i\theta})| = 1$$

for almost all $e^{i\theta} \in \mathbb{T}$. What is implicit in this statement is that *almost all* is with respect to the Lebesgue measure. On the other hand, we also know that $\sigma'(e^{i\theta}) = +\infty$ for almost all $e^{i\theta} \in \mathbb{T}$, which implies

$$\lim_{r \to 1} S(re^{i\theta}) = 0 \qquad (7.20)$$

for almost all $e^{i\theta} \in \mathbb{T}$. However, in the latter statement, almost all is with respect to the measure σ. In particular, if σ is nonnull, at least at one point on $\mathbb{T}$, the radial limit of S is zero.

Exercises

Exercise 7.6.1   Let I be an inner function. Show that I is a Blaschke product if and only if

$$\lim_{r \to 1} \int_{-\pi}^{\pi} \log |I(re^{i\theta})| \, d\theta = 0.$$

Hint: We have $I = B S$, where B is a Blaschke product and S is a singular inner function. Apply Corollary 7.5 and (7.19).
Remark: Compare with Corollary 7.5.

Exercise 7.6.2   Let I be an inner function. Suppose that for each $e^{i\theta} \in \mathbb{T}$, either $\lim_{r \to 1} |I(re^{i\theta})|$ does not exist or it exists but $\lim_{r \to 1} |I(re^{i\theta})| > 0$. Show that I is a Blaschke product.
Hint: Use (7.20).

7.7   The Nevanlinna class

Let h be a positive measurable function with $\log h \in L^1(\mathbb{T})$, and let σ be a signed Borel measure on $\mathbb{T}$. We emphasize that h is not necessarily in any Lebesgue space $L^p(\mathbb{D})$ and σ is not necessarily positive. Let B be any Blaschke product for the unit disc. Let

$$F = B \, S_\sigma \, O_h, \qquad (7.21)$$

where O_h and S_σ are defined by (7.14) and (7.18). Write $\sigma = \sigma^+ - \sigma^-$, where σ^+ and σ^- are positive Borel measures on $\mathbb{T}$. Then

$$
\begin{aligned}
\log |F(re^{i\theta})| \;\le\; & \frac{1}{2\pi} \int_{\mathbb{T}} \frac{1 - r^2}{1 + r^2 - 2r\cos(\theta - t)} \, d\sigma^-(e^{it}) \\
+ \; & \frac{1}{2\pi} \int_{-\pi}^{\pi} \frac{1 - r^2}{1 + r^2 - 2r\cos(\theta - t)} \, \log^+ h(e^{it}) \, dt.
\end{aligned}
$$

Therefore, by Fubini's theorem,

$$
\int_{-\pi}^{\pi} \log^+ |F(re^{i\theta})| \, d\theta \le \sigma^-(\mathbb{T}) + \int_{-\pi}^{\pi} \log^+ h(e^{it}) \, dt
$$

for all $0 \le r < 1$. The integral on the left side is an increasing function of r (Corollary 4.12). Hence, functions defined by (7.21) satisfy

$$
\sup_{0 \le r < 1} \int_{-\pi}^{\pi} \log^+ |F(re^{i\theta})| \, d\theta < \infty. \qquad (7.22)
$$

The Nevanlinna class $\mathcal{N}$ is, by definition, the family of all analytic functions F on the open unit disc which satisfy (7.22). The following result shows that any such function also has a representation of the form (7.21). Note that, according to Theorem 7.9,

$$
\bigcup_{p>0} H^p(\mathbb{D}) \subset \mathcal{N}.
$$

Theorem 7.11 *Let $F \in \mathcal{N}$, $F \not\equiv 0$. Then*

(a) the zeros of F satisfy the Blaschke condition;

(b) for almost all $e^{i\theta} \in \mathbb{T}$,

$$
f(e^{i\theta}) = \lim_{r \to 1} F(re^{i\theta})
$$

exists and is finite;

(c) $\log |f| \in L^1(\mathbb{T})$;

(d) there is a signed measure σ such that

$$
F = B \, S_\sigma \, O_{|f|},
$$

where B is the Blaschke product formed with the zeros of F.

Proof. Fix $\rho < 1$. By Corollary 7.3, for each $z \in D_\rho$, we have

$$F(z) = \gamma \prod_{k=1}^{n} \frac{\rho(z - z_k)}{\rho^2 - \bar{z}_k z} \exp\left\{ \frac{1}{2\pi} \int_{-\pi}^{\pi} \frac{\rho e^{it} + z}{\rho e^{it} - z} \log|F(\rho e^{it})| \, dt \right\},$$

where γ is a constant of modulus one and $z_1, \ldots, z_n$ denote the zeros of F in D_ρ repeated according to their multiplicities. Hence, for each $z \in \mathbb{D}$,

$$F(\rho z) = \gamma \prod_{k=1}^{n} \frac{z - z_k/\rho}{1 - z \bar{z}_k/\rho} \exp\left\{ \frac{1}{2\pi} \int_{-\pi}^{\pi} \frac{e^{it} + z}{e^{it} - z} \log|F(\rho e^{it})| \, dt \right\}.$$

Define

$$\Phi_\rho(z) \;=\; \gamma \prod_{k=1}^{n} \frac{z - z_k/\rho}{1 - z \bar{z}_k/\rho} \exp\left\{ -\frac{1}{2\pi} \int_{-\pi}^{\pi} \frac{e^{it} + z}{e^{it} - z} \log^-|F(\rho e^{it})| \, dt \right\},$$

$$\Psi_\rho(z) \;=\; \exp\left\{ -\frac{1}{2\pi} \int_{-\pi}^{\pi} \frac{e^{it} + z}{e^{it} - z} \log^+|F(\rho e^{it})| \, dt \right\}.$$

Then Φ_ρ and Ψ_ρ are analytic functions in the closed unit ball of $H^\infty(\mathbb{D})$, i.e. $|\Phi_\rho(z)| \leq 1$ and $|\Psi_\rho(z)| \leq 1$ for all $z \in \mathbb{D}$, and

$$F(\rho z) = \frac{\Phi_\rho(z)}{\Psi_\rho(z)}, \qquad (z \in \mathbb{D}).$$

Based on our main assumption on F, there is a constant $C > 0$ such that

$$\Psi_\rho(0) = \exp\left\{ -\frac{1}{2\pi} \int_{-\pi}^{\pi} \log^+|F(\rho e^{it})| \, dt \right\} \geq C \qquad (7.23)$$

for all $\rho < 1$. Now take a sequence converging to 1, say $\rho_n = 1 - 1/n$, $n \geq 1$. Hence, by Montel's theorem, there is a subsequence $(n_k)_{k \geq 1}$ and functions Φ and Ψ in the unit ball of $H^\infty(\mathbb{D})$ such that, as $k \to \infty$,

$$\Phi_{\rho_{n_k}}(z) \longrightarrow \Phi(z) \qquad \text{and} \qquad \Psi_{\rho_{n_k}}(z) \longrightarrow \Psi(z)$$

uniformly on compact subsets of $\mathbb{D}$. Therefore,

$$F(z) = \frac{\Phi(z)}{\Psi(z)}, \qquad (z \in \mathbb{D}).$$

Moreover, by (7.23) and the fact that $\Psi_\rho(z) \neq 0$ for all $z \in \mathbb{D}$, we deduce that $\Psi(z) \neq 0$ for all $z \in \mathbb{D}$. This representation allows us to derive all parts of the theorem.

The zeros of F are the same as the zeros of $\Phi \in H^\infty(\mathbb{D})$. Hence (a) follows from Lemma 7.6. By Fatou's theorem (Lemma 3.10),

$$\varphi(e^{i\theta}) \;=\; \lim_{r \to 1} \Phi(r e^{i\theta}),$$

$$\psi(e^{i\theta}) \;=\; \lim_{r \to 1} \Psi(r e^{i\theta})$$

exist and are finite for almost all $e^{i\theta} \in \mathbb{T}$, and by the uniqueness theorem (Theorem 5.13) the set $\{ e^{i\theta} \in \mathbb{T} : \psi(e^{i\theta}) = 0 \}$ is of Lebesgue measure zero. Hence, part (b) follows. Since $f = \varphi/\psi$, part (c) follows from Theorem 7.9. Finally, as discussed in Section 7.6, there are positive Borel measures σ_1 and σ_2 such that

$$\Phi = B \, S_{\sigma_1} \, O_{|\varphi|} \qquad \text{and} \qquad \Psi = S_{\sigma_2} \, O_{|\psi|},$$

where B is the Blaschke product formed with the zeros of Φ. Hence, we obtain $F = B \, S_{\sigma_1 - \sigma_2} \, O_{|\varphi|/|\psi|} = B \, S_\sigma \, O_{|f|}.$ $\qquad\square$

In the proof of the preceding theorem it was shown that each element $F \in \mathcal{N}$ is the quotient of two bounded analytic functions. On the other hand, suppose that $F = \Phi/\Psi$, where Φ and Ψ are bounded analytic functions on the open unit disc. Without loss of generality, we may assume that $\|\Phi\|_\infty \le 1$ and $\|\Psi\|_\infty \le 1$, and that $\Psi(0) \ne 0$. Hence,

$$\log^+ |F(re^{i\theta})| \le - \log |\Psi(re^{i\theta})|$$

and thus, by Jensen's formula (Theorem 7.2),

$$\int_{-\pi}^\pi \log^+ |F(re^{i\theta})| \, d\theta \le - \int_{-\pi}^\pi \log |\Psi(re^{i\theta})| \, d\theta \le -2\pi \log |\Psi(0)|$$

for all $r < 1$. Therefore, $F \in \mathcal{N}$. We thus obtain another characterization of the Nevanlinna class.

Corollary 7.12 *Let F be analytic on the open unit disc. Then $F \in \mathcal{N}$ if and only if F is the quotient of two bounded analytic functions.*

The measure σ appearing in the canonical decomposition of functions in the Nevanlinna class is not necessarily positive. Hence, we define a subclass as follows:

$$\mathcal{N}^+ = \{ F \in \mathcal{N} : F = B \, S_\sigma \, O_{|f|} \text{ with } \sigma \ge 0 \}.$$

Based on the canonical factorization theorem, we have

$$\bigcup_{p>0} H^p(\mathbb{D}) \subset \mathcal{N}^+.$$

The main advantage of $\mathcal{N}^+$ over $\mathcal{N}$ is the following result.

Theorem 7.13 (Smirnov) *Let $F \in \mathcal{N}^+$, and let $f(e^{i\theta}) = \lim_{r \to 1} F(re^{i\theta})$, wherever the limit exists. Then $F \in H^p(\mathbb{D})$, $0 < p \le \infty$, if and only if $f \in L^p(\mathbb{T})$. In particular, if $F \in H^s(\mathbb{D})$ for some $s \in (0, \infty]$ and $f \in L^p(\mathbb{T})$, $0 < p \le \infty$, then $F \in H^p(\mathbb{D})$.*

Proof. Since $F \in \mathcal{N}^+$ we have the canonical factorization $F = B \, S_\sigma \, O_{|f|}$, where B is the Blaschke product formed with the zeros of F and σ is a positive singular measure. Therefore, $I = I_F = B \, S_\sigma$ is an inner function. Hence, it is enough to show that $O_{|f|} \in H^p(\mathbb{D})$. But, as discussed at the beginning of Section 7.6, $O_{|f|} \in H^p(\mathbb{D})$ if and only if $f \in L^p(\mathbb{T})$. $\qquad\square$

Let

$$F(z) = \exp\left\{\frac{1+z}{1-z}\right\}, \qquad (z \in \mathbb{D}).$$

Clearly $F \in \mathcal{N}$ with unimodular boundary values

$$f(e^{i\theta}) = e^{i\cot(\theta/2)}, \qquad (e^{i\theta} \in \mathbb{T}).$$

However, $F \notin H^\infty(\mathbb{D})$. Hence, in Theorem 7.13 we cannot replace $\mathcal{N}^+$ by $\mathcal{N}$. The following result provides a characterization of the elements of $\mathcal{N}^+$ in the Nevanlinna class $\mathcal{N}$.

Theorem 7.14 *Let $F \in \mathcal{N}$, and let $f(e^{i\theta}) = \lim_{r\to 1} F(re^{i\theta})$, wherever the limit exists. Then $F \in \mathcal{N}^+$ if and only if*

$$\lim_{r\to 1}\int_{-\pi}^{\pi} \log^+ |F(re^{i\theta})|\, d\theta = \int_{-\pi}^{\pi} \log^+ |f(e^{i\theta})|\, d\theta. \qquad (7.24)$$

Proof. Suppose that $F \in \mathcal{N}^+$. Hence $F = B S_\sigma O_{|f|}$, where σ is a positive singular measure. Therefore, $|F| \leq |O_{|f|}|$, which implies

$$\log^+ |F(re^{i\theta})| \leq \frac{1}{2\pi}\int_{-\pi}^{\pi} \frac{1-r^2}{1+r^2-2r\cos(\theta-t)} \log^+ |f|(e^{it})\, dt, \qquad (re^{i\theta} \in \mathbb{D}).$$

Thus, by Fubini's theorem,

$$\int_{-\pi}^{\pi} \log^+ |F(re^{i\theta})|\, d\theta \leq \int_{-\pi}^{\pi} \log^+ |f|(e^{it})\, dt, \qquad (0 \leq r < 1).$$

The quantity on the right side is an increasing function of r (Corollary 4.12), and by Fatou's lemma,

$$\int_{-\pi}^{\pi} \log^+ |f(e^{i\theta})|\, d\theta \leq \liminf_{r\to 1}\int_{-\pi}^{\pi} \log^+ |F(re^{i\theta})|\, d\theta.$$

Hence, (7.24) holds.

Now, suppose that (7.24) holds. Let $F = B S_\sigma O_{|f|}$ be the canonical decomposition of $F \in \mathcal{N}$. Write $G = S_\sigma O_{|f|}$ and denote its boundary values by g. Since $F = BG$, it is enough to prove that $G \in \mathcal{N}^+$. We have $|f| = |g|$ and

$$\log |B(z)| + \log^+ |G(z)| \leq \log^+ |F(z)| \leq \log^+ |G(z)|.$$

Thus, by Corollary 7.5 and our assumption (7.24),

$$\lim_{r\to 1}\int_{-\pi}^{\pi} \log^+ |G(re^{i\theta})|\, d\theta = \int_{-\pi}^{\pi} \log^+ |g(e^{i\theta})|\, d\theta. \qquad (7.25)$$

The main reason we replaced F by G is that it has the representation

$$\log |G(re^{i\theta})| = \frac{1}{2\pi}\int_{\mathbb{T}} \frac{1-r^2}{1+r^2-2r\cos(\theta-t)}\, d\lambda(e^{it}), \qquad (re^{i\theta} \in \mathbb{D}),$$

where
$$d\lambda(e^{it}) = \log |g(e^{it})| \, dt - d\sigma(e^{it}).$$

Hence, by Corollary 2.11 the measures $\log |G(re^{it})| \, dt$ converge in the weak* topology to $d\lambda(e^{it})$. Take any sequence $r_n > 0$, $n \geq 1$, such that $r_n \to 1$. Since the sequences $(\log^+ |G(r_n e^{it})| \, dt)_{n \geq 1}$ and $(\log^- |G(r_n e^{it})| \, dt)_{n \geq 1}$ are uniformly bounded in $\mathcal{M}(\mathbb{T})$, there is a subsequence $(n_k)_{k \geq 1}$ and two positive measures $\lambda_1, \lambda_2 \in \mathcal{M}(\mathbb{T})$ such that

$$\log^+ |G(r_{n_k} e^{it})| \, dt \longrightarrow d\lambda_1(e^{it}) \quad \text{and} \quad \log^- |G(r_{n_k} e^{it})| \, dt \longrightarrow d\lambda_2(e^{it})$$

in the weak* topology. Hence, $\lambda = \lambda_1 - \lambda_2$. Our main task is to show that λ_1 is absolutely continuous with respect to the Lebesgue measure. This fact is equivalent to saying that $\sigma \geq 0$, and thus the theorem would be proved.

Let E be any Borel subset of $\mathbb{T}$. Then, by Fatou's lemma,

$$\int_E \log^+ |g(e^{it})| \, dt \leq \liminf_{r \to 1} \int_E \log^+ |G(re^{it})| \, dt$$

and

$$\int_{\mathbb{T} \setminus E} \log^+ |g(e^{it})| \, dt \leq \liminf_{r \to 1} \int_{\mathbb{T} \setminus E} \log^+ |G(re^{it})| \, dt.$$

If, in any one of the last two inequalities, the strict inequality holds, then we add them up and obtain a strict inequality which contradicts (7.25). Hence

$$\liminf_{r \to 1} \int_E \log^+ |G(re^{it})| \, dt = \int_E \log^+ |g(e^{it})| \, dt$$

for all Borel subsets E of $\mathbb{T}$. In particular, if we take $r = r_{n_k}$, $k \to \infty$, then, by (7.25) and Theorem A.4, the measures $\log^+ |G(r_{n_k} e^{it})| \, dt$ converge to $\log^+ |g(e^{it})| \, dt$ in the weak* topology. Therefore,

$$d\lambda_1(e^{it}) = \log^+ |g(e^{it})| \, dt.$$

$\square$

Exercises

Exercise 7.7.1 Show that
$$\bigcup_{p>0} H^p(\mathbb{D}) \subsetneq \mathcal{N}^+ \subsetneq \mathcal{N}.$$

Exercise 7.7.2 Let $F \in \mathcal{N}$, and let $f(e^{i\theta}) = \lim_{r \to 1} F(re^{i\theta})$, wherever the limit exists. Show that the assumption $f \in L^p(\mathbb{T})$, $0 < p \leq \infty$, is not enough to conclude that $F \in H^p(\mathbb{D})$.

Exercise 7.7.3 Let $F \in \mathcal{N}$. Show that there is a Blaschke product B and singular inner functions S_1, S_2 and an outer function O such that

$$F = B\, S_1\, O/S_2.$$

Exercise 7.7.4 [Generalized Smirnov theorem] Let O_g and O_h be outer functions with $g/h \in L^p(\mathbb{T})$, $0 < p \leq \infty$. Show that

$$\frac{O_g}{O_h} \in H^p(\mathbb{D}).$$

Exercise 7.7.5 Let $u, \tilde{u} \in L^1(\mathbb{T})$. Let $V = Q * u$. Show that $V \in h^1(\mathbb{D})$,

$$\lim_{r \to 1} \|V_r - \tilde{u}\|_1 = 0$$

and

$$V(re^{i\theta}) = \frac{1}{2\pi} \int_{-\pi}^{\pi} \frac{1 - r^2}{1 + r^2 - 2r\cos(\theta - t)}\, \tilde{u}(e^{it})\, dt, \qquad (re^{i\theta} \in \mathbb{D}).$$

Hint: Let $U = P * u$, and let $F = U + iV$. By Kolmogorov's theorem (Corollary 6.2), $F \in H^{1/2}(\mathbb{D})$. Since the boundary values of F are given by $f = u + i\tilde{u} \in L^1(\mathbb{T})$, the Smirnov theorem (Theorem 7.13) ensures that $F \in H^1(\mathbb{D})$.

Exercise 7.7.6 Let $u, \tilde{u} \in L^1(\mathbb{T})$. Show that $u + i\tilde{u} \in H^1(\mathbb{T})$.
Hint: Use Exercise 7.7.5.

Exercise 7.7.7 Let $u, \tilde{u} \in L^1(\mathbb{T})$. Suppose that

$$\int_{-\pi}^{\pi} u(e^{it})\, dt = 0.$$

Show that

$$\tilde{\tilde{u}} = -u.$$

Hint: Use Exercise 7.7.6.

Exercise 7.7.8 Let $u, \tilde{u} \in L^1(\mathbb{T})$. Show that

$$\hat{\tilde{u}}(n) = -i\,\mathrm{sgn}(n)\,\hat{u}(n), \qquad (n \in \mathbb{Z}).$$

Hint: Use Exercise 7.7.5.

7.8 The Hardy and Fejér–Riesz inequalities

Another consequence of Theorem 7.7 is the following representation for the elements of $H^1(\mathbb{D})$. This result by itself is interesting and will lead us to Hardy's inequality.

Lemma 7.15 *Let $F \in H^1(\mathbb{D})$. Then there are $G, K \in H^2(\mathbb{D})$ such that*

$$F = GK$$

and

$$\|F\|_1 = \|G\|_2^2 = \|K\|_2^2.$$

Proof. By Theorem 7.7, $F = B\Phi$, where B is a Blaschke product and $\Phi \in H^1(\mathbb{D})$ is free of zeros on $\mathbb{D}$ with $\|\Phi\|_1 = \|F\|_1$. Let

$$G = B\,\Phi^{\frac{1}{2}} \quad \text{and} \quad K = \Phi^{\frac{1}{2}}.$$

Clearly $GK = F$ and $\|K\|_2^2 = \|\Phi\|_1 = \|F\|_1$. Moreover, again by Theorem 7.7, $\|G\|_2^2 = \|B\,\Phi^{\frac{1}{2}}\|_2^2 = \|\Phi^{\frac{1}{2}}\|_2^2 = \|\Phi\|_1 = \|F\|_1$. $\qquad\square$

An element $F \in H^1(\mathbb{D})$ corresponds to a unique $f \in H^1(\mathbb{T})$ and we have the power series representation

$$F(z) = \sum_{n=0}^{\infty} \hat{f}(n)\, z^n, \qquad (z \in \mathbb{D}).$$

Since $f \in H^1(\mathbb{T}) \subset L^1(\mathbb{T})$, by the Riemann–Lebesgue lemma (Corollary 2.17), we know that $\hat{f}(n) \to 0$ as $n \to \infty$. Hardy's inequality improves this result.

Theorem 7.16 (Hardy's inequality) *Let $F \in H^1(\mathbb{D})$ and let $f \in H^1(\mathbb{T})$ denote its boundary values. Then*

$$\sum_{n=1}^{\infty} \frac{|\hat{f}(n)|}{n} \leq \pi \, \|F\|_1.$$

Proof. By Lemma 7.15, there are $G, K \in H^2(\mathbb{D})$ such that $F = GK$ and $\|F\|_1 = \|G\|_2^2 = \|K\|_2^2$. Moreover, if

$$G(z) = \sum_{n=0}^{\infty} \hat{g}(n)\, z^n, \qquad (z \in \mathbb{D}),$$

and

$$K(z) = \sum_{n=0}^{\infty} \hat{k}(n)\, z^n, \qquad (z \in \mathbb{D}),$$

then we clearly have

$$\hat{f}(n) = \sum_{\substack{j+\ell=n \\ j,\ell \geq 0}} \hat{g}(j)\hat{k}(\ell).$$

By Parseval's identity (Corollary 2.22),

$$\sum_{n=0}^{\infty} |\hat{g}(n)|^2 = \|G\|_2^2 = \|F\|_1$$

and

$$\sum_{n=0}^{\infty} |\hat{k}(n)|^2 = \|K\|_2^2 = \|F\|_1.$$

Hence

$$\mathbb{G}(z) = \sum_{n=0}^{\infty} |\hat{g}(n)| \, z^n, \qquad (z \in \mathbb{D}),$$

and

$$\mathbb{K}(z) = \sum_{n=0}^{\infty} |\hat{k}(n)| \, z^n, \qquad (z \in \mathbb{D}),$$

are also in $H^2(\mathbb{D})$, with $\|\mathbb{G}\|_2^2 = \|\mathbb{K}\|_2^2 = \|F\|_1$. Therefore, if we define

$$\mathbb{F}(z) = \mathbb{G}(z)\,\mathbb{K}(z) = \sum_{n=0}^{\infty} A_n \, z^n, \qquad (z \in \mathbb{D}),$$

then, by Hölder's inequality, $\mathbb{F} \in H^1(\mathbb{D})$ with

$$\|\mathbb{F}\|_1 \leq \|\mathbb{G}\|_2 \, \|\mathbb{K}\|_2 = \|F\|_1$$

and

$$|\hat{f}(n)| \leq \left| \sum_{\substack{j+\ell=n \\ j,\ell \geq 0}} \hat{g}(j)\hat{k}(\ell) \right| \leq \sum_{\substack{j+\ell=n \\ j,\ell \geq 0}} |\hat{g}(j)| \, |\hat{k}(\ell)| = A_n$$

for all $n \geq 0$. Hence, it is enough to prove that

$$\sum_{n=1}^{\infty} \frac{A_n}{n} \leq \pi \, \|\mathbb{F}\|_1.$$

The main advantage of $\mathbb{F}$ over F is that all its coefficients are positive.
    Let

$$\Phi(z) = \sum_{n=1}^{\infty} \frac{z^n}{n} = -\log(1-z), \qquad (z \in \mathbb{D}),$$

where log is the main branch of logarithm. Hence, $\Im\Phi \in h^\infty(\mathbb{D})$ with

$$\|\Im\Phi\|_\infty = \frac{\pi}{2}.$$

Then, by Parseval's identity, we have

$$\frac{1}{2\pi} \int_{-\pi}^{\pi} \mathbb{F}(re^{i\theta}) \, \overline{\Phi(re^{i\theta})} \, d\theta = \sum_{n=1}^{\infty} \frac{A_n}{n} \, r^{2n}$$

and
$$\frac{1}{2\pi} \int_{-\pi}^{\pi} \mathbb{F}(re^{i\theta})\, \Phi(re^{i\theta})\, d\theta = 0.$$

Hence,
$$-\frac{i}{\pi} \int_{-\pi}^{\pi} \mathbb{F}(re^{i\theta})\, \Im\Phi(re^{i\theta})\, d\theta = \sum_{n=1}^{\infty} \frac{A_n}{n}\, r^{2n}.$$

Therefore, for each $0 \le r < 1$,
$$\sum_{n=1}^{\infty} \frac{A_n}{n}\, r^{2n} \le 2\|\mathbb{F}\|_1 \, \|\Im\Phi\|_\infty \le \pi \, \|\mathbb{F}\|_1.$$

Now, let $r \to 1$.   □

Let $F \in H^\infty(\mathbb{D})$. Abusing the notation, let F also denote the restriction of F on the interval $I = (-1,1)$. Then F is clearly bounded on I and $\|F\|_{L^\infty(I)} \le \|F\|_{H^\infty(\mathbb{D})}$. The Fejér–Riesz inequality is a generalization of this simple observation for other $H^p(\mathbb{D})$ classes.

Theorem 7.17 (Fejér–Riesz) *Let $F \in H^p(\mathbb{D})$, $0 < p < \infty$. Then*
$$\left(\int_{-1}^{1} |F(x)|^p \, dx \right)^{\frac{1}{p}} \le \pi^{1/p} \, \|F\|_p.$$

Proof. **Case 1:** $F \in H^2(\mathbb{D})$ and the Taylor coefficients of F are real.

In this case, $F(x)$ is real for all $x \in (-1,1)$. Fix $0 < r < 1$. Let Γ_r be the curve formed with the interval $[-r, r]$ and the semicircle $re^{i\theta}$, $0 \le \theta \le \pi$. Then
$$\int_{\Gamma_r} F(z)\, dz = 0,$$

which implies
$$\int_{-r}^{r} F^2(x)\, dx = -ir \int_{0}^{\pi} F^2(re^{i\theta})\, e^{i\theta}\, d\theta,$$

and thus
$$\int_{-r}^{r} F^2(x)\, dx \le \int_{0}^{\pi} |F(re^{i\theta})|^2 \, d\theta. \tag{7.26}$$

A similar argument shows that
$$\int_{-r}^{r} F^2(x)\, dx \le \int_{-\pi}^{0} |F(re^{i\theta})|^2 \, d\theta. \tag{7.27}$$

To obtain this inequality we should start with the curve $\overline{\Gamma}_r$ formed with the interval $[-r, r]$ and the semicircle $re^{i\theta}$, $-\pi \le \theta \le 0$. Adding (7.26) to (7.27) gives
$$2 \int_{-r}^{r} F^2(x)\, dx \le \int_{-\pi}^{\pi} |F(re^{i\theta})|^2 \, d\theta \le 2\pi\|F\|_2^2.$$

Let $r \to 1$ to obtain

$$\left(\int_{-1}^{1} F^2(x)\, dx \right)^{\frac{1}{2}} \leq \pi^{\frac{1}{2}} \|F\|_2.$$

Case 2: $F \in H^2(\mathbb{D})$.

Write

$$F(z) = \sum_{n=0}^{\infty} (a_n + i b_n)\, z^n = \sum_{n=0}^{\infty} a_n\, z^n + i \sum_{n=0}^{\infty} b_n\, z^n = F_1(z) + i F_2(z),$$

where a_n and b_n are real numbers. Then F_1 and F_2 are real on the interval $(-1, 1)$ and

$$\|F_1\|_2^2 + \|F_2\|_2^2 = \sum_{n=0}^{\infty} |a_n|^2 + \sum_{n=0}^{\infty} |b_n|^2 = \sum_{n=0}^{\infty} |a_n + i b_n|^2 = \|F\|_2^2.$$

Hence, by Case 1,

$$\begin{aligned}
\int_{-1}^{1} |F(x)|^2\, dx &= \int_{-1}^{1} F_1^2(x)\, dx + \int_{-1}^{1} F_2^2(x)\, dx \\
&\leq \pi \|F_1\|_2^2 + \pi \|F_2\|_2^2 = \pi \|F\|_2^2.
\end{aligned}$$

Case 3: $F \in H^p(\mathbb{D})$.

By Theorem 7.7, $F = BG$, where $G \in H^p(\mathbb{D})$, G is zero-free and $\|F\|_p = \|G\|_p$. Let $K = G^{p/2}$. Hence, $K \in H^2(\mathbb{D})$ and

$$\|K\|_2^2 = \|G\|_p^p = \|F\|_p^p.$$

Therefore, by Case 2,

$$\begin{aligned}
\int_{-1}^{1} |F(x)|^p\, dx &\leq \int_{-1}^{1} |G(x)|^p\, dx \\
&= \int_{-1}^{1} |K(x)|^2\, dx \\
&\leq \pi \|K\|_2^2 = \pi \|F\|_p^p.
\end{aligned}$$

$\square$

Exercises

Exercise 7.8.1   Let $F \in H^1(\mathbb{D})$ and let $f \in H^1(\mathbb{T})$ denote its boundary values. Show that

$$\sum_{n=0}^{\infty} \frac{|\hat{f}(n)|}{n+1} \leq \pi \, \|F\|_1.$$

Exercise 7.8.2   Let $G, K \in H^2(\mathbb{D})$ and let

$$F = GK.$$

Show that $F \in H^1(\mathbb{D})$ and

$$\|F\|_1 \leq \|G\|_2 \, \|K\|_2.$$

Exercise 7.8.3   Let $F \in H^1(\mathbb{D})$. Show that there are $G, K \in H^1(\mathbb{D})$ such that

$$F = G + K,$$

G and K are free of zeros on $\mathbb{D}$, and

$$\|G\|_1 \leq \|F\|_1, \qquad\qquad \|K\|_1 \leq \|F\|_1.$$

Hint: If B is a Blaschke product, $B \not\equiv 1$, then $1 + B$ and $1 - B$ are zero-free on $\mathbb{D}$.

Exercise 7.8.4   Let

$$F(z) = \sum_{n=2}^{\infty} \frac{z^n}{\log n}, \qquad (z \in \mathbb{D}).$$

Show that $F \notin H^1(\mathbb{D})$.

Exercise 7.8.5   Give an example to show that the constant $\pi^{1/p}$ in Theorem 7.17 is sharp, i.e. it cannot be replaced by a smaller one.
Hint: Consider $F_\varepsilon(z) = (\varphi'(z))^{1/p}$, where φ is the conformal mapping from $\mathbb{D}$ to the rectangle $(-1, 1) \times (-\varepsilon, \varepsilon)$, and φ maps $(-1, 1)$ to itself.

Chapter 8

Interpolating linear operators

8.1 Operators on Lebesgue spaces

Let $(X, \mathfrak{M}, \mu)$ be a measure space with $\mu \geq 0$, and let $L^p(X)$, $0 < p < \infty$, denote the space of all equivalent classes of measurable functions f such that

$$\|f\|_p = \left(\int_X |f|^p \, d\mu \right)^{\frac{1}{p}} < \infty.$$

The space $L^\infty(X)$ is defined similarly by requiring

$$\|f\|_\infty = \inf_{M > 0} \left\{ M \ : \ \mu\{ \, t \ : \ |f(t)| > M \, \} = 0 \right\} < \infty.$$

The family of Lebesgue spaces $L^p(\mathbb{T})$, $0 < p \leq \infty$, is an important example that we have already seen several times before.

Let $(X, \mathfrak{M}, \mu)$ and $(Y, \mathfrak{N}, \nu)$ be two measure spaces. Let $p, q \in (0, \infty]$. If an operator

$$\Lambda : L^p(X) \longrightarrow L^q(Y)$$

is bounded, i.e. with a suitable constant C,

$$\|\Lambda f\|_q \leq C \, \|f\|_p \tag{8.1}$$

for all $f \in L^p(X)$, we say that Λ is of *type* (p, q) and we denote its norm by

$$\|\Lambda\|_{(p,q)} = \sup_{\substack{f \in L^p(X) \\ f \neq 0}} \frac{\|\Lambda f\|_q}{\|f\|_p}.$$

In other words, $\|\Lambda\|_{(p,q)}$ is the smallest possible constant C in (8.1).

Suppose that the operator Λ is defined on the set

$$L^{p_0}(X) + L^{p_1}(X) = \{\, f + g : f \in L^{p_0}(X) \text{ and } g \in L^{p_1}(X)\}$$

and is of types (p_0, q_0) and (p_1, q_1). More precisely, when we restrict Λ to $L^{p_0}(X)$ and $L^{p_1}(X)$, both operators

$$\Lambda : L^{p_0}(X) \longrightarrow L^{q_0}(Y)$$

and

$$\Lambda : L^{p_1}(X) \longrightarrow L^{q_1}(Y)$$

are well-defined and bounded. Then if p is between p_0 and p_1, we are faced with a natural question. Can we find q between q_0 and q_1 such that

$$\Lambda : L^p(X) \longrightarrow L^q(Y)$$

is also well-defined and bounded? The Riesz–Thorin interpolation theorem (Theorem 8.2) provides an affirmative answer. Let us mention two relevant examples that have already been discussed.

The Fourier transform

$$
\begin{array}{ccc}
\mathcal{F} : L^1(\mathbb{T}) & \longrightarrow & \ell^\infty(\mathbb{Z}) \\
f & \longmapsto & \hat{f}
\end{array}
$$

was defined in Section 1.3. By Lemma 1.1, $\mathcal{F}$ is of type $(1, \infty)$. Moreover, by Parseval's identity (Corollary 2.22), the operator

$$
\begin{array}{ccc}
\mathcal{F} : L^2(\mathbb{T}) & \longrightarrow & \ell^2(\mathbb{Z}) \\
f & \longmapsto & \hat{f}
\end{array}
$$

is well-defined and has type $(2, 2)$. As a matter of fact, it is not difficult to show that

$$\|\mathcal{F}\|_{(1,\infty)} = \|\mathcal{F}\|_{(2,2)} = 1.$$

Therefore, if $p \in [1, 2]$ is given, is there a proper $q \in [2, \infty]$ such that

$$
\begin{array}{ccc}
\mathcal{F} : L^p(\mathbb{T}) & \longrightarrow & \ell^q(\mathbb{Z}) \\
f & \longmapsto & \hat{f}
\end{array}
$$

is well-defined and bounded?

Our second example is the convolution operator. Fix $1 \le r \le \infty$ and fix $f \in L^r(\mathbb{T})$. Let

$$
\begin{array}{ccc}
\Lambda : L^1(\mathbb{T}) & \longrightarrow & L^r(\mathbb{T}) \\
g & \longmapsto & f * g.
\end{array}
$$

By Corollary 1.5, Λ is of type $(1, r)$. On the other hand, the same corollary implies

$$
\begin{array}{ccc}
\Lambda : L^{r'}(\mathbb{T}) & \longrightarrow & L^\infty(\mathbb{T}) \\
g & \longmapsto & f * g
\end{array}
$$

is of type (r', ∞), where $1/r + 1/r' = 1$. Thanks to Young's inequality (Theorem 1.4), we already know that if

$$s \in [1, r'],$$

and we pick

$$p = \frac{sr'}{r' - s} \in [r, \infty],$$

then

$$\Lambda : L^s(\mathbb{T}) \longrightarrow L^p(\mathbb{T})$$
$$g \longmapsto f * g$$

is of type (s, p). However, using the Riesz–Thorin interpolation theorem (Theorem 8.2), we will give a second proof of this fact.

Exercises

Exercise 8.1.1   Let $(X, \mathfrak{M}, \mu)$ be a measure space. Let $0 < p_0 \le p \le p_1 \le \infty$. Show that

$$L^p(X) \subset L^{p_0}(X) + L^{p_1}(X).$$

Exercise 8.1.2   Let $(X, \mathfrak{M}, \mu)$ be a measure space. When is the family $L^p(X)$, $0 < p \le \infty$, monotone?

8.2   Hadamard's three-line theorem

In Section 4.2 we discussed several versions of the maximum principle. In particular, we saw that if F is analytic on a proper unbounded domain Ω, continuous on $\overline{\Omega}$ and $|F(\zeta)| \le 1$ at all boundary points ζ, we still cannot deduce that F is bounded on Ω. However, as Phragmen-Lindelöf observed, under some extra conditions, we can show that the maximum principle still holds even for unbounded domains. There are many types of Phragmen-Lindelöf theorems with different conditions for different domains. To prove the Riesz–Thorin interpolation theorem, we need a very special case due to Hadamard.

Theorem 8.1 (Hadamard's three-line theorem) *Let S be the strip*

$$S = \{ x + iy : 0 < x < 1, \, y \in \mathbb{R} \}.$$

Suppose that F is continuous and bounded on $\overline{S} = [0, 1] \times \mathbb{R}$, and that F is analytic on S. Suppose that

$$|F(iy)| \le m_0$$

and

$$|F(1 + iy)| \le m_1$$

for all $y \in \mathbb{R}$. Then

$$|F(x + iy)| \le m_0^{1-x} \, m_1^x$$

for all $0 < x < 1$ and all $y \in \mathbb{R}$.

Proof. Without loss of generality assume that $m_0, m_1 > 0$. Since otherwise, we can replace them by $m_0 + \varepsilon$ and $m_1 + \varepsilon$, and after proving the inequality with these new constants, we let $\varepsilon \to 0$.

Let

$$G(z) = \frac{F(z)}{m_0^{1-z} \, m_1^z}.$$

The function G is continuous and bounded on $\overline{S} = [0,1] \times \mathbb{R}$, and analytic on S. Moreover,

$$|G(iy)| \le 1$$

and

$$|G(1+iy)| \le 1$$

for all $y \in \mathbb{R}$. If we succeed in proving that $|G(z)| \le 1$ inside the strip, then the required inequality follows immediately. However, this fact is a simple consequence of Corollary 4.5 applied to $\log |G|$. □

Let

$$\|F_x\|_\infty = \sup_{y \in \mathbb{R}} |F(x+iy)|.$$

Then Hadamard's three-line theorem can be rewritten as

$$\log \|F_x\|_\infty \le (1-x) \log \|F_0\|_\infty + x \log \|F_1\|_\infty.$$

In other words, $\log \|F_x\|_\infty$ is a convex function of x.

Exercises

Exercise 8.2.1   [Hadamard's three-circle theorem] Let Ω be the annulus

$$\Omega = \{ z : 0 < r_1 < |z| < r_2 < \infty \}.$$

Suppose that F is continuous on $\overline{\Omega}$, and that F is analytic on Ω. Suppose that

$$|F(r_1 e^{i\theta})| \le m_1$$

and

$$|F(r_2 e^{i\theta})| \le m_2$$

for all θ. Show that

$$|F(re^{i\theta})| \le m_0^{\frac{\log r_2 - \log r}{\log r_2 - \log r_1}} \, m_1^{\frac{\log r - \log r_1}{\log r_2 - \log r_1}}$$

for all $r_1 < r < r_2$ and all θ.

Remark: Hadamard's three-circle theorem says that $\log \|F_r\|_\infty$ is a convex function of $\log r$. Hardy's convexity theorem (Theorem 4.15) is a generalization of this result.

Exercise 8.2.2 Use

$$F_n(z) = \frac{F(z)}{m_0^{1-z} m_1^z} e^{(z^2-1)/n}, \qquad (n \geq 1),$$

and the ordinary maximum principle for analytic functions to give a direct proof of Hadamard's three-line theorem.
Hint: For each fixed n, we have

$$\lim_{\substack{z \to \infty \\ z \in \overline{S}}} F_n(z) = 0.$$

Hence, the maximum principle ensures that $|F_n(z)| \leq 1$ for all $z \in \overline{S}$. Now, let $n \to \infty$.

8.3 The Riesz–Thorin interpolation theorem

Let $(X, \mathfrak{M}, \mu)$ be a measure space. A *simple function* is a finite linear combination

$$\sum_{n=1}^{N} c_n \chi_{A_n},$$

where c_n are complex numbers and $A_n \in \mathfrak{M}$ with $\mu(A_n) < \infty$. Let us recall that

$$\chi_A(x) = \begin{cases} 1 & \text{if } x \in A, \\ 0 & \text{if } x \notin A \end{cases}$$

is the characteristic function of A. Clearly, we can assume that A_n are disjoints. Simple functions belong to all $L^p(X)$ spaces and they are dense in $L^p(X)$ whenever $0 < p < \infty$.

Theorem 8.2 (Riesz–Thorin interpolation theorem) *Let $(X, \mathfrak{M}, \mu)$ and $(Y, \mathfrak{N}, \nu)$ be two measure spaces. Let $p_0, q_0, p_1, q_1 \in [1, \infty]$. Suppose that*

$$\Lambda : L^{p_0}(X) + L^{p_1}(X) \longrightarrow L^{q_0}(Y) + L^{q_1}(Y)$$

is a linear map of types (p_0, q_0) and (p_1, q_1). Let $t \in [0, 1]$ and define

$$\frac{1}{p_t} = \frac{1-t}{p_0} + \frac{t}{p_1} \qquad and \qquad \frac{1}{q_t} = \frac{1-t}{q_0} + \frac{t}{q_1}.$$

Then

$$\Lambda : L^{p_t}(X) \longrightarrow L^{q_t}(Y)$$

is a linear map of type (p_t, q_t), and

$$\|\Lambda\|_{(p_t, q_t)} \leq \|\Lambda\|_{(p_0, q_0)}^{1-t} \|\Lambda\|_{(p_1, q_1)}^t.$$

Remark: We consider $1/\infty = 0$ and $1/0 = \infty$.

Proof. Fix $t \in (0,1)$. Without loss of generality, we also assume that $p_0 \le p_1$. We consider three cases.

Case 1: Suppose that $p_t, q_t \in (1, \infty)$.

This assumption has two advantages. First, $L^{q_t}(Y)$ is the dual of $L^{q_t'}(Y)$, where $1/q_t + 1/q_t' = 1$. Second, simple functions are dense in $L^{p_t}(X)$ and $L^{q_t'}(Y)$.

Let φ be a simple function on X. Since $\varphi \in L^{p_0}(X) + L^{p_1}(X)$, we have $\Lambda\varphi \in L^{q_0}(Y) + L^{q_1}(Y)$, and thus, by duality,

$$\|\Lambda\varphi\|_{q_t} = \sup \left| \int_Y (\Lambda\varphi)\, \psi \, d\nu \right|,$$

where the supremum is taken over all simple functions $\psi \in L^{q_t'}(Y)$ with

$$\|\psi\|_{q_t'} \le 1.$$

We apply Hadamard's three-line theorem (Theorem 8.1) to estimate $\int_Y (\Lambda\varphi)\,\psi\,d\nu$.

Since φ and ψ are simple functions we can write them as

$$\varphi = \sum_m r_m \, e^{i\theta_m} \, \chi_{A_m} \qquad \text{and} \qquad \psi = \sum_n \rho_n \, e^{i\vartheta_n} \, \chi_{B_n},$$

where $r_m, \rho_n > 0$, $A_m \in \mathfrak{M}$ and $B_n \in \mathfrak{N}$, and each sum has finitely many terms. Define

$$\varphi_z = \sum_m r_m^{p_t\left(\frac{1-z}{p_0} + \frac{z}{p_1}\right)} e^{i\theta_m} \, \chi_{A_m} \qquad \text{and} \qquad \psi_z = \sum_n \rho_n^{q_t'\left(\frac{1-z}{q_0'} + \frac{z}{q_1'}\right)} e^{i\vartheta_n} \, \chi_{B_n}.$$

The definition is arranged such that $\varphi_t = \varphi$ and $\psi_t = \psi$. Without loss of generality we can assume the sets $\{A_m\}$ and $\{B_n\}$ are respectively pairwise disjoint. Hence, for all $\alpha \ge 0$, we have

$$|\varphi_{x+iy}|^\alpha = \sum_m r_m^{\alpha p_t\left(\frac{1-x}{p_0} + \frac{x}{p_1}\right)} \chi_{A_m} \qquad \text{and} \qquad |\psi_{x+iy}|^\alpha = \sum_n \rho_n^{\alpha q_t'\left(\frac{1-x}{q_0'} + \frac{x}{q_1'}\right)} \chi_{B_n}.$$

In particular, the identities

$$|\varphi_{iy}|^{p_0} = |\varphi_{1+iy}|^{p_1} = |\varphi|^{p_t} \tag{8.2}$$

and

$$|\psi_{iy}|^{q_0'} = |\psi_{1+iy}|^{q_1'} = |\psi|^{q_t'} \tag{8.3}$$

are essential for us.

Let

$$F(z) = \int_Y (\Lambda \varphi_z)\, \psi_z\, d\nu$$

$$= \sum_m \sum_n e^{i(\theta_m + \vartheta_n)} \left(\int_Y (\Lambda \chi_{A_m})\, \chi_{B_n}\, d\nu \right) r_m^{p_t \left(\frac{1-z}{p_0} + \frac{z}{p_1} \right)} \rho_n^{q_t' \left(\frac{1-z}{q_0} + \frac{z}{q_1} \right)}.$$

Hence F is an entire function, and

$$F(t) = \int_Y (\Lambda \varphi)\, \psi\, d\nu.$$

Let S be the strip

$$S = \{\, x + iy \; : \; 0 < x < 1,\, y \in \mathbb{R} \,\}.$$

Since

$$\left| r_m^{p_t \left(\frac{1-z}{p_0} + \frac{z}{p_1} \right)} \rho_n^{q_t' \left(\frac{1-z}{q_0} + \frac{z}{q_1} \right)} \right| = r_m^{p_t \left(\frac{1-x}{p_0} + \frac{x}{p_1} \right)} \rho_n^{q_t' \left(\frac{1-x}{q_0} + \frac{x}{q_1} \right)},$$

the entire function F is bounded on $\bar{S}$. Moreover, by Hölder's inequality, on the vertical line $\Re z = 0$ we have

$$|F(iy)| = \left| \int_Y (\Lambda \varphi_{iy})\, \psi_{iy}\, d\nu \right|$$

$$\leq \left(\int_Y |\Lambda \varphi_{iy}|^{q_0}\, d\nu \right)^{\frac{1}{q_0}} \left(\int_Y |\psi_{iy}|^{q_0'}\, d\nu \right)^{\frac{1}{q_0'}}$$

$$\leq \|\Lambda\|_{(p_0, q_0)} \left(\int_Y |\varphi_{iy}|^{p_0}\, d\nu \right)^{\frac{1}{p_0}} \left(\int_Y |\psi_{iy}|^{q_0'}\, d\nu \right)^{\frac{1}{q_0'}}.$$

Hence, by (8.2)–(8.3) and the fact that $\|\psi\|_{q_t'} \leq 1$, we obtain

$$|F(iy)| \leq \|\Lambda\|_{(p_0, q_0)} \|\varphi\|_{p_t}^{p_t/p_0}. \tag{8.4}$$

Similarly, on the vertical line $\Re z = 1$ we have

$$|F(1 + iy)| = \left| \int_Y (\Lambda \varphi_{1+iy})\, \psi_{1+iy}\, d\nu \right|$$

$$\leq \left(\int_Y |\Lambda \varphi_{iy}|^{q_1}\, d\nu \right)^{\frac{1}{q_1}} \left(\int_Y |\psi_{iy}|^{q_1'}\, d\nu \right)^{\frac{1}{q_1'}}$$

$$\leq \|\Lambda\|_{(p_1, q_1)} \left(\int_Y |\varphi_{iy}|^{p_1}\, d\nu \right)^{\frac{1}{p_1}} \left(\int_Y |\psi_{iy}|^{q_1'}\, d\nu \right)^{\frac{1}{q_1'}}.$$

Hence, by (8.2)–(8.3), we obtain

$$|F(1 + iy)| \leq \|\Lambda\|_{(p_1, q_1)} \|\varphi\|_{p_t}^{p_t/p_1}. \tag{8.5}$$

Therefore, by Hadamard's three-line theorem (Theorem 8.1) and by (8.4) and (8.5), we have

$$|F(t)| \leq \left(\|\Lambda\|_{(p_0,q_0)} \|\varphi\|_{p_t}^{p_t/p_0} \right)^{1-t} \left(\|\Lambda\|_{(p_1,q_1)} \|\varphi\|_{p_t}^{p_t/p_1} \right)^t$$
$$= \|\Lambda\|_{(p_0,q_0)}^{1-t} \|\Lambda\|_{(p_1,q_1)}^t \|\varphi\|_{p_t}.$$

Thus

$$\left| \int_Y (\Lambda\varphi)\,\psi\,d\nu \right| \leq \|\Lambda\|_{(p_0,q_0)}^{1-t} \|\Lambda\|_{(p_1,q_1)}^t \|\varphi\|_{p_t}.$$

Taking the supremum with respect to all ψ with $\|\psi\|_{q'_t} \leq 1$ gives

$$\|\Lambda\varphi\|_{q_t} \leq \|\Lambda\|_{(p_0,q_0)}^{1-t} \|\Lambda\|_{(p_1,q_1)}^t \|\varphi\|_{p_t}. \tag{8.6}$$

This is the required inequality, but just for simple functions $\varphi \in L^{p_t}(X)$. We use some tools from measure theory to show that (8.6) holds for all elements of $L^{p_t}(X)$.

Let $f \in L^{p_t}(X)$. Then there is a sequence of simple functions $\varphi_n \in L^{p_t}(X)$ such that

$$\lim_{n\to\infty} \|\varphi_n - f\|_{p_t} = 0.$$

If we can show that

$$\lim_{n\to\infty} \|\Lambda\varphi_n - \Lambda f\|_{q_t} = 0,$$

then immediately we see that (8.6) is also fulfilled by f and we are done. This fact is actually our main assumption whenever $p_t = p_0$ or $p_t = p_1$. Hence we assume that $p_0 < p_t < p_1$.

Since $(\varphi_n)_{n\geq 1}$ is convergent in $L^{p_t}(X)$, it is necessarily a Cauchy sequence, and thus, by (8.6), $(\Lambda\varphi_n)_{n\geq 1}$ is a Cauchy sequence in $L^{q_t}(Y)$. Therefore, there is $g \in L^{q_t}(Y)$ such that

$$\lim_{n\to\infty} \|\Lambda\varphi_n - g\|_{q_t} = 0.$$

We show that $\Lambda\varphi_n$ also converges in measure to Λf and this is enough to ensure that $g = \Lambda f$.

Let

$$\varepsilon_n = \|f - \varphi_n\|_{p_t}$$

and let

$$S_n = \{ x \in X : |f(x) - \varphi_n(x)| > \varepsilon_n \}.$$

Clearly we can assume that $\varepsilon_n > 0$. Since otherwise, f would be a simple function, and we saw that (8.6) is fulfilled by such functions. Now, on the one hand, by Chebyshev's inequality,

$$\mu(S_n) = \int_{S_n} d\mu \leq \int_{S_n} \left| \frac{f - \varphi_n}{\varepsilon_n} \right|^{p_t} d\mu \leq \frac{\|f - \varphi_n\|_{p_t}^{p_t}}{\varepsilon_n^{p_t}} = 1,$$

and thus, by Hölder's inequality with $\alpha = p_t/p_0$, we obtain

$$
\begin{aligned}
\|(f - \varphi_n)\chi_{S_n}\|_{p_0}^{p_0} &= \int_X |f - \varphi_n|^{p_0} \chi_{S_n}\, d\mu \\
&\leq \left(\int_X |f - \varphi_n|^{\alpha p_0}\, d\mu \right)^{\frac{1}{\alpha}} \left(\int_X |\chi_{S_n}|^{\alpha'}\, d\mu \right)^{\frac{1}{\alpha'}} \\
&\leq \|f - \varphi_n\|_{p_t}^{p_0}.
\end{aligned}
$$

On the other hand, on $X \setminus S_n$ we have $|f - \varphi_n|/\varepsilon_n \leq 1$, and thus

$$
\int_{X\setminus S_n} \left| \frac{f - \varphi_n}{\varepsilon_n} \right|^{p_1} d\mu \leq \int_{X\setminus S_n} \left| \frac{f - \varphi_n}{\varepsilon_n} \right|^{p_t} d\mu \leq \frac{\|f - \varphi_n\|_{p_t}^{p_t}}{\varepsilon_n^{p_t}} = 1,
$$

which gives

$$
\|(f - \varphi_n)\chi_{X\setminus S_n}\|_{p_1} \leq \|f - \varphi_n\|_{p_t}.
$$

Therefore,

$$
\|(f - \varphi_n)\chi_{S_n}\|_{p_0} \longrightarrow 0 \qquad \text{and} \qquad \|(f - \varphi_n)\chi_{X\setminus S_n}\|_{p_1} \longrightarrow 0
$$

as $n \to \infty$. Since Λ is of type (p_0, q_0) and (p_1, q_1), we conclude

$$
\|\Lambda\big((f - \varphi_n)\chi_{S_n} \big)\|_{q_0} \longrightarrow 0 \qquad \text{and} \qquad \|\Lambda\big((f - \varphi_n)\chi_{X\setminus S_n} \big)\|_{q_1} \longrightarrow 0
$$

as $n \to \infty$. In particular,

$$
\Lambda\big((f - \varphi_n)\chi_{S_n} \big) \longrightarrow 0 \qquad \text{and} \qquad \Lambda\big((f - \varphi_n)\chi_{X\setminus S_n} \big) \longrightarrow 0
$$

in measure as $n \to \infty$. Adding both together gives

$$
\Lambda\varphi_n \longrightarrow \Lambda f
$$

in measure, and we are done.

Case 2: $p_t \in (1, \infty)$, but either $q_t = 1$ or $q_t = \infty$.

In this case, we necessarily have $q_0 = q_1 = q_t$. Let T be a bounded linear functional on $L^{q_t}(Y)$ and let

$$
F(z) = T(\Lambda\varphi_z),
$$

where φ is a simple function on X. Hence

$$
|F(iy)| \leq \|T\|\, \|\Lambda\varphi_{iy}\|_{q_0} \leq \|T\|\, \|\Lambda\|_{(p_0, q_0)}\, \|\varphi\|_{p_t}^{p_t/p_0}
$$

and

$$
|F(1 + iy)| \leq \|T\|\, \|\Lambda\varphi_{1+iy}\|_{q_1} \leq \|T\|\, \|\Lambda\|_{(p_1, q_1)}\, \|\varphi\|_{p_t}^{p_t/p_1}.
$$

Therefore, by Hadamard's three-line theorem (Theorem 8.1), we have

$$
|T(\Lambda\varphi)| = |F(it)| \leq \|T\|\, \|\Lambda\|_{(p_0, q_0)}^{1-t}\, \|\Lambda\|_{(p_1, q_1)}^{t}\, \|\varphi\|_{p_t}.
$$

Finally, the Hahn–Banach theorem implies that

$$\|\Lambda\varphi\|_{q_t} \le \|\Lambda\|_{(p_0,q_0)}^{1-t}\,\|\Lambda\|_{(p_1,q_1)}^{t}\,\|\varphi\|_{p_t}.$$

The rest is exactly as in Case 1.

Case 3: Either $p_t = 1$ or $p_t = \infty$.

In this case, we necessarily have $p_0 = p_1 = p_t$. Let $f \in L^{p_t}(X)$. Then, by Hölder's inequality,

$$\begin{aligned}
\|Tf\|_{q_t} &= \big\|\,|Tf|^{1-t}\,|Tf|^{t}\,\big\|_{q_t} \\
&\le \|Tf\|_{q_0}^{1-t}\,\|Tf\|_{q_1}^{t} \\
&\le \big(\|\Lambda\|_{(p_0,q_0)}\|f\|_{p_0}\big)^{1-t}\big(\|\Lambda\|_{(p_1,q_1)}\|f\|_{p_1}\big)^{t} \\
&= \|\Lambda\|_{(p_0,q_0)}^{1-t}\,\|\Lambda\|_{(p_1,q_1)}^{t}\,\|f\|_{p_1}.
\end{aligned}$$

$\square$

As a very special case, if $r = p_0 = q_0$ and $s = p_1 = q_1$ in the Riesz–Thorin interpolation theorem, and, by assumption, the linear map Λ is of types (r,r) and (s,s), i.e.

$$\|\Lambda f\|_r \le \|\Lambda\|_{(r,r)}\|f\|_r, \qquad (f \in L^r(X)),$$

and

$$\|\Lambda f\|_s \le \|\Lambda\|_{(s,s)}\|f\|_s, \qquad (f \in L^s(X)),$$

then, for each p between r and s with

$$\frac{1}{p} = \frac{1-t}{r} + \frac{t}{s}, \qquad (0 \le t \le 1),$$

the map

$$\Lambda : L^p(X) \longrightarrow L^p(Y)$$

is of type (p,p), i.e.

$$\|\Lambda f\|_p \le \|\Lambda\|_{(p,p)}\|f\|_p, \qquad (f \in L^p(X)),$$

and

$$\|\Lambda\|_{(p,p)} \le \|\Lambda\|_{(r,r)}^{1-t}\,\|\Lambda\|_{(s,s)}^{t}.$$

Exercises

Exercise 8.3.1 First, give a direct proof of Corollary 1.5. Then use this corollary and the Riesz–Thorin interpolation theorem to obtain another proof of Young's inequality (Theorem 1.4).

Exercise 8.3.2 Let $(X, \mathfrak{M}, \mu)$ and $(Y, \mathfrak{N}, \nu)$ be two measure spaces. Let $p_0, q_0, p_1, q_1 \in [1, \infty]$. Suppose that Λ is a linear map defined for all simple functions φ on X such that $\Lambda\varphi$ is a measurable function on Y and

$$\|\Lambda\varphi\|_{q_0} \leq M_0 \|\varphi\|_{p_0}$$

and

$$\|\Lambda\varphi\|_{q_1} \leq M_1 \|\varphi\|_{p_1}.$$

Let $t \in [0, 1]$ and define

$$\frac{1}{p_t} = \frac{1-t}{p_0} + \frac{t}{p_1} \qquad \text{and} \qquad \frac{1}{q_t} = \frac{1-t}{q_0} + \frac{t}{q_1}.$$

Show that

$$\|\Lambda\varphi\|_{q_t} \leq M_0^{1-t} M_1^t \|\varphi\|_{p_t}.$$

Moreover, if $p_t < \infty$, then Λ can be uniquely extended to the whole space $L^{p_t}(\mu)$.

8.4 The Hausdorff–Young theorem

In this section we study the Housdorff–Young theorem about the Fourier transform on $\mathbb{T}$. There is another version with a similar proof about the Fourier transform of functions defined on $\mathbb{R}$.

Theorem 8.3 (The Hausdorff–Young theorem) *Let $f \in L^p(\mathbb{T})$, $1 \leq p \leq 2$. Then $\hat{f}$, the Fourier transform of f, belongs to $\ell^q(\mathbb{Z})$, where q is the conjugate exponent of p, and*

$$\|\hat{f}\|_q \leq \|f\|_p.$$

Proof. By Lemma 1.1,

$$\begin{array}{ccc} \mathcal{F} : L^1(\mathbb{T}) & \longrightarrow & \ell^\infty(\mathbb{Z}) \\ f & \longmapsto & \hat{f} \end{array}$$

is of type $(1, \infty)$ with

$$\|\mathcal{F}\|_{(1,\infty)} \leq 1.$$

By Parseval's identity (Corollary 2.22),

$$\begin{array}{ccc} \mathcal{F} : L^2(\mathbb{T}) & \longrightarrow & \ell^2(\mathbb{Z}) \\ f & \longmapsto & \hat{f} \end{array}$$

is of type $(2, 2)$ with

$$\|\mathcal{F}\|_{(2,2)} \leq 1.$$

(As a matter of fact, we have $\|\mathcal{F}\|_{1,\infty} = \|\mathcal{F}\|_{2,2} = 1$. But we do not need this here.) Hence, with

$$p_0 = 1, \quad q_0 = \infty, \quad p_1 = 2, \quad q_1 = 2,$$

the Riesz–Thorin interpolation theorem (Theorem 8.2) implies that

$$\mathcal{F} : L^{p_t}(\mathbb{T}) \longrightarrow \ell^{q_t}(\mathbb{Z})$$
$$f \longmapsto \hat{f}$$

is well-defined and has type (p_t, q_t) with

$$\|\mathcal{F}\|_{(p_t, q_t)} \leq \|\mathcal{F}\|_{(1,\infty)}^{1-t} \|\mathcal{F}\|_{(2,2)}^{t} \leq 1.$$

Hence

$$\|\hat{f}\|_{q_t} \leq \|f\|_{p_t}.$$

To find the relation between p_t and q_t, note that

$$\frac{1}{p_t} = \frac{1-t}{\infty} + \frac{t}{2} = \frac{t}{2}$$

and

$$\frac{1}{q_t} = \frac{1-t}{1} + \frac{t}{2} = 1 - \frac{t}{2}.$$

Hence $p_t \in [1, 2]$ and

$$\frac{1}{p_t} + \frac{1}{q_t} = 1.$$

$\square$

Corollary 8.4 *Let $f \in H^p(\mathbb{T})$, $1 \leq p \leq 2$. Then $\hat{f}$, the Fourier transform of f, belongs to $\ell^q(\mathbb{Z}^+)$, where q is the conjugate exponent of p, and*

$$\|\hat{f}\|_q \leq \|f\|_p.$$

Exercises

Exercise 8.4.1   Let $1 \leq p \leq 2$, and let q be the conjugate exponent of p. Let $\{a_n\}_{n\in\mathbb{Z}}$ be a sequence in $\ell^p(\mathbb{Z})$. Show that there is an $f \in L^q(\mathbb{T})$ such that

$$\hat{f}(n) = a_n, \qquad (n \in \mathbb{Z}),$$

and

$$\|f\|_q \leq \|\hat{f}\|_p.$$

Hint 1: Use duality and Theorem 8.3.
Hint 2: Give a direct proof using the Riesz–Thorin interpolation theorem.

Exercise 8.4.2   Let $1 \leq p \leq 2$ and let q be the conjugate exponent of p. Let $\{a_n\}_{n \in \mathbb{Z}}$ be a sequence in $\ell^p(\mathbb{Z}^+)$. Show that there is an $f \in H^q(\mathbb{T})$ such that

$$\hat{f}(n) = a_n, \qquad (n \in \mathbb{Z}),$$

and

$$\|f\|_q \leq \|\hat{f}\|_p.$$

Hint: Use duality and Corollary 8.4.

Exercise 8.4.3   Let $1 < p < 2$ and let q be the conjugate exponent of p. According to Theorem 8.3, the map

$$
\begin{array}{ccc}
L^p(\mathbb{T}) & \longrightarrow & \ell^q(\mathbb{Z}) \\
f & \longmapsto & \hat{f}
\end{array}
$$

is well-defined, and by the uniqueness theorem (Corollary 2.13) it is injective. However, show that it is not surjective.

Exercise 8.4.4   Let $1 < p < 2$ and let q be the conjugate exponent of p. According to Corollary 8.4, the map

$$
\begin{array}{ccc}
H^p(\mathbb{T}) & \longrightarrow & \ell^q(\mathbb{Z}^+) \\
f & \longmapsto & \hat{f}
\end{array}
$$

is well-defined, and by the uniqueness theorem (Corollary 2.13) it is injective. However, show that it is not surjective.
Hint: See Exercise 8.4.3.

Exercise 8.4.5   According to Corollary 8.4 and by the Riemann–Lebesgue lemma (2.17), the map

$$
\begin{array}{ccc}
H^1(\mathbb{T}) & \longrightarrow & c_0(\mathbb{Z}^+) \\
f & \longmapsto & \hat{f}
\end{array}
$$

is well-defined, and by the uniqueness theorem (Corollary 2.13) it is injective. However, show that it is not surjective.
Hint: See Exercise 2.5.6.

Exercise 8.4.6   [The generalized Hausdorff–Young theorem]   Let $(X, \mathfrak{M}, \mu)$ be a measure space. Let $(\varphi_n)_{n \in \mathbb{Z}}$ be an orthonormal family in $L^2(\mu)$ such that

$$|\varphi_n(x)| \leq M$$

for all $n \in \mathbb{Z}$ and all $x \in X$. In the first place, show that $\varphi_n \in L^q(\mu)$ for any $2 \le q \le \infty$. Then, for each $f \in L^p(\mu)$, $1 \le p \le 2$, define

$$\hat{f}(n) = \int_X f \overline{\varphi_n} \, d\mu, \qquad (n \in \mathbb{Z}),$$

and $\hat{f} = (\hat{f}(n))_{n \in \mathbb{Z}}$. Show that $\hat{f} \in \ell^q(\mathbb{Z})$, where $1/p + 1/q = 1$ and

$$\|\hat{f}\|_q \le M^{(2-p)/p} \, \|f\|_p.$$

Hint: Prove the inequality for $p = 1$ and $p = 2$ and then use the Riesz–Thorin interpolation theorem (Theorem 8.2).

8.5   An interpolation theorem for Hardy spaces

Let $V, V_1, V_2, \ldots, V_n$ be complex vector spaces. A map

$$\Lambda : V_1 \times V_2 \times \cdots \times V_n \longrightarrow V$$

is called a *multilinear operator* if it is linear with respect to each of its arguments while the others are kept fixed. If these vector spaces are subspaces (not necessarily closed) of certain Lebesgue spaces, we say that Λ is of type $(p_1, p_2, \ldots, p_n, q)$ whenever

$$\|\Lambda(f_1, f_2, \ldots, f_n)\|_q \le C \, \|f_1\|_{p_1} \|f_2\|_{p_2} \cdots \|f_n\|_{p_n}.$$

The following result is another version of the Riesz–Thorin interpolation theorem (Theorem 8.2). More results of this type can be found in [2], [12] and [22].

Lemma 8.5 *Let $(X_i, \mathfrak{M}_i, \mu_i)$, $i = 1, 2, \ldots, n$, and $(Y, \mathfrak{N}, \nu)$ be measure spaces. Let V_i represent the vector space of simple functions on X_i, and let V be the space of measurable functions on Y. Suppose that*

$$\Lambda : V_1 \times V_2 \times \cdots \times V_n \longrightarrow V$$

is a multilinear operator of types $(p_{1,0}, p_{2,0}, \ldots, p_{n,0}, q_0)$ and $(p_{1,1}, p_{2,1}, \ldots, p_{n,1}, q_1)$, where $p_{i,0}, q_0, p_{i,1}, q_1 \in [1, \infty]$, with constants C_0 and C_1. Let $t \in [0,1]$ and define

$$\frac{1}{p_{i,t}} = \frac{1-t}{p_{i,0}} + \frac{t}{p_{i,1}} \qquad \text{and} \qquad \frac{1}{q_t} = \frac{1-t}{q_0} + \frac{t}{q_1}.$$

Then Λ is of type $(p_{1,t}, p_{2,t}, \ldots, p_{n,t}, q_t)$ with constant $C_0^{1-t} C_1^t$, i.e.

$$\|\Lambda(\varphi_1, \varphi_2, \ldots, \varphi_n)\|_{q_t} \le C_0^{1-t} C_1^t \, \|\varphi_1\|_{p_{1,t}} \|\varphi_2\|_{p_{2,t}} \cdots \|\varphi_n\|_{p_{n,t}}$$

for all simple functions $\varphi_1, \varphi_2, \ldots, \varphi_n$. Moreover, if $p_{1,t}, p_{2,t}, \ldots, p_{n,t}$ are all finite, then Λ can be extended to

$$L^{p_{1,t}}(\mu_1) \times L^{p_{2,t}}(\mu_2) \times \cdots \times L^{p_{n,t}}(\mu_n),$$

preserving the last inequality.

Proof. The proof is similar to the proof of the Riesz–Thorin interpolation theorem (Theorem 8.2). We will use the same notations applied there.

Case 1: $p_{1,t}, p_{2,t}, \ldots, p_{n,t} < \infty$ and $q_t > 1$.

Fix simple functions $\varphi_1, \varphi_2, \ldots, \varphi_n$ and ψ with

$$\|\varphi_1\|_{p_{1,t}} = \|\varphi_2\|_{p_{2,t}} = \cdots = \|\varphi_n\|_{p_{n,t}} = \|\psi\|_{q_t'} = 1.$$

For the simple function

$$\varphi_i = \sum_m r_m\, e^{i\theta_m}\, \chi_{A_m}$$

on the measure spaces X_i, we define

$$\varphi_{i,z} = \sum_m r_m^{p_{i,t}\left(\frac{1-z}{p_{i,0}} + \frac{z}{p_{i,1}}\right)} e^{i\theta_m}\, \chi_{A_m}.$$

Similarly, for

$$\psi = \sum_n \rho_n\, e^{i\vartheta_n}\, \chi_{B_n},$$

let

$$\psi_z = \sum_n \rho_n^{q_t'\left(\frac{1-z}{q_0'} + \frac{z}{q_1'}\right)} e^{i\vartheta_n}\, \chi_{B_n}.$$

Let

$$F(z) = \int_Y \Lambda(\varphi_{1,z}, \varphi_{2,z}, \ldots, \varphi_{n,z})\, \psi_z\, d\nu.$$

Hence, F is an entire function, which is bounded on $\overline{S}$ and

$$F(t) = \int_Y \Lambda(\varphi_1, \varphi_2, \ldots, \varphi_n)\, \psi\, d\nu.$$

By Hölder's inequality,

$$
\begin{aligned}
|F(iy)| &\leq \left(\int_Y |\Lambda(\varphi_{1,iy}, \varphi_{2,iy}, \ldots, \varphi_{n,iy})|^{q_0}\, d\nu\right)^{\frac{1}{q_0}} \left(\int_Y |\psi_{iy}|^{q_0'}\, d\nu\right)^{\frac{1}{q_0'}} \\
&\leq C_0\, \|\varphi_{1,iy}\|_{p_{1,0}} \|\varphi_{2,iy}\|_{p_{2,0}} \cdots \|\varphi_{n,iy}\|_{p_{n,0}} \times \|\psi_{iy}\|_{q_0'} \\
&= C_0\, \|\varphi_1\|_{p_{1,t}}^{p_{1,t}/p_{1,0}} \|\varphi_2\|_{p_{2,t}}^{p_{2,t}/p_{2,0}} \cdots \|\varphi_n\|_{p_{n,t}}^{p_{n,t}/p_{n,0}} \times \|\psi\|_{q_t'}^{q_t'/q_0'} = C_0
\end{aligned}
$$

and

$$
\begin{aligned}
|F(1+iy)| &\leq \left(\int_Y |\Lambda(\varphi_{1,1+iy}, \varphi_{2,1+iy}, \ldots, \varphi_{n,1+iy})|^{q_1}\, d\nu\right)^{\frac{1}{q_1}} \left(\int_Y |\psi_{1+iy}|^{q_1'}\, d\nu\right)^{\frac{1}{q_1'}} \\
&\leq C_1\, \|\varphi_{1,1+iy}\|_{p_{1,1}} \|\varphi_{2,1+iy}\|_{p_{2,1}} \cdots \|\varphi_{n,1+iy}\|_{p_{n,1}} \times \|\psi_{1+iy}\|_{q_1'} \\
&= C_1\, \|\varphi_1\|_{p_{1,t}}^{p_{1,t}/p_{1,1}} \|\varphi_2\|_{p_{2,t}}^{p_{2,t}/p_{2,1}} \cdots \|\varphi_n\|_{p_{n,t}}^{p_{n,t}/p_{n,1}} \times \|\psi\|_{q_t'}^{q_t'/q_1'} = C_1,
\end{aligned}
$$

for all $y \in \mathbb{R}$. Hence, by Hadamard's three-line theorem (Theorem 8.1),

$$|F(t)| = \left| \int_Y \Lambda(\varphi_1, \varphi_2, \ldots, \varphi_n) \, \psi \, d\nu \right| \leq C_0^{1-t} C_1^t.$$

Taking the supremum with respect to ψ with $\|\psi\|_{q_t'} = 1$ gives

$$\left(\int_Y |\Lambda(\varphi_1, \varphi_2, \ldots, \varphi_n)|^{q_t} \, d\nu \right)^{\frac{1}{q_t}} \leq C_0^{1-t} C_1^t,$$

which is equivalent to the required result.

Case 2: Any of $p_{1,t} = \infty$, $p_{2,t} = \infty$, $\ldots$, $p_{n,t} = \infty$, or $q_t = 1$ happens.

If, for example, $p_{1,t} = \infty$, then we necessarily have $p_{1,t} = p_{1,0} = p_{1,1} = \infty$. Similarly, $q_t = 1$ implies $q_t = q_0 = q_1 = 1$. If any of these possibilities happens, in the definition of $F(z)$ remove z from the corresponding function, e.g.

$$F(z) = \int_Y \Lambda(\varphi_1, \varphi_{2,z}, \ldots, \varphi_{n,z}) \, \psi \, d\nu$$

corresponds to the case $p_{1,t} = \infty$, $q_t = 1$ and $p_{2,t}, \ldots, p_{n,t} < \infty$. Then follow the same procedure as in Case 1.

It remains to show that if $p_{1,t}, p_{2,t}, \ldots, p_{n,t}$ are all finite, then Λ can be extended by continuity to

$$L^{p_{1,t}}(\mu_1) \times L^{p_{2,t}}(\mu_2) \times \cdots \times L^{p_{n,t}}(\mu_n),$$

preserving the inequality

$$\|\Lambda(f_1, f_2, \ldots, f_n)\|_{q_t} \leq C_0^{1-t} C_1^t \|f_1\|_{p_{1,t}} \|f_2\|_{p_{2,t}} \cdots \|f_n\|_{p_{n,t}}.$$

The proof is based on the following simple observation. Let φ_i and ϕ_i be simple functions on X_i. Let

$$\Delta = \Lambda(\phi_1, \phi_2, \ldots, \phi_n) - \Lambda(\varphi_1, \varphi_2, \ldots, \varphi_n).$$

To estimate Δ, write

$$
\begin{aligned}
\Delta \quad = \quad & \Lambda(\phi_1, \phi_2, \ldots, \phi_n) - \Lambda(\varphi_1, \phi_2, \ldots, \phi_n) \\
+ \quad & \Lambda(\varphi_1, \phi_2, \ldots, \phi_n) - \Lambda(\varphi_1, \varphi_2, \ldots, \phi_n) \\
& \quad\vdots \\
+ \quad & \Lambda(\varphi_1, \ldots, \varphi_{n-1}, \phi_n) - \Lambda(\varphi_1, \varphi_2, \ldots, \varphi_n).
\end{aligned}
$$

Hence,

$$\|\Delta\|_{q_t} \leq C_0^{1-t} C_1^t \left(\max_i \{ \|\phi_i\|_{p_{i,t}}, \|\varphi_i\|_{p_{i,t}} \} \right)^{n-1} \left(\sum_{i=1}^n \|\phi_i - \varphi_i\|_{p_{i,t}} \right).$$

Therefore, by continuity, Λ can be extended to $L^{p_{1,t}}(\mu_1) \times L^{p_{2,t}}(\mu_2) \times \cdots \times L^{p_{n,t}}(\mu_n)$.

$\square$

The Riesz–Thorin interpolation theorem (Theorem 8.2) shows that linear operators between Lebesgue spaces can be interpolated. Using Lemma 8.5, we show that linear operators on Hardy spaces enjoy a similar property.

Theorem 8.6 *Let $(Y, \mathfrak{N}, \nu)$ be a measure space. Suppose that*

$$\Lambda : H^{p_0}(\mathbb{D}) \longrightarrow L^{q_0}(Y) \qquad and \qquad \Lambda : H^{p_1}(\mathbb{D}) \longrightarrow L^{q_1}(Y),$$

where $p_0, p_1 \in (0, \infty)$ and $q_0, q_1 \in [1, \infty]$, are bounded with constants C_0 and C_1. Let $t \in [0, 1]$ and define

$$\frac{1}{p_t} = \frac{1-t}{p_0} + \frac{t}{p_1} \qquad and \qquad \frac{1}{q_t} = \frac{1-t}{q_0} + \frac{t}{q_1}.$$

Then

$$\Lambda : H^{p_t}(\mathbb{D}) \longrightarrow L^{q_t}(Y)$$

is bounded and

$$\|\Lambda(F)\|_{q_t} \leq C\, C_0^{1-t} C_1^t \,\|F\|_{p_t}, \qquad (F \in H^{p_t}(\mathbb{D})),$$

where $C = C(p_0, p_1)$ is an absolute constant.

Proof. Without loss of generality, assume that $p_0 \leq p_1$. By assumption,

$$\begin{align}
\|\Lambda(F)\|_{q_0} &\leq C_0 \|F\|_{p_0}, \qquad (F \in H^{p_0}(\mathbb{D})), \tag{8.7}\\
\|\Lambda(F)\|_{q_1} &\leq C_1 \|F\|_{p_1}, \qquad (F \in H^{p_1}(\mathbb{D})). \tag{8.8}
\end{align}$$

For a function $\varphi \in L^p(\mathbb{T})$, $1 < p < \infty$, define

$$\Phi(z) = \frac{1}{2\pi} \int_{-\pi}^{\pi} \frac{e^{it} + z}{e^{it} - z}\, \varphi(e^{it})\, dt, \qquad (z \in \mathbb{D}). \tag{8.9}$$

Then, by Corollary 6.7, $\Phi \in H^p(\mathbb{D})$ and

$$\|\Phi\|_p \leq A_p \|\varphi\|_p, \tag{8.10}$$

where A_p is a constant just depending on p. Fix an integer n such that $np_0 > 1$. Let $\varphi_1, \varphi_2, \ldots, \varphi_n$ be functions in $L^{np_0}(\mathbb{T})$. By (8.10), $\Phi_i \in H^{np_0}(\mathbb{D})$ and thus, by Hölder's inequality, $\Phi_1 \Phi_2 \cdots \Phi_n \in H^{p_0}(\mathbb{D})$. Hence, we can define

$$\Upsilon(\varphi_1, \varphi_2, \ldots, \varphi_n) = \Lambda(\Phi_1 \Phi_2 \cdots \Phi_n) \tag{8.11}$$

on $L^{np_0}(\mathbb{T}) \times \cdots \times L^{np_0}(\mathbb{T})$.

Clearly, Υ is linear with respect to each argument. Moreover, by (8.7)–(8.8) and by Hölder's inequality,

$$\|\Upsilon(\varphi_1, \varphi_2, \ldots, \varphi_n)\|_{q_0} \leq C_0 \|\Phi_1 \Phi_2 \cdots \Phi_n\|_{p_0} \leq C_0 \|\Phi_1\|_{np_0} \|\Phi_2\|_{np_0} \cdots \|\Phi_n\|_{np_0}$$

and

$$\|\Upsilon(\varphi_1, \varphi_2, \ldots, \varphi_n)\|_{q_1} \leq C_1 \|\Phi_1 \Phi_2 \cdots \Phi_n\|_{p_1} \leq C_1 \|\Phi_1\|_{np_1} \|\Phi_2\|_{np_1} \cdots \|\Phi_n\|_{np_1}.$$

Hence, by (8.10),

$$\|\Upsilon(\varphi_1,\varphi_2,\ldots,\varphi_n)\|_{q_0} \le C_0\,A_{np_0}^n\,\|\varphi_1\|_{np_0}\|\varphi_2\|_{np_0}\cdots\|\varphi_n\|_{np_0},$$
$$\|\Upsilon(\varphi_1,\varphi_2,\ldots,\varphi_n)\|_{q_1} \le C_1\,A_{np_1}^n\,\|\varphi_1\|_{np_1}\|\varphi_2\|_{np_1}\cdots\|\varphi_n\|_{np_1}.$$

Therefore, by Lemma 8.5,

$$\|\Upsilon(\varphi_1,\ldots,\varphi_n)\|_{q_t} \le C_0^{1-t}C_1^t\,(A_{np_0}^{1-t}A_{np_1}^t)^n\,\|\varphi_1\|_{np_t}\cdots\|\varphi_n\|_{np_t}. \tag{8.12}$$

Let $F \in H^{p_t}(\mathbb{D})$. Without loss of generality, assume that $F(0) > 0$. Since otherwise, either multiply F by a constant of modulus one or replace it by $F+\varepsilon$ if $F(0) = 0$. By Theorem 7.7, $F = B\Phi^n$, where B is a Blaschke product and $\Phi \in H^{np_t}(\mathbb{D})$ with $\|B\Phi\|_{np_t}^{np_t} = \|\Phi\|_{np_t}^{np_t} = \|F\|_{p_t}^{p_t}$ and $B(0) > 0$, $\Phi(0) > 0$. Let

$$\Phi_1 = B\Phi \qquad \text{and} \qquad \Phi_2 = \Phi_3 = \cdots = \Phi_n = \Phi.$$

Since $np_t > 1$, each Φ_i has a representation of the form (8.9) with $\varphi_i = \Re\Phi_i$. Hence, by (8.11) and (8.12),

$$\|\Lambda(F)\|_{q_t} \le C_0^{1-t}C_1^t\,(A_{np_0}^{1-t}A_{np_1}^t)^n\,\|\varphi_1\|_{np_t}\cdots\|\varphi_n\|_{np_t}.$$

But

$$\|\varphi_i\|_{np_t} \le \|\Phi_i\|_{np_t} = \|F\|_{p_t}^{1/n}.$$

Therefore,

$$\|\Lambda(F)\|_{q_t} \le C\,C_0^{1-t}C_1^t\,\|F\|_{p_t}, \qquad (F \in H^{p_t}(\mathbb{D})),$$

where $C = (A_{np_0}^{1-t}A_{np_1}^t)^n$. $\qquad\square$

Using the well-known isometry between $H^p(\mathbb{D})$ and $H^p(\mathbb{T})$, we can replace $\mathbb{D}$ by $\mathbb{T}$ in the preceding theorem.

Exercises

Exercise 8.5.1 Let P be an analytic polynomial. Show that there is a finite Blaschke product B and an analytic polynomial Q with no zeros in the open unit disc $\mathbb{D}$ such that
$$P = BQ.$$

Exercise 8.5.2 Let $(Y,\mathfrak{N},\nu)$ be a measure space, let V be the space of measurable functions on Y, and let $\mathcal{P}$ denote the space of analytic polynomials on $\mathbb{T}$. Suppose that
$$\Lambda : \mathcal{P} \longrightarrow V$$
is a linear operator of types (p_0,q_0) and (p_1,q_1), where $p_0,p_1 \in (0,\infty)$ and $q_0,q_1 \in [1,\infty]$, with constants C_0 and C_1. Let $t \in [0,1]$ and define
$$\frac{1}{p_t} = \frac{1-t}{p_0} + \frac{t}{p_1} \qquad \text{and} \qquad \frac{1}{q_t} = \frac{1-t}{q_0} + \frac{t}{q_1}.$$

Show that Λ is of type (p_t, q_t) and

$$\|\Lambda(P)\|_{q_t} \le C\, C_0^{1-t}\, C_1^t\, \|P\|_{p_t}, \qquad (P \in \mathcal{P}),$$

where $C = C(p_0, p_1)$ is an absolute constant. Moreover, Λ can be extended to $H^{p_t}(\mathbb{T})$, preserving the last inequality.

Hint: Modify the proof of Theorem 8.6. Moreover, the ideas used at the end of the proof of Lemma 8.5 and the factorization given in Exercise 8.5.1 might be useful.

8.6   The Hardy–Littlewood inequality

The Riemann–Lebesgue lemma (Corollary 2.17) says that the Fourier coefficients of a function in $L^1(\mathbb{T})$ tend to zero as $|n|$ grows. Parseval's identity (Corollary 2.22) and Hardy's inequality (Theorem 7.16) provide further information if we consider smaller classes of functions. An easy argument based on Theorem 8.6 enables us to generalize these two results.

Let $Y = \{0, 1, 2, \dots\}$ be equipped with the measure

$$\mu(n) = \frac{1}{(n+1)^2}, \qquad (n \ge 0).$$

Hence, $L^p(Y)$, $0 < p < \infty$, consists of all complex sequences $(a_n)_{n \ge 0}$ such that

$$\| (a_n)_{n \ge 0} \|_p = \left(\sum_{n=0}^{\infty} \frac{|a_n|^p}{(n+1)^2} \right)^{\frac{1}{p}} < \infty.$$

According to Hardy's inequality, the operator

$$\Lambda : H^1(\mathbb{T}) \longrightarrow L^1(Y)$$

$$f \longmapsto ((n+1)\hat{f}(n))_{n \ge 0}$$

is of type $(1,1)$ with $\|\Lambda\|_{(1,1)} \le \pi$.

On the other hand, by Parseval's identity, the operator

$$\Lambda : H^2(\mathbb{T}) \longrightarrow L^2(Y)$$

$$f \longmapsto ((n+1)\hat{f}(n))_{n \ge 0}$$

is of type $(2,2)$ with $\|\Lambda\|_{(2,2)} = 1$. To apply Theorem 8.6, note that

$$p_0 = q_0 = 1, \qquad\qquad p_1 = q_1 = 2,$$

and thus $p = p_t = q_t \in [1, 2]$, and that

$$\| \Lambda f \|_p = \left(\sum_{n=0}^{\infty} \frac{|\hat{f}(n)|^p}{(n+1)^{2-p}} \right)^{\frac{1}{p}}, \qquad (f \in H^p(\mathbb{T})).$$

Hence we immediately obtain the following result.

Theorem 8.7 (Hardy–Littlewood) *Let $f \in H^p(\mathbb{T})$, $1 \le p \le 2$. Then*

$$\left(\sum_{n=0}^{\infty} \frac{|\hat{f}(n)|^p}{(n+1)^{2-p}} \right)^{\frac{1}{p}} \le C_p \, \|f\|_p,$$

where C_p is a constant just depending on p.

The Hardy–Littlewood inequality is valid even if $0 < p < 1$. But we do not discuss this case here.

Exercises

Exercise 8.6.1   Let $2 \le q < \infty$ and let $(a_n)_{n \ge 0}$ be a sequence of complex numbers such that

$$\sum_{n=0}^{\infty} (n+1)^{q-2} \, |a_n|^q < \infty.$$

Show that $F(z) = \sum_{n=0}^{\infty} a_n z^n$ is in $H^q(\mathbb{D})$ and

$$\|F\|_q \le C_q \left(\sum_{n=0}^{\infty} (n+1)^{q-2} \, |a_n|^q \right)^{\frac{1}{q}},$$

where C_q is a constant just depending on q.
Hint: Use duality and Theorem 8.7.

Exercise 8.6.2   Let $f \in L^p(\mathbb{T})$, $1 < p \le 2$. Show that

$$\left(\sum_{n=-\infty}^{\infty} \frac{|\hat{f}(n)|^p}{(|n|+1)^{2-p}} \right)^{\frac{1}{p}} \le C_p \, \|f\|_p,$$

where C_p is a constant just depending on p.
Hint: Use Theorem 8.7 and Exercise 6.3.4.

Exercise 8.6.3   Let $2 \le q < \infty$. Let $(a_n)_{n \in \mathbb{Z}}$ be a sequence of complex numbers such that

$$\sum_{n=-\infty}^{\infty} (|n|+1)^{q-2} \, |a_n|^q < \infty.$$

Show that there is an $f \in L^q(\mathbb{T})$ such that $\hat{f}(n) = a_n$, $n \in \mathbb{Z}$, and

$$\|f\|_q \le C_q \left(\sum_{n=-\infty}^{\infty} (|n|+1)^{q-2} \, |a_n|^q \right)^{\frac{1}{q}},$$

where C_q is a constant just depending on q.
Hint: Use Exercise 8.6.2 and duality.

Chapter 9

The Fourier transform

9.1   Lebesgue spaces on the real line

From now on we study upper half plane analogues of the results obtained before for the unit disc. Since the real line $\mathbb{R}$ is not compact and the Lebesgue measure is not finite on $\mathbb{R}$, we face several difficulties. Using a conformal mapping between the upper half plane $\mathbb{C}_+$ and the open unit disc $\mathbb{D}$, which also establishes a correspondence between $\mathbb{R}$ and $\mathbb{T} \setminus \{-1\}$, some of the preceding representation theorems can be rewritten for the upper half plane. Nevertheless, working with the Poisson kernel for $\mathbb{C}_+$ is somehow easier than its counterpart for $\mathbb{D}$. That is why we mostly provide direct proofs in the following.

Let f be a measurable function on $\mathbb{R}$ and let

$$\|f\|_p = \left(\int_{-\infty}^{\infty} |f(t)|^p \, dt \right)^{\frac{1}{p}}, \qquad (0 < p < \infty),$$

and

$$\|f\|_\infty = \inf_{M>0} \left\{ M \, : \, |\{ t \, : \, |f(t)| > M \}| = 0 \right\}.$$

The *Lebesgue spaces* $L^p(\mathbb{R})$, $0 < p \leq \infty$, are defined by

$$L^p(\mathbb{R}) = \{ f \, : \, \|f\|_p < \infty \}.$$

For $1 \leq p \leq \infty$, $L^p(\mathbb{R})$ is a Banach space and $L^2(\mathbb{R})$ equipped with the inner product

$$\langle f, g \rangle = \int_{-\infty}^{\infty} f(t) \, \overline{g(t)} \, dt$$

is a Hilbert space. It is important to note that, contrary to the case of the unit circle, the Lebesgue spaces on $\mathbb{R}$ do not form a chain. In other words, for each $p, q \in (0, \infty]$, $p \neq q$, we have

$$L^p(\mathbb{R}) \setminus L^q(\mathbb{R}) \neq \emptyset.$$

We will have to cope with this fact later on.

The space of all continuous functions on $\mathbb{R}$ is denoted by $\mathcal{C}(\mathbb{R})$. An element of $\mathcal{C}(\mathbb{R})$ is not necessarily bounded or uniformly continuous on $\mathbb{R}$. For most applications it is enough to consider the smaller space

$$\mathcal{C}_0(\mathbb{R}) = \{\, f \,:\, f \text{ continuous on } \mathbb{R} \text{ and } \lim_{|t|\to\infty} f(t) = 0 \,\}.$$

Elements of $\mathcal{C}_0(\mathbb{R})$ are certainly bounded and uniformly continuous on $\mathbb{R}$. However, the simple example $f(t) = \sin t$ shows that the inverse is not true. Since elements of $\mathcal{C}_0(\mathbb{R})$ are bounded, $\mathcal{C}_0(\mathbb{R})$ can be considered as a subspace of $L^\infty(\mathbb{R})$. Indeed, in this case, we have

$$\|f\|_\infty = \max_{t\in\mathbb{R}} |f(t)|$$

and the maximum is attained. Another important subclass of $\mathcal{C}_0(\mathbb{R})$ is $\mathcal{C}_c(\mathbb{R})$, which consists of all continuous functions of compact support, i.e. there is an $M = M(f)$ such that

$$f(t) = 0$$

for all $|t| \geq M$. The smooth subspaces $\mathcal{C}^n(\mathbb{R})$, $\mathcal{C}_c^n(\mathbb{R})$, $\mathcal{C}^\infty(\mathbb{R})$ and $\mathcal{C}_c^\infty(\mathbb{R})$ are defined similarly. In spectral analysis of Hardy spaces, we will also need

$$L^p(\mathbb{R}^+) = \{\, f \in L^p(\mathbb{R}) \,:\, f(t) = 0 \text{ for almost all } t \leq 0 \,\}$$

and

$$\mathcal{C}_0(\mathbb{R}^+) = \{\, f \in \mathcal{C}_0(\mathbb{R}) \,:\, f(t) = 0 \text{ for all } t \leq 0 \,\}.$$

The space of all *Borel measures* on $\mathbb{R}$ is denoted by $\mathcal{M}(\mathbb{R})$. This class equipped with the norm

$$\|\mu\| = |\mu|(\mathbb{R}),$$

where $|\mu|$ is the total variation of μ, is a Banach space. Each function $f \in L^1(\mathbb{R})$ corresponds uniquely to the measure

$$d\mu(t) = f(t)\, dt$$

and thus we can consider $L^1(\mathbb{R})$ as a subspace of $\mathcal{M}(\mathbb{R})$. Note that $\|\mu\| = \|f\|_1$. By a celebrated theorem of F. Riesz, the dual of $\mathcal{C}_0(\mathbb{R})$ is $\mathcal{M}(\mathbb{R})$.

Exercises

Exercise 9.1.1 Let $f : \mathbb{R} \longrightarrow \mathbb{C}$ and define its translations by

$$f_\tau(t) = f(t - \tau), \qquad (\tau \in \mathbb{R}).$$

Show that

$$\lim_{\tau\to 0} \|f_\tau - f\|_X = 0$$

if $f \in X$ with $X = L^p(\mathbb{R})$, $1 \le p < \infty$, or $X = C_0(\mathbb{R})$. Provide an example to show that this property does not hold if $X = L^\infty(\mathbb{R})$. However, show that

$$\|f_\tau\|_X = \|f\|_X, \qquad (\tau \in \mathbb{R}),$$

in all spaces mentioned above.

Exercise 9.1.2 Show that $C_0(\mathbb{R})$ is closed in $L^\infty(\mathbb{R})$.

Exercise 9.1.3 Show that each element of $C_0(\mathbb{R})$ is uniformly continuous on $\mathbb{R}$. More generally, suppose that $f \in C(\mathbb{R})$ and that

$$L_1 = \lim_{t \to +\infty} f(t)$$

and

$$L_2 = \lim_{t \to -\infty} f(t)$$

exist (we do not assume that $L_1 = L_2$). Show that f is uniformly continuous on $\mathbb{R}$.

Exercise 9.1.4 Show that $C_c(\mathbb{R})$ is dense in $L^p(\mathbb{R})$, $1 \le p < \infty$. Can you show that $C_c^\infty(\mathbb{R})$ is dense in $L^p(\mathbb{R})$, $1 \le p < \infty$?

Exercise 9.1.5 Show that $C_c(\mathbb{R})$ is dense in $C_0(\mathbb{R})$. Can you show that $C_c^\infty(\mathbb{R})$ is dense in $C_0(\mathbb{R})$?
Remark: We will develop certain techniques later on which might be useful for the second part of this question and the previous one. See Sections 9.4 and 10.4.

Exercise 9.1.6 Show that $L^p(\mathbb{R}^+)$ is a closed subspace of $L^p(\mathbb{R})$. Similarly, show that $C_0(\mathbb{R}^+)$ is a closed subspace of $C_0(\mathbb{R})$.

9.2 The Fourier transform on $L^1(\mathbb{R})$

The *Fourier transform* of $f \in L^1(\mathbb{R})$ is defined by

$$\hat{f}(t) = \int_{-\infty}^{\infty} f(\tau)\, e^{-i\,2\pi t\tau}\, d\tau, \qquad (t \in \mathbb{R}).$$

The *Fourier integral* of f is given formally by

$$\int_{-\infty}^{\infty} \hat{f}(\tau)\, e^{i\,2\pi t\tau}\, d\tau, \qquad (t \in \mathbb{R}).$$

One of our tasks is to study the convergence of this integral. The Fourier transform of a measure $\mu \in \mathcal{M}(\mathbb{R})$ is defined similarly by

$$\hat{\mu}(t) = \int_{\mathbb{R}} e^{-i\,2\pi t\tau}\, d\mu(\tau), \qquad (t \in \mathbb{R}).$$

However, if we consider $L^1(\mathbb{R})$ as a subspace of $\mathcal{M}(\mathbb{R})$, the two definitions of Fourier transform are consistent.

No wonder our most important example is the *Poisson kernel* for the upper half plane, which is defined by

$$P_y(t) = \frac{1}{\pi} \frac{y}{t^2 + y^2}, \qquad (y > 0). \tag{9.1}$$

(See Figure 9.1.)

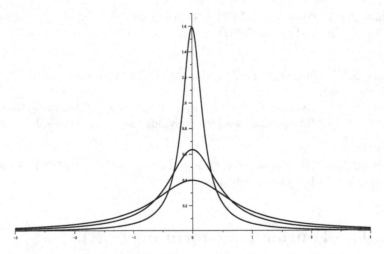

Fig. 9.1. The Poisson kernel $P_y(t)$ for $y = 0.8,\ 0.5,\ 0.2$.

For a fixed $x + iy \in \mathbb{C}_+$, let us find the Fourier transform of

$$f(t) = \frac{1}{\pi} \frac{y}{(x - t)^2 + y^2}.$$

Since $P_y(x)$ is an even function of x, we have

$$
\begin{aligned}
\hat{f}(t) &= \int_{-\infty}^{\infty} P_y(x - \tau)\, e^{-i\, 2\pi t\tau}\, d\tau \\
&= \int_{-\infty}^{\infty} P_y(\tau)\, e^{-i\, 2\pi t(x-\tau)}\, d\tau \\
&= e^{-i\, 2\pi x t} \int_{-\infty}^{\infty} P_y(\tau)\, e^{i\, 2\pi t\tau}\, d\tau \\
&= e^{-i\, 2\pi x t} \int_{-\infty}^{\infty} P_y(\tau)\, e^{-i\, 2\pi t\tau}\, d\tau
\end{aligned}
$$

and thus

$$
\hat{f}(t) = e^{-i\, 2\pi x t} \int_{-\infty}^{\infty} P_y(\tau)\, e^{i\, 2\pi |t|\tau}\, d\tau.
$$

Let $R > y$, and let Γ_R be the positively oriented curve formed with the interval $[-R, R]$ and the semicircle $\{\, Re^{i\theta} : 0 \le \theta \le \pi \,\}$. (See Figure 9.2.)

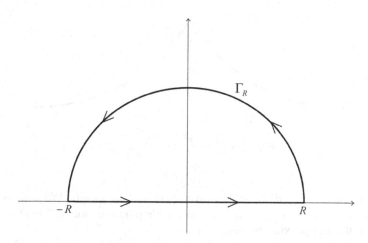

Fig. 9.2. The curve Γ_R.

Then

$$
\begin{aligned}
\hat{f}(t) &= e^{-i\, 2\pi x t} \lim_{R \to \infty} \int_{-R}^{R} \frac{y}{\pi(\tau^2 + y^2)}\, e^{i\, 2\pi |t|\tau}\, d\tau \\
&= e^{-i\, 2\pi x t} \lim_{R \to \infty} \int_{\Gamma_R} \frac{y}{\pi(w^2 + y^2)}\, e^{i\, 2\pi |t|w}\, dw.
\end{aligned}
$$

The only pole of our integrand

$$
F(w) = \frac{y}{\pi(w^2 + y^2)}\, e^{i\, 2\pi |t|w}
$$

inside Γ_R is iy. Moreover, the residue of F at this point is $\frac{1}{2\pi i}\, e^{-2\pi y|t|}$. Therefore, by the residue theorem,

$$\hat{f}(t) = e^{-i\, 2\pi x t - 2\pi y|t|}.$$

In particular, the Fourier transform of the Poisson kernel

$$P_y(t) = \frac{1}{\pi}\, \frac{y}{t^2 + y^2}$$

is

$$\widehat{P}_y(t) = e^{-2\pi y|t|}.$$

(See Figure 9.3.)

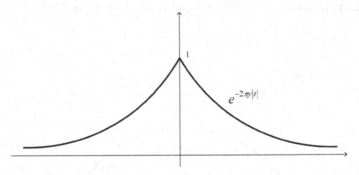

Fig. 9.3. The spectrum of P_y.

Note that, for each fixed $t \in \mathbb{R}$, we have $\widehat{P}_y(t) \to 1$ as $y \to 0$. We will see other families of functions having this behavior. This phenomenon will lead us to the definition of an approximate identity on $\mathbb{R}$.

We need the Fourier transform of four functions, which are gathered in Table 9.1. Since in each case our function is exponential, it is easy to calculate its Fourier transform. In all cases, $z = x + iy \in \mathbb{C}_+$. Let us also remember that

$$\operatorname{sgn}(t) = \begin{cases} 1 & \text{if } t > 0, \\ 0 & \text{if } t = 0, \\ -1 & \text{if } t < 0. \end{cases}$$

In Table 9.1, it is enough to verify the first and second lines. The third and last lines are linear combinations of the first two lines.

Table 9.1. Examples of the Fourier transform.

$f(t)$	$\hat{f}(t)$
$e^{i\,2\pi x\,t - 2\pi y\,|t|}$	$\dfrac{1}{\pi}\dfrac{y}{(x-t)^2 + y^2}$
$-i\,\mathrm{sgn}(t)\,e^{i\,2\pi x\,t - 2\pi y\,|t|}$	$\dfrac{1}{\pi}\dfrac{x-t}{(x-t)^2 + y^2}$
$\begin{cases} e^{i\,2\pi z\,t} & \text{if } t > 0 \\ 0 & \text{if } t < 0 \end{cases}$	$\dfrac{1}{2\pi i}\dfrac{1}{t-z}$
$\begin{cases} 0 & \text{if } t > 0 \\ e^{i\,2\pi \bar{z}\,t} & \text{if } t < 0 \end{cases}$	$-\dfrac{1}{2\pi i}\dfrac{1}{t-\bar{z}}$

Lemma 9.1 *Let $\mu \in \mathcal{M}(\mathbb{R})$. Then $\hat{\mu} \in \mathcal{C}(\mathbb{R}) \cap L^\infty(\mathbb{R})$ and*

$$\|\hat{\mu}\|_\infty \leq \|\mu\|.$$

In particular, for each $f \in L^1(\mathbb{R})$,

$$\|\hat{f}\|_\infty \leq \|f\|_1.$$

Proof. For all $t \in \mathbb{R}$, we have

$$
\begin{aligned}
|\hat{\mu}(t)| &= \left| \int_{\mathbb{R}} e^{-i\,2\pi t\tau}\, d\mu(\tau) \right| \\
&\leq \int_{\mathbb{R}} |e^{-i\,2\pi t\tau}|\, d|\mu|(\tau) \\
&= \int_{\mathbb{R}} d|\mu|(\tau) = |\mu|(\mathbb{R}) = \|\mu\|,
\end{aligned}
$$

which is equivalent to $\|\hat{\mu}\|_\infty \leq \|\mu\|$. The second inequality is a special case of the first one if we consider $d\mu(t) = f(t)\,dt$ and note that $\|\mu\| = \|f\|_1$.

To show that $\hat{\mu}$ is (uniformly) continuous on $\mathbb{R}$, let $t, t' \in \mathbb{R}$. Then

$$
\begin{aligned}
|\hat{\mu}(t) - \hat{\mu}(t')| &= \left| \int_{\mathbb{R}} \left(e^{-i\,2\pi t\tau} - e^{-i\,2\pi t'\tau} \right) d\mu(\tau) \right| \\
&\leq \int_{\mathbb{R}} |e^{-i\,2\pi t\tau} - e^{-i\,2\pi t'\tau}| \, d|\mu|(\tau) \\
&= \int_{\mathbb{R}} |1 - e^{i\,2\pi(t-t')\tau}| \, d|\mu|(\tau).
\end{aligned}
$$

The integrand is bounded,

$$
|1 - e^{i\,2\pi(t-t')\tau}| \leq 2,
$$

and $|\mu|$ is a finite positive Borel measure on $\mathbb{R}$. The required result now follows from the dominated convergence theorem.   □

Exercises

Exercise 9.2.1   Let $z = x + iy \in \mathbb{C}_+$. Show that

$$
P_y(x - t) = \frac{1}{\pi} \frac{y}{(x-t)^2 + y^2} = \Re\left(\frac{i}{\pi} \frac{1}{z - t} \right).
$$

Exercise 9.2.2   Let $f \in L^1(\mathbb{R})$ and let $g(t) = e^{i2\pi\tau_0 t} f(t - \tau_1)$, where τ_0, τ_1 are fixed real constants. Evaluate $\hat{g}$ in terms of $\hat{f}$.

Exercise 9.2.3   Let $\mu \in \mathcal{M}(\mathbb{R})$ and let

$$
\lambda(E) = \overline{\mu(-E)}.
$$

Evaluate $\hat{\lambda}$ in terms of $\hat{\mu}$.

Exercise 9.2.4   Let $f \in L^1(\mathbb{R})$ and let $g(t) = -2\pi i t f(t)$. Suppose that $g \in L^1(\mathbb{R})$. Show that $\hat{f} \in C^1(\mathbb{R})$ and

$$
\frac{d\hat{f}}{dt} = \hat{g}.
$$

What can we say if $t^n f(t) \in L^1(\mathbb{R})$?

Exercise 9.2.5   Let $f \in L^1(\mathbb{R})$ and define

$$
F(t) = \int_{-\infty}^t f(\tau) \, d\tau, \qquad (t \in \mathbb{R}).
$$

Suppose that $F \in L^1(\mathbb{R})$. Evaluate $\hat{F}$ in terms of $\hat{f}$.

Exercise 9.2.6   [Gauss–Weierstrass kernel] Let $\varepsilon > 0$ and let

$$\mathbf{G}_\varepsilon(t) = e^{-\pi\varepsilon t^2}, \qquad (t \in \mathbb{R}).$$

Show that

$$\widehat{\mathbf{G}}_\varepsilon(t) = \frac{1}{\sqrt{\varepsilon}} e^{-\pi t^2/\varepsilon}, \qquad (t \in \mathbb{R}).$$

Remark: Fix $x \in \mathbb{R}$ and $\varepsilon > 0$. A small modification of the preceding result shows that the Fourier transform of

$$f(t) = e^{i\,2\pi x t - \pi\varepsilon t^2}, \qquad (t \in \mathbb{R}),$$

is

$$\hat{f}(t) = \frac{1}{\sqrt{\varepsilon}} e^{-\pi(x-t)^2/\varepsilon}, \qquad (t \in \mathbb{R}).$$

(See Figure 9.4.)

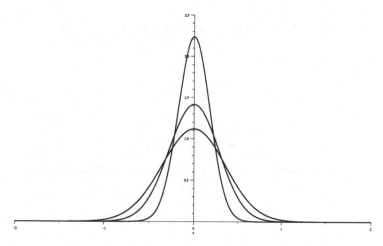

Fig. 9.4. The spectrum of $\mathbf{G}_\varepsilon$ for $\varepsilon = 0.8,\ 0.5,\ 0.2$.

Exercise 9.2.7   [Fejér's kernel] Fix $\lambda > 0$ and let

$$\mathbf{K}_\lambda(t) = \lambda \left(\frac{\sin(\pi\lambda t)}{\pi\lambda t} \right)^2, \qquad (t \in \mathbb{R}).$$

(See Figure 9.5.) Show that

$$\widehat{\mathbf{K}}_\lambda(t) = \max\left\{ 1 - \frac{|t|}{\lambda},\, 0 \right\}, \qquad (t \in \mathbb{R}).$$

(See Figure 9.6.)

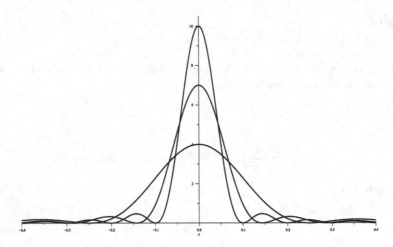

Fig. 9.5. The Fejér kernel $\mathbf{K}_\lambda(t)$ for $\lambda = 4, 7, 10$.

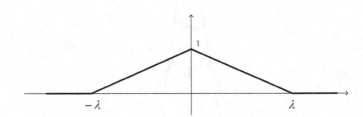

Fig. 9.6. The spectrum of $\mathbf{K}_\lambda$.

Exercise 9.2.8   Fix $x \in \mathbb{R}$ and $\lambda > 0$. Let

$$f(t) = \begin{cases} \left(1 - \frac{|t|}{\lambda}\right) e^{i \, 2\pi x t} & \text{if } |t| \le \lambda, \\ \\ 0 & \text{if } |t| \ge \lambda. \end{cases}$$

Show that

$$\hat{f}(t) = \lambda \left(\frac{\sin(\pi \lambda (x - t))}{\pi \lambda (x - t)} \right)^2, \qquad (t \in \mathbb{R}).$$

Hint: Use Exercises 9.2.2 and 9.2.7.

Exercise 9.2.9 Let $f \in \mathcal{C}^1(\mathbb{R})$ and suppose that $f, f' \in L^1(\mathbb{R})$. Show that

$$\widehat{f'}(t) = 2\pi i t \widehat{f}(t), \qquad (t \in \mathbb{R}).$$

What can we say if $f \in \mathcal{C}^n(\mathbb{R})$ and $f, f', \dots, f^{(n)} \in L^1(\mathbb{R})$?

Exercise 9.2.10 [de la Vallée Poussin's kernel] Fix $\lambda > 0$ and let

$$\mathbf{V}_\lambda(t) = 2\mathbf{K}_{2\lambda}(t) - \mathbf{K}_\lambda(t), \qquad (t \in \mathbb{R}).$$

(See Figure 9.7.) Show that

$$\widehat{\mathbf{V}}_\lambda(t) = \begin{cases} 1 & \text{if} & |t| \leq \lambda, \\[2mm] 2 - \frac{|t|}{\lambda} & \text{if} & \lambda \leq |t| \leq 2\lambda, \\[2mm] 0 & \text{if} & |t| \geq 2\lambda. \end{cases}$$

(See Figure 9.8.)

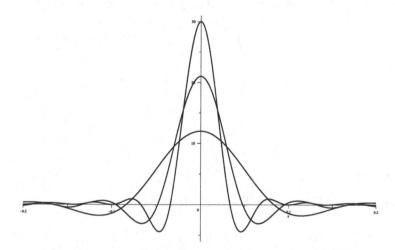

Fig. 9.7. The de la Vallée Poussin kernel $\mathbf{V}_\lambda(t)$ for $\lambda = 4, 7, 10$.

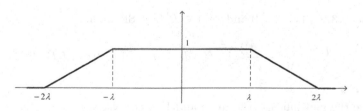

Fig. 9.8. The spectrum of $\mathbf{V}_\lambda$.

9.3   The multiplication formula on $L^1(\mathbb{R})$

The multiplication formula is an easy consequence of Fubini's theorem. However, it has profound implications in the spectral synthesis of harmonic functions in the upper half plane and also in extending the definition of the Fourier transform to other $L^p(\mathbb{R})$ spaces.

Lemma 9.2 (Multiplication formula) *Let $f \in L^1(\mathbb{R})$ and let $\mu \in \mathcal{M}(\mathbb{R})$. Then*

$$\int_{\mathbb{R}} \hat{f}(t)\, d\mu(t) = \int_{-\infty}^{\infty} f(t)\, \hat{\mu}(t)\, dt.$$

In particular, for $f, g \in L^1(\mathbb{R})$,

$$\int_{-\infty}^{\infty} \hat{f}(t)\, g(t)\, dt = \int_{-\infty}^{\infty} f(t)\, \hat{g}(t)\, dt.$$

Proof. By Fubini's theorem, we have

$$
\begin{aligned}
\int_{\mathbb{R}} \hat{f}(t)\, d\mu(t) &= \int_{\mathbb{R}} \left(\int_{-\infty}^{\infty} f(\tau)\, e^{-i\,2\pi t\tau}\, d\tau \right) d\mu(t) \\
&= \int_{-\infty}^{\infty} f(\tau) \left(\int_{\mathbb{R}} e^{-i\,2\pi\tau t}\, d\mu(t) \right) d\tau \\
&= \int_{-\infty}^{\infty} f(\tau)\, \hat{\mu}(\tau)\, d\tau.
\end{aligned}
$$

$\square$

Exercises

Exercise 9.3.1   Let $\varepsilon > 0$ and let $f \in L^1(\mathbb{R})$. Show that

$$\int_{-\infty}^{\infty} \frac{1}{\sqrt{\varepsilon}}\, e^{-\pi(x-t)^2/\varepsilon}\, f(t)\, dt = \int_{-\infty}^{\infty} e^{-\pi\varepsilon t^2}\, \hat{f}(t)\, e^{i\,2\pi x t}\, dt$$

for all $x \in \mathbb{R}$.
Hint: Use the multiplication formula and Exercise 9.2.6.

Exercise 9.3.2   Let $\lambda > 0$ and let $f \in L^1(\mathbb{R})$. Show that

$$\int_{-\infty}^{\infty} \lambda \left(\frac{\sin(\pi\lambda(x-t))}{\pi\lambda(x-t)} \right)^2 f(t)\, dt = \int_{-\lambda}^{\lambda} \left(1 - |t|/\lambda \right) \hat{f}(t)\, e^{i\,2\pi x t}\, dt$$

for all $x \in \mathbb{R}$.
Hint: Use the multiplication formula and Exercise 9.2.8.

9.4 Convolution on $\mathbb{R}$

As in the unit circle, we define the convolution of two functions $f, g \in L^1(\mathbb{R})$ by

$$(f * g)(t) = \int_{-\infty}^{\infty} f(\tau)\, g(t - \tau)\, d\tau.$$

By Fubini's theorem, we see that

$$
\begin{aligned}
\int_{-\infty}^{\infty} \left(\int_{-\infty}^{\infty} |f(\tau)\, g(t - \tau)|\, d\tau \right) dt &= \int_{-\infty}^{\infty} |f(\tau)| \left(\int_{-\infty}^{\infty} |g(t - \tau)|\, dt \right) d\tau \\
&= \left(\int_{-\infty}^{\infty} |f(\tau)|\, d\tau \right) \left(\int_{-\infty}^{\infty} |g(t)|\, dt \right) \\
&= \|f\|_1 \|g\|_1 < \infty.
\end{aligned}
$$

Hence, $(f * g)(t)$ is well-defined for almost all $t \in \mathbb{R}$, and moreover $f * g \in L^1(\mathbb{R})$ with

$$\|f * g\|_1 \leq \|f\|_1 \|g\|_1. \tag{9.2}$$

Lemma 9.3 *Let* $f, g \in L^1(\mathbb{R})$. *Then*

$$\widehat{f * g} = \hat{f}\, \hat{g}.$$

Proof. By Fubini's theorem, for each $t \in \mathbb{R}$, we have

$$
\begin{aligned}
\widehat{f * g}(t) &= \int_{-\infty}^{\infty} (f * g)(\tau)\, e^{-i\, 2\pi t \tau}\, d\tau \\
&= \int_{-\infty}^{\infty} \left(\int_{-\infty}^{\infty} f(s)\, g(\tau - s)\, ds \right) e^{-i\, 2\pi t \tau}\, d\tau \\
&= \int_{-\infty}^{\infty} f(s) \left(\int_{-\infty}^{\infty} g(\tau - s)\, e^{-i\, 2\pi t \tau}\, d\tau \right) ds \\
&= \int_{-\infty}^{\infty} f(s) \left(\int_{-\infty}^{\infty} g(\tau)\, e^{-i\, 2\pi t (\tau + s)}\, d\tau \right) ds \\
&= \left(\int_{-\infty}^{\infty} f(s)\, e^{-i\, 2\pi t s}\, ds \right) \left(\int_{-\infty}^{\infty} g(\tau)\, e^{-i\, 2\pi t \tau}\, d\tau \right) \\
&= \hat{f}(t)\, \hat{g}(t).
\end{aligned}
$$

$\square$

Using Riesz's theorem, we can also define the convolution of two measures $\mu, \nu \in \mathcal{M}(\mathbb{R})$ by the relation

$$\int_{\mathbb{R}} \int_{\mathbb{R}} \varphi(t + \tau)\, d\mu(t)\, d\nu(\tau) = \int_{\mathbb{R}} \varphi(t)\, d(\mu * \nu)(t), \tag{9.3}$$

where $\varphi \in C_0(\mathbb{R})$. It is rather easy to see that

$$\|\mu * \nu\| \leq \|\mu\|\, \|\nu\|.$$

But, since the exponential functions $\varphi_x(t) = e^{-i\,2\pi x t}$ are not in $C_0(\mathbb{R})$, it is slightly more difficult to show that

$$\widehat{\mu * \nu} = \hat{\mu}\,\hat{\nu}.$$

However, we do not need this result in the following.

If $\mu \in \mathcal{M}(\mathbb{R})$ and ν is absolutely continuous with respect to the Lebesgue measure, i.e. $d\nu(t) = f(t)\,dt$, where $f \in L^1(\mathbb{R})$, then, for each $\varphi \in C_0(\mathbb{R})$, we have

$$
\begin{aligned}
\int_{\mathbb{R}} \varphi(t)\, d(\mu * \nu)(t) &= \int_{\mathbb{R}}\int_{\mathbb{R}} \varphi(\tau + s)\, d\mu(\tau)\, d\nu(s) \\
&= \int_{\mathbb{R}}\left(\int_{\mathbb{R}} \varphi(\tau + s)\, f(s)\, ds \right) d\mu(\tau) \\
&= \int_{\mathbb{R}}\left(\int_{\mathbb{R}} \varphi(t)\, f(t - \tau)\, dt \right) d\mu(\tau) \\
&= \int_{\mathbb{R}} \varphi(t)\left(\int_{\mathbb{R}} f(t - \tau)\, d\mu(\tau) \right) dt.
\end{aligned}
$$

Therefore, $\mu * \nu$ is also absolutely continuous with respect to the Lebesgue measure, and according to the previous calculation we may write

$$d(\mu * \nu)(t) = (\mu * f)(t)\,dt,$$

where

$$(\mu * f)(t) = \int_{\mathbb{R}} f(t - \tau)\, d\mu(\tau) \tag{9.4}$$

for almost all $t \in \mathbb{R}$.

Exercises

Exercise 9.4.1   Let $f, g \in L^1(\mathbb{R})$. Show that

$$\int_{-\infty}^{\infty}\int_{-\infty}^{\infty} \varphi(t + \tau)\, f(t)\, g(\tau)\, dt\, d\tau = \int_{-\infty}^{\infty} \varphi(t)\, (f * g)(t)\, dt$$

for all $\varphi \in C_0(\mathbb{R})$.

Remark: This exercise shows that the generalized definition (9.3) is consistent with the old one whenever both measures are absolutely continuous with respect to the Lebesgue measure.

Exercise 9.4.2   Let

$$f(t) = \begin{cases} 1 & \text{if } |t| \leq 1, \\ 0 & \text{if } |t| > 1. \end{cases}$$

Evaluate $f * f$.

9.5 Young's inequality

On the unit circle the space $L^1(\mathbb{T})$ contains all Lebesgue spaces $L^p(\mathbb{T})$, $1 \leq p \leq \infty$. As a simple consequence, after defining $f * g$ over $L^1(\mathbb{T})$, it was used even if f and g were from different subclasses $L^p(\mathbb{T})$. Due to the fact that dt is not a finite measure on $\mathbb{R}$, Lebesgue spaces $L^p(\mathbb{R})$, $1 \leq p \leq \infty$, do not form a chain and thus $f * g$ is not yet defined if either f or g does not belong to $L^1(\mathbb{R})$. However, Young's theorem assures us that if $f \in L^r(\mathbb{R})$ and $g \in L^s(\mathbb{R})$, for certain values of r and s, then $f * g$ is a well-defined measurable function on $\mathbb{R}$. We start with a special case of this result and then use the Riesz–Thorin interpolation theorem (Theorem 8.2) to prove the general case.

Lemma 9.4 *Let $f \in L^p(\mathbb{R})$, $1 \leq p \leq \infty$, and let $g \in L^1(\mathbb{R})$. Then $(f * g)(t)$ is well-defined for almost all $t \in \mathbb{R}$, $f * g \in L^p(\mathbb{R})$ and*

$$\|f * g\|_p \leq \|f\|_p \, \|g\|_1.$$

Proof. The case $p = 1$ was studied at the beginning of Section 9.4. If $p = \infty$, the result is an immediate consequence of Hölder's inequality. As a matter of fact, in this case $(f * g)(t)$ is well-defined for all $t \in \mathbb{R}$.

Hence, suppose that $1 < p < \infty$. Let q be the conjugate exponent of p. Then, by Hölder's inequality,

$$
\begin{aligned}
\int_{-\infty}^{\infty} |f(\tau)\, g(t - \tau)| \, d\tau &= \int_{-\infty}^{\infty} \left(|f(\tau)| \, |g(t - \tau)|^{\frac{1}{p}} \right) |g(t - \tau)|^{\frac{1}{q}} \, d\tau \\
&\leq \left(\int_{-\infty}^{\infty} |f(\tau)|^p \, |g(t - \tau)| \, d\tau \right)^{\frac{1}{p}} \left(\int_{-\infty}^{\infty} |g(t - \tau)| \, d\tau \right)^{\frac{1}{q}} \\
&= \left(\int_{-\infty}^{\infty} |f(\tau)|^p \, |g(t - \tau)| \, d\tau \right)^{\frac{1}{p}} \|g\|_1^{\frac{1}{q}}.
\end{aligned}
$$

Hence, by Fubini's theorem,

$$
\begin{aligned}
\int_{-\infty}^{\infty} \left(\int_{-\infty}^{\infty} |f(\tau)\, g(t - \tau)| \, d\tau \right)^p dt &\leq \|g\|_1^{\frac{p}{q}} \int_{-\infty}^{\infty} \left(\int_{-\infty}^{\infty} |f(\tau)|^p \, |g(t - \tau)| \, d\tau \right) dt \\
&\leq \|g\|_1^{\frac{p}{q}} \int_{-\infty}^{\infty} |f(\tau)|^p \left(\int_{-\infty}^{\infty} |g(t - \tau)| \, dt \right) d\tau \\
&= \|g\|_1^{\frac{p}{q}} \times \|f\|_p^p \, \|g\|_1 = \|f\|_p^p \, \|g\|_1^p.
\end{aligned}
$$

Hence

$$\left\{ \int_{-\infty}^{\infty} \left(\int_{-\infty}^{\infty} |f(\tau)\, g(t - \tau)| \, d\tau \right)^p dt \right\}^{\frac{1}{p}} \leq \|f\|_p \, \|g\|_1,$$

and this inequality ensures that $(f * g)(t)$ is well-defined for almost all $t \in \mathbb{R}$, $f * g \in L^p(\mathbb{R})$ and that

$$\|f * g\|_p \leq \|f\|_p \, \|g\|_1.$$

$\square$

Lemma 9.5 *Let $1 \leq p \leq \infty$ and let q be its conjugate exponent. Let $f \in L^p(\mathbb{R})$ and let $g \in L^q(\mathbb{R})$. Then $f * g$ is well-defined at all points of $\mathbb{R}$ and $f * g$ is a bounded uniformly continuous function on the real line with*

$$\|f * g\|_\infty \leq \|f\|_p \, \|g\|_q.$$

Proof. First, suppose that $1 < p < \infty$. Then, by Hölder's inequality, we have

$$\int_{-\infty}^{\infty} |f(\tau)\, g(t - \tau)|\, d\tau \leq \left(\int_{-\infty}^{\infty} |f(\tau)|^p \, d\tau \right)^{\frac{1}{p}} \left(\int_{-\infty}^{\infty} |g(t - \tau)|^q \, d\tau \right)^{\frac{1}{q}},$$

and thus

$$\int_{-\infty}^{\infty} |f(\tau)\, g(t - \tau)|\, d\tau \leq \|f\|_p \, \|g\|_q$$

for all $t \in \mathbb{R}$. Hence, $f * g$ is well-defined at all points of $\mathbb{R}$ and

$$\|f * g\|_\infty \leq \|f\|_p \, \|g\|_q.$$

To show that $f * g$ is uniformly continuous on the real line, note that

$$
\begin{aligned}
|(f * g)(t) - (f * g)(t')| &\leq \int_{-\infty}^{\infty} |f(\tau)| \, |g(t - \tau) - g(t' - \tau)| \, d\tau \\
&= \int_{-\infty}^{\infty} |f(-\tau)| \, |g(t + \tau) - g(t' + \tau)| \, d\tau \\
&\leq \|f\|_p \, \|g_t - g_{t'}\|_q,
\end{aligned}
$$

where $g_t(\tau) = g(\tau + t)$. But the translation operator is continuous on $L^q(\mathbb{R})$, $1 \leq q < \infty$. The verification of this fact is similar to the one given for $L^q(\mathbb{T})$ classes, except that on $\mathbb{R}$ we should exploit continuous functions of compact support. Fix $\varepsilon > 0$ and pick $\varphi \in C_c(\mathbb{R})$ such that

$$\|g - \varphi\|_q < \varepsilon.$$

Let $\mathrm{supp}\, \varphi \subset [-M, M]$. Hence, for all $t, t' \in \mathbb{R}$ with $|t - t'| < 1$,

$$
\begin{aligned}
\|g_t - g_{t'}\|_q &\leq \|g_t - \varphi_t\|_q + \|\varphi_t - \varphi_{t'}\|_q + \|\varphi_{t'} - g_{t'}\|_q \\
&\leq 2\|g - \varphi\|_q + (2M + 2)^{1/q} \, \|\varphi - \varphi_{(t-t')}\|_\infty.
\end{aligned}
$$

Since φ is uniformly continuous on $\mathbb{R}$, there is a $\delta < 1$ such that $|s - s'| < \delta$ implies $|\varphi(s) - \varphi(s')| < \varepsilon/(2M + 2)^{1/q}$. Therefore,

$$\|g_t - g_{t'}\|_q \leq 3\varepsilon$$

provided that $|t - t'| < \delta$. Hence, we have

$$|(f * g)(t) - (f * g)(t')| \leq 3\varepsilon \, \|f\|_p$$

if $|t - t'| < \delta$. A small modification of this proof also works for $p = 1$ and $p = \infty$. $\qquad \square$

Now we have all the necessary ingredients to apply the Riesz–Thorin interpolation theorem (Theorem 8.2) in order to prove Young's inequality on the real line.

Theorem 9.6 (Young's inequality) *Let $f \in L^r(\mathbb{R})$ and let $g \in L^s(\mathbb{R})$ with*

$$1 \leq r, s \leq \infty$$

and

$$\frac{1}{r} + \frac{1}{s} \geq 1.$$

Then $f * g$ is well-defined almost everywhere on $\mathbb{R}$ and $f * g \in L^p(\mathbb{R})$, where

$$\frac{1}{p} = \frac{1}{r} + \frac{1}{s} - 1$$

and

$$\|f * g\|_p \leq \|f\|_r \, \|g\|_s.$$

Proof. Fix $f \in L^r(\mathbb{R})$. By Lemma 9.4, the convolution operator

$$\begin{aligned} \Lambda : L^1(\mathbb{R}) &\longrightarrow L^r(\mathbb{R}) \\ g &\longmapsto f * g \end{aligned}$$

is of type $(1, r)$ with

$$\|\Lambda\|_{(1,r)} \leq \|f\|_r.$$

On the other hand, Lemma 9.5 implies that

$$\begin{aligned} \Lambda : L^{r'}(\mathbb{R}) &\longrightarrow L^\infty(\mathbb{R}) \\ g &\longmapsto f * g \end{aligned}$$

is of type (r', ∞) with

$$\|\Lambda\|_{(r',\infty)} \leq \|f\|_r,$$

where $1/r + 1/r' = 1$. Hence, by the Riesz–Thorin interpolation theorem (Theorem 8.2) with

$$p_0 = 1, \quad q_0 = r, \quad p_1 = r', \quad q_1 = \infty,$$

we conclude that

$$\begin{aligned} \Lambda : L^{p_t}(\mathbb{R}) &\longrightarrow L^{q_t}(\mathbb{R}) \\ g &\longmapsto f * g \end{aligned}$$

is well-defined and has type (p_t, q_t) with

$$\|\Lambda\|_{(p_t,q_t)} \leq \|\Lambda\|_{(1,r)}^{1-t} \, \|\Lambda\|_{(r',\infty)}^{t} \leq \|f\|_r.$$

Hence,

$$\|f * g\|_{q_t} \leq \|f\|_r \|g\|_{p_t}.$$

To find the relation between p_t and q_t, note that

$$\frac{1}{p_t} = \frac{1-t}{1} + \frac{t}{r'} = 1 - \frac{t}{r}$$

and

$$\frac{1}{q_t} = \frac{1-t}{r} + \frac{t}{\infty} = \frac{1}{r} - \frac{t}{r}.$$

Hence,

$$\frac{1}{q_t} = \frac{1}{r} + \frac{1}{p_t} - 1.$$

□

Exercises

Exercise 9.5.1 Write

$$|f(\tau)\, g(t-\tau)| = \left(|g(t-\tau)|^{1-\frac{s}{p}} \right) \left(|f(\tau)|^{1-\frac{r}{p}} \right) \left(|f(\tau)|^{\frac{r}{p}} \, |g(t-\tau)|^{\frac{s}{p}} \right).$$

Now, modify the proof of Theorem 1.4 to give another proof of Young's inequality on the real line.

Exercise 9.5.2 Let $1 \le p \le \infty$ and let q be its conjugate exponent. Let $f \in L^p(\mathbb{R})$ and let $g \in L^q(\mathbb{R})$. Show that

$$f * g \in \mathcal{C}_0(\mathbb{R}).$$

Chapter 10

Poisson integrals

10.1 An application of the multiplication formula on $L^1(\mathbb{R})$

Given a family of functions $\{F_y\}_{y>0}$ on the real line $\mathbb{R}$, we define

$$F(x+iy) = F_y(x), \qquad (x+iy \in \mathbb{C}_+),$$

and thus deal with *one* function defined in the upper half plane $\mathbb{C}_+$. On the other hand, if F as a function on $\mathbb{C}_+$ is given, we can use the preceding identity as the definition of the family $\{F_y\}_{y>0}$. This dual interpretation will be encountered many times in what follows. The Poisson kernel

$$P_y(x) = P(x+iy) = \frac{1}{\pi}\frac{y}{x^2+y^2}$$

is a prototype of this phenomenon.

According to Lemma 9.4, if $u \in L^p(\mathbb{R})$, $1 \le p \le \infty$, then $P_y * u$ is well-defined almost everywhere on $\mathbb{R}$ and

$$\|P_y * u\|_p \le \|u\|_p.$$

Looking at $(P_y * u)(x)$ as a function defined in the upper half plane, we show that it represents a harmonic function there. Moreover, using the multiplication formula, we obtain another integral representation for $(P_y * u)(x)$ in terms of the Fourier transform of u. We also consider some other convolutions $K * u$, e.g. when K is the conjugate Poisson kernel. Let us start with a general case.

225

Theorem 10.1 *Let* $\mu \in \mathcal{M}(\mathbb{R})$. *Then we have*

$$\frac{1}{\pi} \int_{\mathbb{R}} \frac{y}{(x-t)^2 + y^2}\, d\mu(t) \ = \ \int_{-\infty}^{\infty} e^{-2\pi y|t|}\, \hat{\mu}(t)\, e^{i\,2\pi xt}\, dt,$$

$$\frac{1}{\pi} \int_{\mathbb{R}} \frac{x-t}{(x-t)^2 + y^2}\, d\mu(t) \ = \ \int_{-\infty}^{\infty} -i\, \mathrm{sgn}(t)\, e^{-2\pi y|t|}\, \hat{\mu}(t)\, e^{i\,2\pi xt}\, dt,$$

$$\frac{1}{2\pi i} \int_{\mathbb{R}} \frac{d\mu(t)}{t-z} \ = \ \int_{0}^{\infty} \hat{\mu}(t)\, e^{i\,2\pi zt}\, dt,$$

$$\frac{1}{2\pi i} \int_{\mathbb{R}} \frac{d\mu(t)}{t-\bar{z}} \ = \ -\int_{-\infty}^{0} \hat{\mu}(t)\, e^{i\,2\pi \bar{z}t}\, dt,$$

for all $z = x + iy \in \mathbb{C}_{+}$. *Each integral is absolutely convergent on compact subsets of* $\mathbb{C}_{+}$. *Moreover, each line represents a harmonic function on* $\mathbb{C}_{+}$. *The third line gives an analytic function.*

Proof. All the required identities are special cases of the multiplication formula (Lemma 9.2) applied to the Fourier transforms given in Table 9.1.

In Theorem 2.1, by applying the Laplace operator directly, we showed that the given function U is harmonic on the unit disc. This method applies here too. However, we provide a more instructive proof. Without loss of generality, assume that μ is real. Hence,

$$\begin{aligned}
U(z) \ &= \ \frac{1}{\pi} \int_{\mathbb{R}} \frac{y}{(t-x)^2 + y^2}\, d\mu(t) \\
&= \ \frac{1}{\pi} \int_{\mathbb{R}} \Re\left\{\frac{i}{z-t}\right\} d\mu(t) \\
&= \ \Re\left\{\frac{i}{\pi} \int_{\mathbb{R}} \frac{1}{z-t}\, d\mu(t)\right\}.
\end{aligned}$$

So U is the real part of an analytic function and thus it is harmonic in the upper half plane. Since

$$\left| e^{-2\pi y|t|}\, \hat{\mu}(t)\, e^{i2\pi xt} \right| \leq \|\mu\|\, e^{-2\pi y|t|},$$

by the dominated convergence theorem, the integral is absolutely and uniformly convergent on compact subsets of $\mathbb{C}_{+}$.

The second integral is the imaginary part of the preceding analytic function and thus it is harmonic too. The last two lines are linear combinations of the first two and, moreover, the third integral represents an analytic function. $\square$

In Theorem 10.1, if the measure μ is absolutely continuous with respect to the Lebesgue measure, i.e.

$$d\mu(t) = u(t)\, dt,$$

where $u \in L^1(\mathbb{R})$, then we obtain the following corollary. One of our main goals is to show that this result also holds even if $u \in L^p(\mathbb{R})$, $1 \leq p \leq 2$. The great task is to define $\hat{u}$ whenever $1 < p \leq 2$.

Corollary 10.2 *Let $u \in L^1(\mathbb{R})$. Then we have*

$$\frac{1}{\pi} \int_{-\infty}^{\infty} \frac{y}{(x-t)^2 + y^2} \, u(t) \, dt = \int_{-\infty}^{\infty} e^{-2\pi y|t|} \, \hat{u}(t) \, e^{i\, 2\pi x t} \, dt,$$

$$\frac{1}{\pi} \int_{-\infty}^{\infty} \frac{x-t}{(x-t)^2 + y^2} \, u(t) \, dt = \int_{-\infty}^{\infty} -i \, \mathrm{sgn}(t) \, e^{-2\pi y|t|} \, \hat{u}(t) \, e^{i\, 2\pi x t} \, dt,$$

$$\frac{1}{2\pi i} \int_{-\infty}^{\infty} \frac{u(t)}{t - z} \, dt = \int_0^{\infty} \hat{u}(t) \, e^{i\, 2\pi z t} \, dt,$$

$$\frac{1}{2\pi i} \int_{-\infty}^{\infty} \frac{u(t)}{t - \bar{z}} \, dt = -\int_{-\infty}^0 \hat{u}(t) \, e^{i\, 2\pi \bar{z} t} \, dt,$$

for all $z = x + iy \in \mathbb{C}_+$. Each integral is absolutely convergent on compact subsets of $\mathbb{C}_+$. Moreover, each line represents a harmonic function on $\mathbb{C}_+$. The third line gives an analytic function.

10.2 The conjugate Poisson kernel

Let $z = x + iy \in \mathbb{C}_+$. Since

$$\frac{y}{(t-x)^2 + y^2} = O(1/t^2)$$

as $|t| \to \infty$, the function

$$U(z) = \frac{1}{\pi} \int_{\mathbb{R}} \frac{y}{(t-x)^2 + y^2} \, d\mu(t)$$

is well-defined whenever μ is a Borel measure on $\mathbb{R}$ such that

$$\int_{\mathbb{R}} \frac{d|\mu|(t)}{1 + t^2} < \infty. \tag{10.1}$$

Clearly, (10.1) is fulfilled whenever $\mu \in \mathcal{M}(\mathbb{R})$. Moreover, U represents a harmonic function in the upper half plane. The verification of this fact is exactly as given in the proof of Theorem 10.1. In particular, if $d\mu(t) = u(t) \, dt$, where $u \in L^p(\mathbb{R})$, $1 \le p \le \infty$, then

$$U(z) = \frac{1}{\pi} \int_{-\infty}^{\infty} \frac{y}{(x-t)^2 + y^2} \, u(t) \, dt$$

is well-defined and represents a harmonic function in $\mathbb{C}_+$. Moreover, by Lemma 9.4,

$$\|U_y\|_p \le \|u\|_p$$

for all $y > 0$.

We already saw that

$$\Re \frac{i}{z - t} = \frac{y}{(x-t)^2 + y^2},$$

and since

$$\Im \frac{i}{z-t} = \frac{x-t}{(x-t)^2 + y^2},$$

we are tempted to say that the harmonic conjugate of U is given by

$$V(z) = \frac{1}{\pi} \int_{\mathbb{R}} \frac{x-t}{(x-t)^2 + y^2}\, d\mu(t). \qquad (10.2)$$

However, since

$$\frac{x-t}{(x-t)^2 + y^2} = O(1/|t|)$$

as $|t| \to \infty$, (10.2) is well-defined whenever

$$\int_{-\infty}^{\infty} \frac{d|\mu|(t)}{1 + |t|} < \infty.$$

For example, if $u \in L^p(\mathbb{R})$, $1 \leq p < \infty$, then

$$\int_{-\infty}^{\infty} \frac{|u(t)|}{1 + |t|}\, dt < \infty,$$

and thus the harmonic conjugate of

$$U(z) = \frac{1}{\pi} \int_{-\infty}^{\infty} \frac{y}{(x-t)^2 + y^2}\, u(t)\, dt$$

is given by

$$V(z) = \frac{1}{\pi} \int_{-\infty}^{\infty} \frac{x-t}{(x-t)^2 + y^2}\, u(t)\, dt. \qquad (10.3)$$

For the general case, where μ satisfies (10.1), simply note that

$$\Re\left(\frac{i}{z-t} + \frac{it}{1+t^2} \right) = \frac{y}{(x-t)^2 + y^2},$$

$$\Im\left(\frac{i}{z-t} + \frac{it}{1+t^2} \right) = \frac{x-t}{(x-t)^2 + y^2} + \frac{t}{1+t^2}$$

and that, as $|t| \to \infty$,

$$\frac{x-t}{(x-t)^2 + y^2} + \frac{t}{1+t^2} = O(1/t^2).$$

Therefore, assuming (10.1), the harmonic conjugate of

$$U(z) = \frac{1}{\pi} \int_{\mathbb{R}} \frac{y}{(t-x)^2 + y^2}\, d\mu(t)$$

is well-defined by the formula

$$V(z) = \frac{1}{\pi} \int_{-\infty}^{\infty} \left(\frac{x-t}{(x-t)^2 + y^2} + \frac{t}{1+t^2} \right) d\mu(t). \qquad (10.4)$$

The family

$$Q_y(t) = \frac{1}{\pi} \frac{t}{t^2 + y^2}, \qquad (y \geq 0), \qquad (10.5)$$

is called the *conjugate Poisson kernel* for the upper half plane. (See Figure 10.1.) However, as we will see, this kernel is more difficult to handle than the Poisson kernel.

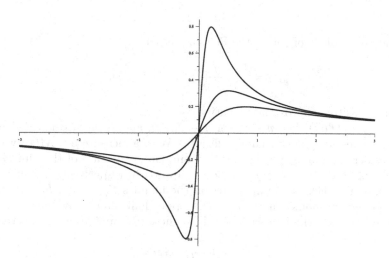

Fig. 10.1. The conjugate Poisson kernel $Q_y(t)$ for $y = 0.8, 0.5, 0.2$.

10.3 Approximate identities on $\mathbb{R}$

Let $\{\Phi_\iota\}$ be a family of integrable functions on $\mathbb{R}$. In our examples below, we will use y, ε and λ instead of ι and in all cases they range over the interval $(0, \infty)$. However, it should be noted that y and ε tend to zero while λ tends to infinity. Therefore, depending on the example, in the following $\lim_\iota$ means $\lim_{y \to 0}$, $\lim_{\varepsilon \to 0}$ or $\lim_{\lambda \to \infty}$. Similarly, $\iota \succ \iota_0$ means $y < y_0$, $\varepsilon < \varepsilon_0$ or $\lambda > \lambda_0$.

The family $\{\Phi_\iota\}$ of integrable functions is an *approximate identity* on $\mathbb{R}$ if it satisfies the following properties:

(a) for all ι,

$$C_\Phi = \sup_\iota \int_{-\infty}^\infty |\Phi_\iota(t)| \, dt < \infty;$$

(b) for all ι,

$$\int_{-\infty}^\infty \Phi_\iota(t) \, dt = 1;$$

(c) for each fixed $\delta > 0$,

$$\lim_{\iota} \int_{|t|>\delta} |\Phi_\iota(t)|\, dt = 0.$$

If, for all ι and for all $t \in \mathbb{R}$,

$$\Phi_\iota(t) \geq 0,$$

we say that $\{\Phi_\iota\}$ is a *positive approximate identity* on $\mathbb{R}$. In this case, (a) follows from (b) with

$$C_\Phi = 1.$$

Our main example is of course the Poisson kernel

$$P_y(t) = \frac{1}{\pi}\frac{y}{t^2 + y^2}, \qquad (t \in \mathbb{R}),$$

which is a positive approximate identity.

Given $f \in L^p(\mathbb{R})$, $1 \leq p \leq \infty$, or $f \in C_0(\mathbb{R})$, and an approximate identity $\{\Phi_\iota\}$ on $\mathbb{R}$, we form the new family $\{\Phi_\iota * f\}$. With a measure $\mu \in \mathcal{M}(\mathbb{R})$ we can also consider the family $\{\Phi_\iota * \mu\}$. Then, similar to the case of the unit circle, we will explore the way $\Phi_\iota * f$ and $\Phi_\iota * \mu$ approach, respectively, f and μ as ι grows. In particular, we are interested in the families $P_y * f$ and $P_y * \mu$, which also represent harmonic functions in the upper half plane. By Theorem 10.1 and Corollary 10.2, the Fourier integrals of these two families are respectively

$$\int_{-\infty}^{\infty} e^{-2\pi y|t|}\, \hat{f}(t)\, e^{i\,2\pi x t}\, dt$$

and

$$\int_{-\infty}^{\infty} e^{-2\pi y|t|}\, \hat{\mu}(t)\, e^{i\,2\pi x t}\, dt,$$

which are weighted Fourier integrals of f and μ.

Exercises

Exercise 10.3.1 Let $f \in L^1(\mathbb{R})$ and define

$$\Phi_\lambda(t) = \lambda\, f(\lambda t), \qquad (t \in \mathbb{R}),$$

where $\lambda > 0$. Evaluate $\hat{\Phi}_\lambda$ in terms of $\hat{f}$. For each fixed $t \in \mathbb{R}$, what is the behavior of $\hat{\Phi}_\lambda(t)$ if $\lambda \to \infty$?

Exercise 10.3.2 Let $f \in L^1(\mathbb{R})$ with

$$\int_{-\infty}^{\infty} f(t)\, dt = 1.$$

Let $\lambda > 0$ and define

$$\Phi_\lambda(t) = \lambda\, f(\lambda t), \qquad (t \in \mathbb{R}).$$

Show that Φ_λ is an approximate identity on $\mathbb{R}$. Under what condition is Φ_λ a positive approximate identity?

Exercise 10.3.3 [Gauss–Weierstrass kernel] Let $\varepsilon > 0$ and let

$$\mathbf{G}_\varepsilon(t) = \frac{1}{\sqrt{\varepsilon}}\, e^{-\pi t^2/\varepsilon}, \qquad (t \in \mathbb{R}).$$

Show that $\{\mathbf{G}_\varepsilon\}$ is a positive approximate identity on $\mathbb{R}$.

Exercise 10.3.4 Let p be a polynomial of degree n, and let $f(t) = p(t)\, e^{-t^2/2}$. Show that $\hat{f}(t) = q(t)\, e^{-t^2/2}$, where q is a polynomial of degree n. Moreover, show that q is odd (even) if p is odd (even).

Exercise 10.3.5 [Fejér kernel] Let $\lambda > 0$ and let

$$\mathbf{K}_\lambda(t) = \lambda \left(\frac{\sin(\pi \lambda t)}{\pi \lambda t} \right)^2, \qquad (t \in \mathbb{R}).$$

Show that $\mathbf{K}_\lambda$ is a positive approximate identity on $\mathbb{R}$.

Exercise 10.3.6 [de la Vallée Poussin's kernel] Let $\lambda > 0$ and let

$$\mathbf{V}_\lambda(t) = 2\mathbf{K}_{2\lambda}(t) - \mathbf{K}_\lambda(t), \qquad (t \in \mathbb{R}).$$

Show that $\mathbf{V}_\lambda$ is an approximate identity on $\mathbb{R}$.

Exercise 10.3.7 Show that

$$\int_{-\infty}^{\infty} \mathbf{K}_\lambda^2(t)\, dt = \frac{2\lambda}{3}.$$

Exercise 10.3.8 [Jackson's kernel] Let $\lambda > 0$ and let

$$\mathbf{J}_\lambda(t) = \frac{\mathbf{K}_\lambda^2(t)}{\|\mathbf{K}_\lambda\|_2^2} = \frac{3\lambda}{2} \left(\frac{\sin(\pi \lambda t)}{\pi \lambda t} \right)^4, \qquad (t \in \mathbb{R}).$$

(See Figure 10.2.) Show that $\mathbf{J}_\lambda$ is a positive approximate identity on $\mathbb{R}$.

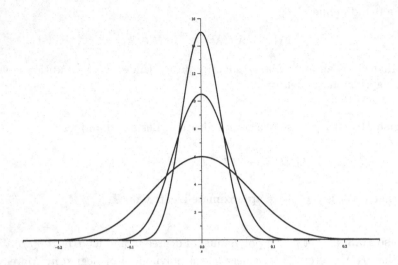

Fig. 10.2. The Jackson kernel $\mathbf{J}_\lambda(t)$ for $\lambda = 4, 7, 10$.

10.4 Uniform convergence and pointwise convergence

To study the convergence of $\Phi_\iota * f$ toward f, we start with the simple case $f \in \mathcal{C}_0(\mathbb{R})$. Let us recall that the elements of $\mathcal{C}_0(\mathbb{R})$ are uniformly continuous and bounded on $\mathbb{R}$.

Theorem 10.3 *Let* Φ_ι *be an approximate identity on* $\mathbb{R}$, *and let* $f \in \mathcal{C}_0(\mathbb{R})$. *Then, for all* ι, $\Phi_\iota * f \in \mathcal{C}_0(\mathbb{R})$ *with*

$$\|\Phi_\iota * f\|_\infty \leq C_\Phi \, \|f\|_\infty.$$

Moreover, $\Phi_\iota * f$ *converges uniformly to* f *on* $\mathbb{R}$, *i.e.*

$$\lim_\iota \|\Phi_\iota * f - f\|_\infty = 0.$$

Proof. Fix ι. By Lemma 9.5, $\Phi_\iota * f$ is continuous on $\mathbb{R}$. Since $\lim_{|t| \to \infty} f(t) = 0$, there is an $M > 0$ such that

$$|f(t)| < \varepsilon$$

for all $|t| \geq M$. Choose $M' = M'(\iota)$ such that

$$\int_{|t| \geq M'} |\Phi_\iota(t)| \, dt < \varepsilon.$$

Hence, for $|t| \geq M + M'$,

$$
\begin{aligned}
|(\Phi_\iota * f)(t)| &= \left| \int_{-\infty}^{\infty} \Phi_\iota(t - \tau)\, f(\tau)\, d\tau \right| \\
&\leq \left(\int_{|\tau| \leq M} + \int_{|\tau| \geq M} \right) |\Phi_\iota(t - \tau)|\, |f(\tau)|\, d\tau \\
&\leq \|f\|_\infty \int_{|s| \geq M'} |\Phi_\iota(s)|\, ds + \varepsilon \int_{-\infty}^{\infty} |\Phi_\iota(s)|\, ds \\
&\leq \varepsilon \left(\|f\|_\infty + C_\Phi \right).
\end{aligned}
$$

The previous two facts show that $\Phi_\iota * f \in \mathcal{C}_0(\mathbb{R})$. Moreover, by Lemma 9.5,

$$
\|\Phi_\iota * f\|_\infty \leq \|\Phi_\iota\|_1 \|f\|_\infty \leq C_\Phi \|f\|_\infty.
$$

To show that $\Phi_\iota * f$ converges uniformly to f on $\mathbb{R}$, given $\varepsilon > 0$, take $\delta = \delta(\varepsilon) > 0$ such that $|f(t - \tau) - f(t)| < \varepsilon$, for all $|\tau| < \delta$ and for *all* $t \in \mathbb{R}$. Therefore, we have

$$
\begin{aligned}
|(\Phi_\iota * f)(t) - f(t)| &= \left| \int_{-\infty}^{\infty} \Phi_\iota(\tau)\, \left(f(t - \tau) - f(t) \right) d\tau \right| \\
&\leq \left(\int_{-\infty}^{-\delta} + \int_{-\delta}^{\delta} + \int_{\delta}^{\infty} \right) |\Phi_\iota(\tau)|\, |f(t - \tau) - f(t)|\, d\tau \\
&\leq 2\|f\|_\infty \int_{-\infty}^{-\delta} |\Phi_\iota(\tau)|\, d\tau + \varepsilon \int_{-\delta}^{\delta} |\Phi_\iota(\tau)|\, d\tau + 2\|f\|_\infty \int_{\delta}^{\infty} |\Phi_\iota(\tau)|\, d\tau \\
&\leq 2\|f\|_\infty \int_{|\tau| > \delta} |\Phi_\iota(\tau)|\, d\tau + \varepsilon\, C_\Phi.
\end{aligned}
$$

Pick $\iota(\varepsilon, \delta) = \iota(\varepsilon)$ so large that

$$
\int_{|\tau| > \delta} |\Phi_\iota(\tau)|\, d\tau < \varepsilon,
$$

whenever $\iota \succ \iota(\varepsilon)$. Thus, for all $\iota \succ \iota(\varepsilon)$ and for all $t \in \mathbb{R}$,

$$
|(\Phi_\iota * f)(t) - f(t)| < (2\|f\|_\infty + C_\Phi)\, \varepsilon.
$$

$\qquad\qquad\qquad\qquad\qquad\qquad\qquad\qquad\qquad\qquad\qquad\qquad\qquad\qquad\quad \square$

As a special case, by using the Poisson kernel we can extend an element of $\mathcal{C}_0(\mathbb{R})$ to the upper half plane $\mathbb{C}_+$. The outcome is a function which is continuous on the closed upper half plane $\overline{\mathbb{C}}_+$ and harmonic on the open half plane $\mathbb{C}_+$. Moreover, due to special properties of the Poisson kernel, our function tends uniformly to zero as the argument goes to infinity in $\overline{\mathbb{C}}_+$.

Corollary 10.4 *Let* $u \in \mathcal{C}_0(\mathbb{R})$, *and let*

$$
U(x+iy) =
\begin{cases}
\dfrac{1}{\pi} \displaystyle\int_{-\infty}^{\infty} \dfrac{y}{(x-t)^2 + y^2}\, u(t)\, dt & \text{if } y > 0, \\[4mm]
\qquad\qquad u(x) & \text{if } y = 0.
\end{cases}
$$

Then

(a) U *is harmonic on* $\mathbb{C}_+$,

(b) U *is continuous on* $\overline{\mathbb{C}}_+$,

(c) for each $y \geq 0$, $U_y \in \mathcal{C}_0(\mathbb{R})$ *and* $\|U_y\|_\infty \leq \|u\|_\infty$,

(d) as $y \to 0$, U_y *converges uniformly to* u *on* $\mathbb{R}$,

(e) given $\varepsilon > 0$, *there is* $R > 0$ *such that*

$$
|U(x+iy)| < \varepsilon,
$$

whenever $|x+iy| \geq R$, $y \geq 0$.

Proof. In Theorem 10.1 we show that U is a harmonic function on $\mathbb{C}_+$. Knowing that the Poisson kernel is an approximate identity on $\mathbb{R}$, properties (b), (c) and (d) are a direct consequence of Theorem 10.3.

To prove (e), first suppose that u has a compact support, say $[-M, M]$. Fix $\varepsilon > 0$. Then

$$
\begin{aligned}
|U(x+iy)| &= \left| \frac{1}{\pi} \int_{-\infty}^{\infty} \frac{y}{(x-t)^2 + y^2}\, u(t)\, dt \right| \\[2mm]
&\leq \frac{1}{\pi} \int_{-M}^{M} \frac{y}{(x-t)^2 + y^2}\, |u(t)|\, dt \\[2mm]
&\leq \frac{2M\,\|u\|_\infty}{\pi\, y}, \qquad (x+iy \in \mathbb{C}_+).
\end{aligned}
$$

Hence, pick $y_0 > 0$ such that

$$
|U(x+iy)| < \varepsilon
$$

for all $y \geq y_0$ (no restriction on x). On the other hand, for $0 < y \leq y_0$ and $|x| \geq M + 2y_0\|u\|_\infty/(\pi\varepsilon)$, we have

$$
\begin{aligned}
|U(x+iy)| &\leq \frac{1}{\pi} \int_{|t|\leq M} \frac{y}{(x-t)^2 + y^2}\, |u(t)|\, dt \\[2mm]
&< \frac{\|u\|_\infty}{\pi} \int_{|\tau|\geq 2y_0\|u\|_\infty/(\pi\varepsilon)} \frac{y}{\tau^2}\, d\tau \leq \varepsilon.
\end{aligned}
$$

Hence it is enough to take

$$
R = y_0 + M + \frac{2y_0\|u\|_\infty}{\pi\varepsilon}.
$$

For the general case, take $v \in \mathcal{C}_c(\mathbb{R})$ such that $\|u - v\|_\infty < \varepsilon$ (remember that $\mathcal{C}_c(\mathbb{R})$ is dense in $\mathcal{C}_0(\mathbb{R})$). Then, by part (c),

$$
\begin{aligned}
|U(x + iy)| &= |P_y * (u - v)(x) + (P_y * v)(x)| \\
&\leq \|u - v\|_\infty + |(P_y * v)(x)| \\
&< \varepsilon + |(P_y * v)(x)|, \qquad (x + iy \in \mathbb{C}_+).
\end{aligned}
$$

As we saw in the previous paragraph, there is an R such that $|(P_y * v)(x)| < \varepsilon$ if $|x + iy| \geq R$ and $y \geq 0$. Hence

$$|U(x + iy)| < 2\varepsilon$$

for all $|x + iy| \geq R$ and $y \geq 0$. □

A small modification of the proof of Theorem 10.3 provides a local version for functions which are continuous at a fixed point.

Theorem 10.5 *Let Φ_ι be an approximate identity on $\mathbb{R}$, and let $f \in L^\infty(\mathbb{R})$. Suppose that f is continuous at $t_0 \in \mathbb{R}$. Then, given $\varepsilon > 0$, there exist $\iota(\varepsilon, t_0)$ and $\delta = \delta(\varepsilon, t_0) > 0$ such that*

$$|(\Phi_\iota * f)(t) - f(t_0)| < \varepsilon,$$

whenever $\iota \succ \iota(\varepsilon, t_0)$ and $|t - t_0| < \delta$. In particular,

$$\lim_\iota (\Phi_\iota * f)(t_0) = f(t_0).$$

Proof. Given $\varepsilon > 0$, there exists $\delta = \delta(\varepsilon, t_0) > 0$ such that

$$|f(\eta) - f(t_0)| < \varepsilon$$

if $|\eta - t_0| < 2\delta$. Therefore, for all t with $|t - t_0| < \delta$,

$$
\begin{aligned}
|(\Phi_\iota * f)(t) - f(t_0)| &= \left| \int_{-\infty}^{\infty} \Phi_\iota(\tau) \left(f(t - \tau) - f(t_0) \right) d\tau \right| \\
&\leq \left(\int_{-\infty}^{-\delta} + \int_{-\delta}^{\delta} + \int_{\delta}^{\infty} \right) |\Phi_\iota(\tau)| \, |f(t - \tau) - f(t_0)| \, d\tau \\
&\leq \int_{|\tau| > \delta} |\Phi_\iota(\tau)| \left(|f(t - \tau)| + |f(t_0)| \right) d\tau + \varepsilon \int_{-\delta}^{\delta} |\Phi_\iota(\tau)| \, d\tau \\
&\leq 2\|f\|_\infty \int_{|\tau| > \delta} |\Phi_\iota(\tau)| \, d\tau + \varepsilon \, C_\Phi.
\end{aligned}
$$

Pick $\iota(\varepsilon, t_0)$ so large that

$$\int_{|\tau| > \delta} |\Phi_\iota(\tau)| \, d\tau < \varepsilon$$

whenever $\iota \succ \iota(\varepsilon, t_0)$. Thus, for $\iota \succ \iota(\varepsilon, t_0)$ and for $|t - t_0| < \delta$,

$$|(\Phi_\iota * f)(t) - f(t_0)| < (2\|f\|_\infty + C_\Phi) \, \varepsilon.$$

□

If $u \in L^\infty(\mathbb{R})$ then the following result is a special case of Theorem 10.5. However, due to special properties of the Poisson kernel, we are able to relax the condition $u \in L^\infty(\mathbb{R})$ slightly.

Corollary 10.6 *Suppose that $u \in L^1(dt/(1+t^2))$, i.e.*

$$\int_{-\infty}^{\infty} \frac{|u(t)|}{1+t^2}\, dt < \infty, \qquad\qquad (10.6)$$

and that u is continuous at $t_0 \in \mathbb{R}$. Let

$$U(x+iy) = \frac{1}{\pi} \int_{-\infty}^{\infty} \frac{y}{(x-t)^2 + y^2}\, u(t)\, dt, \qquad (x+iy \in \mathbb{C}_+).$$

Then U is harmonic on $\mathbb{C}_+$ and besides

$$\lim_{\substack{z \to t_0 \\ z \in \mathbb{C}_+}} U(z) = u(t_0).$$

Remark: The condition (10.6) is satisfied if $u \in L^p(\mathbb{R})$, $1 \le p \le \infty$.

Proof. Without loss of generality, suppose that $t_0 = 0$. Since u is continuous at the origin, it is necessarily bounded in a neighborhood of the origin, say $[-2\delta, 2\delta]$. Let

$$\varphi(t) = \begin{cases} 1 & \text{if} \quad |t| \le \delta, \\ 2 - \frac{|t|}{\delta} & \text{if} \quad \delta \le |t| \le 2\delta, \\ 0 & \text{if} \quad |t| \ge 2\delta. \end{cases}$$

Now we write $u = u_1 + u_2$, where $u_1 = u\varphi$ and $u_2 = u(1-\varphi)$. The function u_1 is bounded on $\mathbb{R}$ and continuous at the origin. Hence, by Theorem 10.5,

$$\lim_{\substack{z \to t_0 \\ z \in \mathbb{C}_+}} (P * u_1)(z) = u_1(0) = u(0).$$

Since $U = P * u_1 + P * u_2$, it is enough to show that

$$\lim_{\substack{z \to t_0 \\ z \in \mathbb{C}_+}} (P * u_2)(z) = 0.$$

But, for $|x| < \delta/2$ and $y > 0$, we have

$$\begin{aligned}
|(P * u_2)(x+iy)| &= \left| \int_{|t| \ge \delta} \frac{y}{(x-t)^2 + y^2} u_2(t)\, dt \right| \\
&\le \int_{|t| \ge \delta} \frac{y}{(|t| - \delta/2)^2} |u(t)|\, dt \\
&\le 4y \int_{|t| \ge \delta} \frac{|u(t)|}{t^2}\, dt \\
&\le \left(4(1+1/\delta^2) \int_{-\infty}^{\infty} \frac{|u(t)|}{1+t^2}\, dt \right) y,
\end{aligned}$$

which shows that $(P * u_2)(z) \to 0$, as $z \to 0$. $\qquad\qquad\qquad\qquad$ $\square$

Exercises

Exercise 10.4.1 Let $f \in C_0(\mathbb{R})$, and let

$$f_\lambda(t) = \lambda \int_{-\infty}^{\infty} \left(\frac{\sin(\pi\lambda(t-\tau))}{\pi\lambda(t-\tau)} \right)^2 f(\tau)\, d\tau, \qquad (t \in \mathbb{R}).$$

Show that, for each $\lambda > 0$,

$$f_\lambda \in C_0^\infty(\mathbb{R})$$

with

$$\|f_\lambda\|_\infty \le \|f\|_\infty$$

and

$$\lim_{\lambda \to \infty} \|f_\lambda - f\|_\infty = 0.$$

Hint: Use Exercise 10.3.5 and Theorem 10.3.

Exercise 10.4.2 Let $f \in C_0(\mathbb{R})$, and let

$$f_\varepsilon(t) = \frac{1}{\sqrt{\varepsilon}} \int_{-\infty}^{\infty} e^{-\pi(t-\tau)^2/\varepsilon}\, f(\tau)\, d\tau, \qquad (t \in \mathbb{R}).$$

Show that, for each $\varepsilon > 0$,

$$f_\varepsilon \in C_0^\infty(\mathbb{R})$$

with

$$\|f_\varepsilon\|_\infty \le \|f\|_\infty$$

and

$$\lim_{\varepsilon \to 0} \|f_\varepsilon - f\|_\infty = 0.$$

Hint: Use Exercise 10.3.3 and Theorem 10.3.

Exercise 10.4.3 Let Φ_ι be an approximate identity on $\mathbb{R}$, and let f be uniformly continuous and bounded on $\mathbb{R}$. Show that, for each ι, $\Phi_\iota * f$ is also uniformly continuous and bounded on $\mathbb{R}$, and moreover $\Phi_\iota * f$ converges uniformly to f on $\mathbb{R}$.
Remark: If $f(x) = \sin x$, then $(P_y * f)(x) = e^{-y}\sin x$. Clearly, for each $y > 0$, $P_y * f$ is uniformly continuous and bounded on $\mathbb{R}$, and $P_y * f$ converges uniformly to f on $\mathbb{R}$, as $y \to 0$. However, for all $y > 0$, $P_y * f \notin C_0(\mathbb{R})$. Compare with Corollary 10.4.

10.5 Weak* convergence of measures

According to (9.4), $\phi * \mu$, the convolution of an integrable function ϕ and a Borel measure μ, is a well-defined integrable function. We now explore the relation between $\phi * \mu$ and μ where ϕ ranges over the elements of an approximate identity.

Theorem 10.7 *Let Φ_ι be an approximate identity on $\mathbb{R}$, and let $\mu \in \mathcal{M}(\mathbb{R})$. Then, for all ι, $\Phi_\iota * \mu \in L^1(\mathbb{R})$ with*

$$\|\Phi_\iota * \mu\|_1 \le C_\Phi \, \|\mu\|$$

and

$$\|\mu\| \le \sup_\iota \|\Phi_\iota * \mu\|_1.$$

Moreover, the measures $d\mu_\iota(t) = (\Phi_\iota * \mu)(t)\, dt$ converge to $d\mu(t)$ in the weak topology, i.e.*

$$\lim_\iota \int_{-\infty}^{\infty} \varphi(t)\, (\Phi_\iota * \mu)(t)\, dt = \int_{\mathbb{R}} \varphi(t)\, d\mu(t)$$

for all $\varphi \in \mathcal{C}_0(\mathbb{R})$.

Proof. The verification of the first and second assertions is similar to the proof of Lemma 9.5. Indeed, according to (9.4), for each ι, we have

$$|(\Phi_\iota * \mu)(t)| \le \int_{\mathbb{R}} |\Phi_\iota(t - \tau)|\, d|\mu|(\tau)$$

almost everywhere on $\mathbb{R}$, and thus, by Fubini's theorem,

$$
\begin{aligned}
\int_{-\infty}^{\infty} |(\Phi_\iota * \mu)(t)|\, dt &\le \int_{-\infty}^{\infty} \left(\int_{\mathbb{R}} |\Phi_\iota(t - \tau)|\, d|\mu|(\tau) \right) dt \\
&= \int_{\mathbb{R}} \left(\int_{-\infty}^{\infty} |\Phi_\iota(t - \tau)|\, dt \right) d|\mu|(\tau) \\
&\le C_\Phi \int_{\mathbb{R}} d|\mu|(\tau) = C_\Phi \, \|\mu\|.
\end{aligned}
$$

To show the weak* convergence, let $\varphi \in \mathcal{C}_0(\mathbb{R})$, and let $\psi(t) = \varphi(-t)$. Then, $\psi \in \mathcal{C}_0(\mathbb{R})$ and by Fubini's theorem,

$$
\begin{aligned}
\int_{-\infty}^{\infty} \varphi(t)\, (\Phi_\iota * \mu)(t)\, dt &= \int_{-\infty}^{\infty} \varphi(t) \left(\int_{\mathbb{R}} \Phi_\iota(t - \tau)\, d\mu(\tau) \right) dt \\
&= \int_{\mathbb{R}} \left(\int_{-\infty}^{\infty} \Phi_\iota(t - \tau)\, \varphi(t)\, dt \right) d\mu(\tau) \\
&= \int_{\mathbb{R}} \left(\int_{-\infty}^{\infty} \Phi_\iota(-\tau - t)\, \varphi(-t)\, dt \right) d\mu(\tau) \\
&= \int_{\mathbb{R}} \left(\int_{-\infty}^{\infty} \Phi_\iota(-\tau - t)\, \psi(t)\, dt \right) d\mu(\tau) \\
&= \int_{\mathbb{R}} (\Phi_\iota * \psi)(-\tau)\, d\mu(\tau).
\end{aligned}
$$

Theorem 10.3 ensures that $(\Phi_\iota * \psi)(-\tau)$ converges uniformly to $\psi(-\tau)$ on $\mathbb{R}$. Since $|\mu|$ is a finite Borel measure on $\mathbb{R}$, we thus have

$$
\begin{aligned}
\lim_\iota \int_{-\infty}^{\infty} \varphi(t)\,(\Phi_\iota * \mu)(t)\,dt &= \lim_\iota \int_{\mathbb{R}} (\Phi_\iota * \psi)(-\tau)\,d\mu(\tau) \\
&= \int_{\mathbb{R}} \psi(-\tau)\,d\mu(\tau) \\
&= \int_{\mathbb{R}} \varphi(\tau)\,d\mu(\tau).
\end{aligned}
$$

Since

$$
\left| \int_{-\infty}^{\infty} \varphi(t)\,(\Phi_\iota * \mu)(t)\,dt \right| \le \left(\sup_\iota \|\Phi_\iota * \mu\|_1 \right) \|\varphi\|_\infty,
$$

and $(\Phi_\iota * \mu)(t)\,dt$ converges to $d\mu(t)$ in the weak* topology, we thus have

$$
\left| \int_{\mathbb{R}} \varphi(t)\,d\mu(t) \right| \le \left(\sup_\iota \|\Phi_\iota * \mu\|_1 \right) \|\varphi\|_\infty,
$$

for all $\varphi \in C_0(\mathbb{R})$. Hence, by the Riesz representation theorem,

$$
\|\mu\| \le \sup_\iota \|\Phi_\iota * \mu\|_1.
$$

$\square$

If $\{\Phi_\iota\}$ is a positive approximate identity on $\mathbb{R}$, then $C_\Phi = 1$ and we obtain

$$
\|\mu\| = \sup_\iota \|\Phi_\iota * \mu\|_1.
$$

We can even show that sup can be replaced by lim in the preceding identity.

As a special case, by Theorem 10.1, the Poisson kernel extends μ to a harmonic function U on $\mathbb{C}_+$ such that the measures $U(t + iy)\,dt$ are uniformly bounded and converge to $d\mu(t)$, as $y \to 0$, in the weak* topology. This observation leads us to the uniqueness theorem, which is an essential result in harmonic analysis.

Corollary 10.8 *Let $\mu \in \mathcal{M}(\mathbb{R})$, and let*

$$
U(x + iy) = \frac{1}{\pi} \int_{-\infty}^{\infty} \frac{y}{(x - t)^2 + y^2}\,d\mu(t), \qquad (x + iy \in \mathbb{C}_+). \qquad (10.7)
$$

Then U is harmonic on $\mathbb{C}_+$, and

$$
\|\mu\| = \sup_{y>0} \|U_y\|_1 = \lim_{y \to 0} \|U_y\|_1.
$$

Moreover, the measures $d\mu_y(t) = U(t + iy)\,dt$ converge to $d\mu(t)$, as $y \to 0$, in the weak* topology, i.e.

$$
\lim_{y \to 0} \int_{\mathbb{R}} \varphi(t)\,d\mu_y(t) = \int_{\mathbb{R}} \varphi(t)\,d\mu(t)
$$

for all $\varphi \in C_0(\mathbb{R})$.

We give two versions of the uniqueness theorem. However, both results are a direct consequence of the relation $\|\mu\| = \sup_{y>0} \|U_y\|_1$, where U is given by (10.7).

Corollary 10.9 (Uniqueness theorem) *Let $\mu \in \mathcal{M}(\mathbb{R})$. Suppose that*

$$\frac{1}{\pi} \int_{-\infty}^{\infty} \frac{y}{(x-t)^2 + y^2} \, d\mu(t) = 0$$

for all $x + iy \in \mathbb{C}_+$. Then $\mu = 0$.

Corollary 10.10 (Uniqueness theorem) *Let $\mu \in \mathcal{M}(\mathbb{R})$. Suppose that*

$$\hat{\mu}(t) = 0$$

for all $t \in \mathbb{R}$. Then $\mu = 0$. In particular, if $f \in L^1(\mathbb{R})$ and $\hat{f}(t) = 0$, for all $t \in \mathbb{R}$, then we have $f = 0$.

Proof. Define U by (10.7) and note that, by Theorem 10.1,

$$U(x + iy) = \int_{-\infty}^{\infty} e^{-2\pi y|t|} \, \hat{\mu}(t) \, e^{i \, 2\pi x t} \, dt = 0$$

for all $x + iy \in \mathbb{C}_+$. $\square$

The uniqueness theorem combined with Lemma 9.1 says that the map

$$\begin{array}{ccc} \mathcal{M}(\mathbb{R}) & \longrightarrow & \mathcal{C}(\mathbb{R}) \cap L^\infty(\mathbb{R}) \\ \mu & \longmapsto & \hat{\mu} \end{array}$$

is one-to-one.

Exercises

Exercise 10.5.1   Let $\mu \in \mathcal{M}(\mathbb{R})$ and let $y_0 > 0$. Suppose that

$$\frac{1}{\pi} \int_{-\infty}^{\infty} \frac{y_0}{(x-t)^2 + y_0^2} \, d\mu(t) = 0$$

for all $x \in \mathbb{R}$. Show that $\mu = 0$.

Exercise 10.5.2   Let $\mu \in \mathcal{M}(\mathbb{R})$. Suppose that

$$\frac{1}{\pi} \int_{-\infty}^{\infty} \frac{x-t}{(x-t)^2 + y^2} \, d\mu(t) = 0$$

for all $x + iy \in \mathbb{C}_+$. Can we conclude that $\mu = 0$? What if our assumption only holds on the horizontal line $y = y_0$?

Exercise 10.5.3 Let $\mu \in \mathcal{M}(\mathbb{R})$, and let

$$f_\lambda(t) = \lambda \int_{-\infty}^{\infty} \left(\frac{\sin(\pi\lambda\,(t-\tau))}{\pi\lambda\,(t-\tau)} \right)^2 d\mu(\tau), \qquad (t \in \mathbb{R}).$$

Show that, for each $\lambda > 0$,

$$f_\lambda \in L^1(\mathbb{R})$$

with

$$\|f_\lambda\|_1 \le \|\mu\|,$$

and the measures $f_\lambda(t)\,dt$ converge to $d\mu(t)$ in the weak* topology, i.e.

$$\lim_{\lambda\to\infty} \int_{-\infty}^{\infty} \varphi(t)\,f_\lambda(t)\,dt = \int_{\mathbb{R}} \varphi(t)\,d\mu(t)$$

for all $\varphi \in C_0(\mathbb{R})$.
Hint: Use Exercise 10.3.5 and Theorem 10.7.

Exercise 10.5.4 Let $\mu \in \mathcal{M}(\mathbb{R})$, and let

$$f_\varepsilon(t) = \frac{1}{\sqrt{\varepsilon}} \int_{-\infty}^{\infty} e^{-\pi(t-\tau)^2/\varepsilon}\,d\mu(\tau), \qquad (t \in \mathbb{R}).$$

Show that, for each $\varepsilon > 0$,

$$f_\varepsilon \in L^1(\mathbb{R})$$

with

$$\|f_\varepsilon\|_1 \le \|\mu\|,$$

and the measures $f_\varepsilon(t)\,dt$ converge to $d\mu(t)$ in the weak* topology, i.e.

$$\lim_{\varepsilon\to 0} \int_{-\infty}^{\infty} \varphi(t)\,f_\varepsilon(t)\,dt = \int_{\mathbb{R}} \varphi(t)\,d\mu(t)$$

for all $\varphi \in C_0(\mathbb{R})$.
Hint: Use Exercise 10.3.3 and Theorem 10.7.

10.6 Convergence in norm

In this section we use the fact that $C_c(\mathbb{R})$, the space of continuous functions of compact support, is dense in $L^p(\mathbb{R})$, $1 \le p < \infty$. This assertion is not true if $p = \infty$ and that is why the following result does not hold for the elements of $L^\infty(\mathbb{R})$.

Theorem 10.11 *Let Φ_ι be an approximate identity on $\mathbb{R}$, and let $f \in L^p(\mathbb{R})$, $1 \le p < \infty$. Then, for all ι, $\Phi_\iota * f \in L^p(\mathbb{R})$ and*

$$\|\Phi_\iota * f\|_p \le C_\Phi \|f\|_p.$$

Moreover,

$$\lim_\iota \|\Phi_\iota * f - f\|_p = 0.$$

Proof. The first two assertions are proved in Lemma 9.4. To prove the last identity, we have

$$(\Phi_\iota * f)(t) - f(t) = \int_{-\infty}^{\infty} \Phi_\iota(\tau)\,(f(t - \tau) - f(t))\,d\tau, \qquad (t \in \mathbb{R}).$$

Hence, by Minkowki's inequality (see Section A.6),

$$\|\Phi_\iota * f - f\|_p \le \int_{-\infty}^{\infty} |\Phi_\iota(\tau)|\,\|f_\tau - f\|_p\,d\tau,$$

where $f_\tau(t) = f(t - \tau)$. It is enough to show that

$$\lim_{\tau \to 0} \|f_\tau - f\|_p = 0,$$

since then we use the same technique applied in the proof of Theorem 10.5 to deduce that $\|\Phi_\iota * f - f\|_p \to 0$. To do so, given $\varepsilon > 0$, pick $\varphi \in C_c(\mathbb{R})$ such that $\|f - \varphi\|_p < \varepsilon$. Hence,

$$
\begin{aligned}
\|f_\tau - f\|_p &\le \|f_\tau - \varphi_\tau\|_p + \|\varphi_\tau - \varphi\|_p + \|\varphi - f\|_p \\
&= 2\|\varphi - f\|_p + +\|\varphi_\tau - \varphi\|_p \\
&\le 2\varepsilon + \|\varphi_\tau - \varphi\|_p.
\end{aligned}
$$

As we saw in the proof of Lemma 9.5, $\|\varphi_\tau - \varphi\|_p \to 0$, as $\tau \to 0$. Hence, there is $\delta > 0$ such that, for $|\tau| < \delta$,

$$\|f_\tau - f\|_p \le 3\varepsilon.$$

$\qquad\qquad\qquad\qquad\qquad\qquad\qquad\qquad\qquad\qquad\qquad\qquad\qquad\qquad\qquad\qquad\qquad\qquad\square$

As in the preceding cases, we use the Poisson kernel, to extend $u \in L^p(\mathbb{R})$ to the harmonic function $U = P * u$ on $\mathbb{C}_+$ whose integral means $\|U_y\|_p$ are uniformly bounded and, $y \to 0$, U_y converges to u in $L^p(\mathbb{R})$.

Corollary 10.12 *Let $u \in L^p(\mathbb{R})$, $1 \le p < \infty$, and let*

$$U(x + iy) = \frac{1}{\pi} \int_{-\infty}^{\infty} \frac{y}{(x - t)^2 + y^2}\,u(t)\,dt, \qquad (x + iy \in \mathbb{C}_+).$$

Then U is harmonic on $\mathbb{C}_+$,

$$\sup_{y > 0} \|U_y\|_p = \lim_{y \to 0} \|U_y\|_p = \|u\|_p$$

and

$$\lim_{y \to 0} \|U_y - u\|_p = 0.$$

Exercises

Exercise 10.6.1 Let $f \in L^p(\mathbb{R})$, $1 \le p < \infty$, and let

$$f_\lambda(t) = \lambda \int_{-\infty}^{\infty} \left(\frac{\sin(\pi\lambda(t-\tau))}{\pi\lambda(t-\tau)} \right)^2 f(\tau) \, d\tau, \qquad (t \in \mathbb{R}).$$

Show that, for each $\lambda > 0$,

$$f_\lambda \in L^p(\mathbb{R})$$

with

$$\|f_\lambda\|_p \le \|f\|_p$$

and

$$\lim_{\lambda \to \infty} \|f_\lambda - f\|_p = 0.$$

Hint: Use Exercise 10.3.5 and Theorem 10.11.

Exercise 10.6.2 Let $f \in L^p(\mathbb{R})$, $1 \le p < \infty$, and let

$$f_\varepsilon(t) = \frac{1}{\sqrt{\varepsilon}} \int_{-\infty}^{\infty} e^{-\pi(t-\tau)^2/\varepsilon} f(\tau) \, d\tau, \qquad (t \in \mathbb{R}).$$

Show that, for each $\varepsilon > 0$,

$$f_\varepsilon \in L^p(\mathbb{R})$$

with

$$\|f_\varepsilon\|_p \le \|f\|_p$$

and

$$\lim_{\varepsilon \to 0} \|f_\varepsilon - f\|_p = 0.$$

Hint: Use Exercise 10.3.3 and Theorem 10.11.

10.7 Weak* convergence of bounded functions

Since $\mathcal{C}_c(\mathbb{R})$ is not dense in $L^\infty(\mathbb{R})$, the results of Section 10.6 are not entirely valid if $p = \infty$. Nevertheless, a slightly weaker version also holds in this case.

Theorem 10.13 *Let Φ_ι be an approximate identity on $\mathbb{R}$, and let $f \in L^\infty(\mathbb{R})$. Then, for all ι, $\Phi_\iota * f \in L^\infty(\mathbb{R}) \cap C(\mathbb{R})$ with*

$$\|\Phi_\iota * f\|_\infty \le C_\Phi \|f\|_\infty$$

and

$$\|f\|_\infty \le \sup_\iota \|\Phi_\iota * f\|_\infty.$$

Moreover, $\Phi_\iota * f$ converges to f in the weak* topology, i.e.

$$\lim_\iota \int_{-\infty}^{\infty} \varphi(t) \, (\Phi_\iota * f)(t) \, dt = \int_{-\infty}^{\infty} \varphi(t) \, f(t) \, dt$$

for all $\varphi \in L^1(\mathbb{R})$.

Proof. The first two assertions are proved in Lemma 9.5. Let $\varphi \in L^1(\mathbb{R})$, and let $\psi(t) = \varphi(-t)$. Then, by Fubini's theorem,

$$
\begin{aligned}
\int_{-\infty}^{\infty} \varphi(t)\,(\Phi_\iota * f)(t)\,dt &= \int_{-\infty}^{\infty} \varphi(t) \left(\int_{-\infty}^{\infty} \Phi_\iota(t-\tau)\,f(\tau)\,d\tau \right) dt \\
&= \int_{-\infty}^{\infty} \left(\int_{-\infty}^{\infty} \Phi_\iota(t-\tau)\,\varphi(t)\,dt \right) f(\tau)\,d\tau \\
&= \int_{-\infty}^{\infty} \left(\int_{-\infty}^{\infty} \Phi_\iota(-\tau-t)\,\varphi(-t)\,dt \right) f(\tau)\,d\tau \\
&= \int_{-\infty}^{\infty} \left(\int_{-\infty}^{\infty} \Phi_\iota(-\tau-t)\,\psi(t)\,dt \right) f(\tau)\,d\tau \\
&= \int_{-\infty}^{\infty} (\Phi_\iota * \psi)(-\tau)\,f(\tau)\,d\tau.
\end{aligned}
$$

Theorem 10.11 ensures that $(\Phi_\iota * \psi)(-\tau)$ converges in $L^1(\mathbb{R})$ to $\psi(-\tau)$. Since f is a bounded function, we thus have

$$
\begin{aligned}
\lim_\iota \int_{-\infty}^{\infty} \varphi(t)\,(\Phi_\iota * f)(t)\,dt &= \lim_\iota \int_{-\infty}^{\infty} (\Phi_\iota * \psi)(-\tau)\,f(\tau)\,d\tau \\
&= \int_{-\infty}^{\infty} \psi(-\tau)\,f(\tau)\,d\tau \\
&= \int_{-\infty}^{\infty} \varphi(\tau)\,f(\tau)\,d\tau.
\end{aligned}
$$

Since

$$
\left| \int_{-\infty}^{\infty} \varphi(t)\,(\Phi_\iota * f)(t)\,dt \right| \leq (\sup_\iota \|\Phi_\iota * f\|_\infty)\,\|\varphi\|_1,
$$

and $\Phi_\iota * f$ converges to f in the weak* topology,

$$
\left| \int_{-\infty}^{\infty} \varphi(\tau)\,f(\tau)\,d\tau \right| \leq (\sup_\iota \|\Phi_\iota * f\|_\infty)\,\|\varphi\|_1
$$

for all $\varphi \in L^1(\mathbb{R})$. Hence, by Riesz's theorem,

$$
\|f\|_\infty \leq \sup_\iota \|\Phi_\iota * f\|_\infty.
$$

$\square$

If $\{\Phi_\iota\}$ is a positive approximate identity, then $C_\Phi = 1$ and we have

$$
\|f\|_\infty = \sup_\iota \|\Phi_\iota * f\|_\infty = \lim_\iota \|\Phi_\iota * f\|_\infty.
$$

Corollary 10.14 *Let $u \in L^\infty(\mathbb{R})$, and let*

$$
U(x+iy) = \frac{1}{\pi} \int_{-\infty}^{\infty} \frac{y}{(x-t)^2 + y^2}\,u(t)\,dt, \qquad (x+iy \in \mathbb{C}_+).
$$

Then U is bounded and harmonic on $\mathbb{C}_+$, *and*

$$\sup_{y>0} \|U_y\|_\infty = \lim_{y\to 0} \|U_y\|_\infty = \|u\|_\infty.$$

Moreover, as $y \to 0$, U_y *converges to* u *in the weak* topology, i.e.*

$$\lim_{y\to 0} \int_{-\infty}^{\infty} \varphi(t)\, U(t+iy)\, dt = \int_{-\infty}^{\infty} \varphi(t)\, u(t)\, dt$$

for all $\varphi \in L^1(\mathbb{R})$.

Corollary 10.15 *Let* Φ *be continuous and bounded on* $\overline{\mathbb{C}_+}$ *and subharmonic on* $\mathbb{C}_+$. *Then*

$$\Phi(x+iy) \le \frac{1}{\pi} \int_{-\infty}^{\infty} \frac{y}{(x-t)^2+y^2}\, \Phi(t)\, dt, \qquad (x+iy \in \mathbb{C}_+).$$

Proof. Let

$$U(x+iy) = \frac{1}{\pi} \int_{-\infty}^{\infty} \frac{y}{(x-t)^2+y^2}\, \Phi(t)\, dt, \qquad (x+iy \in \mathbb{C}_+).$$

By Corollaries 10.6 and 10.14, U is continuous and bounded on $\overline{\mathbb{C}_+}$ and harmonic on $\mathbb{C}_+$ with

$$\lim_{\substack{z\to t \\ z\in\mathbb{C}_+}} U(z) = \Phi(t), \qquad (t \in \mathbb{R}).$$

Let $\Psi = \Phi - U$. Then Ψ is bounded and subharmonic on $\mathbb{C}_+$, and

$$\lim_{\substack{z\to t \\ z\in\mathbb{C}_+}} \Psi(z) = 0, \qquad (t \in \mathbb{R}).$$

Hence, by Corollary 4.5, $\Psi(z) \le 0$ for all $z \in \mathbb{C}_+$. $\qquad\square$

Exercises

Exercise 10.7.1 Let $f \in L^\infty(\mathbb{R})$, and let

$$f_\lambda(t) = \lambda \int_{-\infty}^{\infty} \left(\frac{\sin(\pi\lambda\,(t-\tau))}{\pi\lambda\,(t-\tau)} \right)^2 f(\tau)\, d\tau, \qquad (t \in \mathbb{R}).$$

Show that, for each $\lambda > 0$,

$$f_\lambda \in L^\infty(\mathbb{R})$$

with

$$\|f_\lambda\|_\infty \le \|f\|_\infty$$

and, as $\lambda \to \infty$, f_λ converges to f in the weak* topology, i.e.

$$\lim_{\lambda \to \infty} \int_{-\infty}^{\infty} \varphi(t)\, f_\lambda(t)\, dt = \int_{-\infty}^{\infty} \varphi(t)\, f(t)\, dt$$

for all $\varphi \in L^1(\mathbb{R})$.
Hint: Use Exercise 10.3.5 and Theorem 10.13.

Exercise 10.7.2 Let $f \in L^\infty(\mathbb{R})$, and let

$$f_\varepsilon(t) = \frac{1}{\sqrt{\varepsilon}} \int_{-\infty}^{\infty} e^{-\pi(t-\tau)^2/\varepsilon} f(\tau)\, d\tau, \qquad (t \in \mathbb{R}).$$

Show that, for each $\varepsilon > 0$,

$$f_\varepsilon \in L^\infty(\mathbb{R})$$

with

$$\|f_\varepsilon\|_\infty \le \|f\|_\infty$$

and, as $\varepsilon \to 0$, f_ε converges to f in the weak* topology, i.e.

$$\lim_{\varepsilon \to 0} \int_{-\infty}^{\infty} \varphi(t)\, f_\varepsilon(t)\, dt = \int_{-\infty}^{\infty} \varphi(t)\, f(t)\, dt$$

for all $\varphi \in L^1(\mathbb{R})$.
Hint: Use Exercise 10.3.3 and Theorem 10.13.

Chapter 11

Harmonic functions in the upper half plane

11.1 Hardy spaces on $\mathbb{C}_+$

The family of all complex harmonic functions in the upper half plane is denoted by $h(\mathbb{C}_+)$, and $H(\mathbb{C}_+)$ is its subset containing all analytic functions on $\mathbb{C}_+$. Let $U \in h(\mathbb{C}_+)$, and write

$$\| U \|_p = \sup_{0<y<\infty} \left(\int_{-\infty}^{\infty} |U(x+iy)|^p \, dx \right)^{\frac{1}{p}}, \qquad (0 < p < \infty),$$

and

$$\| U \|_\infty = \sup_{z \in \mathbb{C}_+} |U(z)|.$$

Then, for $0 < p \le \infty$, we define the harmonic family

$$h^p(\mathbb{C}_+) = \{ U \in h(\mathbb{C}_+) : \| U \|_p < \infty \}$$

and the analytic subfamily

$$H^p(\mathbb{C}_+) = \{ F \in H(\mathbb{C}_+) : \| F \|_p < \infty \}.$$

These are Hardy spaces of the upper half plane. We will also need $h(\overline{\mathbb{C}}_+)$, the family of all functions U such that U is harmonic in some open half plane containing $\overline{\mathbb{C}}_+$.

It is straightforward to see that $h^p(\mathbb{C}_+)$ and $H^p(\mathbb{C}_+)$, $1 \le p \le \infty$, are normed vector spaces. We will see that $h^1(\mathbb{C}_+)$ and $h^p(\mathbb{C}_+)$, $1 < p \le \infty$, are actually Banach spaces respectively isomorphic to $\mathcal{M}(\mathbb{R})$ and $L^p(\mathbb{R})$, $1 < p \le \infty$. Similarly, $H^p(\mathbb{C}_+)$, $1 \le p \le \infty$, is a Banach space isomorphic to a closed subspace of $L^p(\mathbb{R})$ denoted by $H^p(\mathbb{R})$.

Using these new notations, Corollaries 10.8, 10.12 and 10.14 can be rewritten as follows:

$$u \in L^p(\mathbb{R}) \Longrightarrow P * u \in h^p(\mathbb{C}_+),$$

for each $1 \le p \le \infty$, and

$$\mu \in M(\mathbb{R}) \Longrightarrow P * \mu \in h^1(\mathbb{C}_+).$$

In this chapter, we start with a harmonic function in $h^p(\mathbb{C}_+)$, $1 \le p \le \infty$, and show that it is representable as $P * u$ or $P * \mu$.

11.2 Poisson representation for semidiscs

To obtain the Poisson representation for $h^p(\mathbb{C}_+)$ classes, we need the Poisson integral formula for semidiscs. This formula recovers a harmonic function inside the open semidisc

$$S_R = \{\, x + iy : y > 0 \text{ and } x^2 + y^2 < R \,\}$$

from its boundary values on the semicircle

$$C_R = \{\, x + iy : y \ge 0 \text{ and } x^2 + y^2 = R \,\}$$

and on the line segment $[-R, R]$.

Lemma 11.1 *Let U be harmonic on a domain containing the closed semidisc $\overline{S}_R$. Then*

$$
\begin{aligned}
U(z) \;=\; & \frac{1}{\pi} \int_{-R}^{R} \left(\frac{y}{(t-x)^2 + y^2} - \frac{xyt + y(R^2 + xt)}{(R^2 + xt)^2 + (yt)^2} \right) U(t)\, dt \\
& + \frac{1}{\pi} \int_0^\pi \left(\frac{R^2 - r^2}{R^2 + r^2 - 2rR\cos(\theta - \Theta)} \right. \\
& \qquad\qquad \left. - \frac{R^2 - r^2}{R^2 + r^2 - 2rR\cos(\theta + \Theta)} \right) U(Re^{i\Theta})\, d\Theta,
\end{aligned}
$$

for all $z = x + iy = re^{i\theta} \in S_R$.

Proof. Without loss of generality, assume that U is real. Let V be a harmonic conjugate of U, and let $F = U + iV$. Then, by the Cauchy integral formula, for each $z \in S_R$, we have

$$
\begin{aligned}
I_1 &= \frac{1}{i\,2\pi} \int_{\partial S_R} \frac{F(\zeta)}{\zeta - z}\, d\zeta = F(z), \\
I_2 &= \frac{1}{2\pi i} \int_{\partial S_R} \frac{F(\zeta)}{\zeta - \bar{z}}\, d\zeta = 0, \\
I_3 &= \frac{1}{2\pi i} \int_{\partial S_R} \frac{F(\zeta)}{\zeta - R^2/\bar{z}}\, d\zeta = 0, \\
I_4 &= \frac{1}{2\pi i} \int_{\partial S_R} \frac{F(\zeta)}{\zeta - R^2/z}\, d\zeta = 0,
\end{aligned}
$$

where ∂S_R represents the boundary of S_R. Hence,

$$F(z) = I_1 - I_2 - I_3 + I_4.$$

Taking the real part of both sides gives the desired formula. □

Exercises

Exercise 11.2.1   Let Ω be a domain in $\mathbb{C}$ and fix $z_0 \in \Omega$. A function $G(z, z_0)$ defined on $\Omega \setminus \{z_0\}$ and satisfying the following four properties is called the *Green function* for Ω corresponding to the point z_0:

(i) $G(z, z_0) > 0$, for all $z \in \Omega \setminus \{z_0\}$;

(ii) G is harmonic on $\Omega \setminus \{z_0\}$;

(iii) $G(z, z_0) + \log|z - z_0|$ has a removable singularity at z_0 (and thus it represents a harmonic function on Ω);

(iv) $G(z, z_0) \to 0$ as z tends to any boundary point of Ω.

First, verify that if G exists then it is unique. Then show that the Green function of the disc $\Omega = D_R$ is

$$G(z, z_0) = \log\left| \frac{R^2 - \bar{z}_0 z}{R(z - z_0)} \right|,$$

and for the semidisc S_R, the Green function is given by

$$G(z, z_0) = \log\left| \frac{R^2 - \bar{z}_0 z}{R(z - z_0)} \frac{R(z - \bar{z}_0)}{R^2 - z_0 z} \right|. \qquad (11.1)$$

Hint: To prove uniqueness, use the maximum principle.

Exercise 11.2.2   [Green's formula] Let Ω be a domain in $\mathbb{C}$. Suppose that U is harmonic on a domain containing $\overline{\Omega}$. Show that, for all $z_0 \in \Omega$,

$$U(z_0) = -\frac{1}{2\pi} \int_{\partial\Omega} U(\zeta) \frac{\partial G}{\partial n}(\zeta, z_0) \, ds(\zeta).$$

Remark: $\partial G / \partial n$ represents the derivative of G in the direction of the outward normal vector n.

Exercise 11.2.3   Use Exercise 11.2.2 and (11.1) to provide another proof for Theorem 11.1.

11.3   Poisson representation of $h(\overline{\mathbb{C}}_+)$ functions

In studying the boundary values of harmonic functions in the upper half plane $\mathbb{C}_+$, the *best* possible assumption for the behavior on $\mathbb{R}$ is to assume that our function is actually defined on a half plane containing $\overline{\mathbb{C}}_+$. This class was already denoted by $h(\overline{\mathbb{C}}_+)$.

Theorem 11.2 *Let $U \in h(\overline{\mathbb{C}}_+)$ and suppose that U is bounded on $\overline{\mathbb{C}}_+$. Then*

$$U(x+iy) = \frac{1}{\pi} \int_{-\infty}^{\infty} \frac{y}{(x-t)^2 + y^2} U(t)\, dt, \qquad (x+iy \in \mathbb{C}_+). \qquad (11.2)$$

Proof. Since $U \in h(\overline{\mathbb{C}}_+)$, by definition, there exists $Y < 0$ such that U is harmonic on the half plane $\{\Im z > Y\}$. Fix $z = x + iy = re^{i\theta} \in \mathbb{C}_+$. By Lemma 11.1, for each $R > r$, we have

$$U(z) = \frac{1}{\pi} \int_{-R}^{R} \left\{ \frac{y}{(x-t)^2 + y^2} - \frac{(R^2 y/r^2)}{((R^2 x/r^2) - t)^2 + (R^2 y/r^2)^2} \right\} U(t)\, dt$$

$$- \frac{1}{\pi} \int_{0}^{\pi} \left\{ \frac{2rR(R^2 - r^2)\sin\theta \sin\Theta}{\left((R^2 + r^2)\cos\Theta - 2Rr\cos\theta\right)^2 + \left((R^2 - r^2)\sin\Theta\right)^2} \right\} U(Re^{i\Theta})\, d\Theta.$$

But, as $R \to \infty$,

$$\frac{(R^2 y/r^2)}{((R^2 x/r^2) - t)^2 + (R^2 y/r^2)^2} = O(1/R^2)$$

and

$$\frac{2rR(R^2 - r^2)\sin\theta \sin\Theta}{\left((R^2 + r^2)\cos\Theta - 2Rr\cos\theta\right)^2 + \left((R^2 - r^2)\sin\Theta\right)^2} = O(1/R).$$

Let $R \to \infty$. Hence, by the dominated convergence theorem, we obtain

$$U(z) = \frac{1}{\pi} \int_{-\infty}^{\infty} \frac{y}{(x-t)^2 + y^2} U(t)\, dt, \qquad (z = x + iy \in \mathbb{C}_+).$$

$\square$

To extend the previous representation theorem for other classes, we need to show that each element of $h^p(\mathbb{C}_+)$ is bounded if we stay away from the real line. Since the disc $\{|z| \leq r < 1\}$ is compact and continuous functions are automatically bounded on compact sets, the need for such a result was not felt in studying $h^p(\mathbb{D})$ spaces.

Lemma 11.3 *Let $U \in h^p(\mathbb{C}_+)$, $1 \leq p < \infty$. Then*

$$|U(x+iy)| \leq (2/\pi)^{1/p} \frac{\|U\|_p}{y^{1/p}}, \qquad (x+iy \in \mathbb{C}_+).$$

Proof. Fix $z = x + iy \in \mathbb{C}_+$ and let $0 < \rho < y$. Then, for each $r < \rho$, by Theorem 3.4, we have

$$U(z) = \frac{1}{2\pi} \int_{-\pi}^{\pi} U(z + re^{it}) \, dt.$$

Hence, by Hölder's inequality,

$$|U(z)|^p \leq \frac{1}{2\pi} \int_{-\pi}^{\pi} |U(z + re^{it})|^p \, dt. \tag{11.3}$$

(As a matter of fact, we just showed that $|U|^p$ is subharmonic.) Integrating both sides with respect to $r dr$ from 0 to ρ gives

$$|U(z)|^p \, \rho^2/2 \leq \frac{1}{2\pi} \int_0^{\rho} \int_{-\pi}^{\pi} |U(z + re^{it})|^p \, r dr dt.$$

Since the disc $D(z, \rho)$ is a subset of the strip $(-\infty, \infty) \times [y - \rho, y + \rho]$, we thus have

$$
\begin{aligned}
|U(z)|^p \, \rho^2/2 &\leq \frac{1}{2\pi} \int_{y-\rho}^{y+\rho} \left(\int_{-\infty}^{\infty} |U(x' + iy')|^p \, dx' \right) dy' \\
&= \frac{1}{2\pi} \int_{y-\rho}^{y+\rho} \|U_{y'}\|_p^p \, dy' \leq \frac{\rho \, \|U\|_p^p}{\pi}.
\end{aligned}
$$

Therefore,

$$|U(z)|^p \leq \frac{2 \, \|U\|_p^p}{\pi \rho}.$$

Now, let $\rho \to y$. $\qquad \square$

The following result is obvious if $p = \infty$. For other values of p, it is a direct consequence of the preceding lemma.

Corollary 11.4 *Let $U \in h^p(\mathbb{C}_+)$, $1 \leq p \leq \infty$. Let $\beta > 0$ and define*

$$U_\beta(z) = U(z + i\beta).$$

Then $U_\beta \in h(\overline{\mathbb{C}}_+)$ and U_β is bounded on $\overline{\mathbb{C}}_+$.

The main ingredient in the proof of Lemma 11.3 is the inequality (11.3), i.e. the fact that $|U|^p$, $1 \leq p < \infty$, is a subharmonic function. If F is analytic then $|F|^p$ is subharmonic for each $0 < p < \infty$. In other words, the inequality

$$|F(z)|^p \leq \frac{1}{2\pi} \int_{-\pi}^{\pi} |F(z + re^{it})|^p \, dt$$

holds for all $0 < p < \infty$. Therefore, Lemma 11.3 can be generalized slightly for analytic functions.

Lemma 11.5 *Let $F \in H^p(\mathbb{C}_+)$, $0 < p < \infty$. Then*

$$|F(x + iy)| \leq (2/\pi)^{1/p} \, \frac{\|F\|_p}{y^{1/p}}, \qquad (x + iy \in \mathbb{C}_+).$$

11.4   Poisson representation of $h^p(\mathbb{C}_+)$ functions $(1 \leq p \leq \infty)$

Using Theorem 11.2 and Corollary 11.4 we are now able to obtain the Poisson representation for $h^p(\mathbb{C}_+)$ classes, $1 \leq p \leq \infty$.

Theorem 11.6 *Let $U \in h^p(\mathbb{C}_+), 1 < p \leq \infty$. Then there exists a unique $u \in L^p(\mathbb{R})$ such that*

$$U(x + iy) = \frac{1}{\pi} \int_{-\infty}^{\infty} \frac{y}{(x-t)^2 + y^2} u(t) \, dt, \qquad (x + iy \in \mathbb{C}_+),$$

and

$$\|U\|_p = \|u\|_p.$$

Remark: Compare with Corollaries 10.6 and 10.14.

Proof. The uniqueness is a consequence of Corollary 10.9. Let $\beta > 0$ and define

$$U_\beta(z) = U(z + i\beta).$$

Then, by Corollary 11.4, U_β is bounded on $\overline{\mathbb{C}}_+$ and $U_\beta \in h(\overline{\mathbb{C}}_+)$. Hence, by Theorem 11.2,

$$U(x + iy + i\beta) = \frac{1}{\pi} \int_{-\infty}^{\infty} \frac{y}{(x-t)^2 + y^2} U(t + i\beta) \, dt$$

for all $x + iy \in \mathbb{C}_+$. Let $\beta \to 0$. The rest of the proof is exactly like that given for Theorem 3.5. $\qquad\square$

A small modification of the preceding proof, as explained similarly before Theorem 3.7 for the case of the open unit disc, yields the Poisson representation for $h^1(\mathbb{C}_+)$ functions.

Theorem 11.7 *Let $U \in h^1(\mathbb{C}_+)$. Then there exists a unique $\mu \in \mathcal{M}(\mathbb{R})$ such that*

$$U(x + iy) = \frac{1}{\pi} \int_{-\infty}^{\infty} \frac{y}{(x-t)^2 + y^2} \, d\mu(t), \qquad (x + iy \in \mathbb{C}_+),$$

and

$$\|U\|_1 = \|\mu\|.$$

Remark: Compare with Corollary 10.8.

Exercises

Exercise 11.4.1   Let $U \in h^p(\mathbb{C}_+)$, $1 \le p < \infty$. Show that the subharmonic function $|U|^p$ has a harmonic majorant.
Hint: Apply Theorem 11.7.

Exercise 11.4.2   Find $U \in h(\mathbb{C}_+)$ such that the subharmonic function $|U|^p$ has a harmonic majorant, but $U \notin h^p(\mathbb{C}_+)$.
Hint: Consider $U(x + iy) = y \notin h^1(\mathbb{C}_+)$.
Remark: Compare with Exercises 4.1.7 and 11.4.1.

11.5   A correspondence between $\overline{\mathbb{C}}_+$ and $\overline{\mathbb{D}}$

The conformal mapping

$$w = \frac{i - z}{i + z},$$

or equivalently

$$z = i \frac{1 - w}{1 + w},$$

gives the correspondence

$$\mathbb{C}_+ \quad \longleftrightarrow \quad \mathbb{D}$$
$$z \quad \longleftrightarrow \quad w$$

between the points $z = x + iy$ of the upper half plane and $w = re^{i\theta}$ of the unit disc. Moreover, on the boundary, the formulas

$$e^{i\tau} = \frac{i - t}{i + t}$$

and

$$t = i \frac{1 - e^{i\tau}}{1 + e^{i\tau}}$$

provide the correspondence

$$\mathbb{R} \quad \longleftrightarrow \quad \mathbb{T} \setminus \{-1\}$$
$$t \quad \longleftrightarrow \quad e^{i\tau}$$

between $\mathbb{R}$ and $\mathbb{T} \setminus \{-1\}$. The fact that -1 is not included is essential. Using the last two relations, we are able to establish the correspondence

$$\mathcal{B}(\mathbb{R}) \quad \longleftrightarrow \quad \mathcal{B}(\mathbb{T} \setminus \{-1\})$$
$$A \quad \longleftrightarrow \quad B$$

between the Borel subsets of $\mathbb{R}$ and the Borel subsets of $\mathbb{T} \setminus \{-1\}$ by defining

$$A = \left\{ t : \frac{i - t}{i + t} \in B \right\},$$

or equivalently

$$B = \left\{ e^{i\tau} : i\frac{1 - e^{i\tau}}{1 + e^{i\tau}} \in A \right\}.$$

Based on this observation, we make a connection between Borel measures defined on $\mathbb{R}$ and Borel measures defined on $\mathbb{T} \setminus \{-1\}$.

Let ν be a positive Borel measure on $\mathbb{T} \setminus \{-1\}$. Define its pullback λ on $\mathbb{R}$ by

$$\lambda(A) = \nu(B), \tag{11.4}$$

where A is an arbitrary Borel subset of $\mathbb{R}$, and B is related to A as explained above. Then λ is a positive Borel measure on $\mathbb{R}$, and the identity $\lambda(A) = \nu(B)$ can be rewritten as

$$\int_{\mathbb{R}} \chi_A(t)\, d\lambda(t) = \int_{\mathbb{T} \setminus \{-1\}} \chi_A\left(i\frac{1 - e^{i\tau}}{1 + e^{i\tau}} \right) d\nu(e^{i\tau}),$$

or

$$\int_{\mathbb{T} \setminus \{-1\}} \chi_B(e^{i\tau})\, d\nu(e^{i\tau}) = \int_{\mathbb{R}} \chi_B\left(\frac{i - t}{i + t} \right) d\lambda(t).$$

Taking positive linear combinations shows that both identities are valid for positive measurable step functions. Hence, by a standard limiting process, we obtain

$$\int_{\mathbb{R}} \varphi(t)\, d\lambda(t) = \int_{\mathbb{T} \setminus \{-1\}} \varphi\left(i\frac{1 - e^{i\tau}}{1 + e^{i\tau}} \right) d\nu(e^{i\tau}) \tag{11.5}$$

and

$$\int_{\mathbb{T} \setminus \{-1\}} \psi(e^{i\tau})\, d\nu(e^{i\tau}) = \int_{\mathbb{R}} \psi\left(\frac{i - t}{i + t} \right) d\lambda(t) \tag{11.6}$$

for all positive measurable functions φ and ψ.

We started with a measure ν on $\mathbb{T} \setminus \{-1\}$ and then defined λ on $\mathbb{R}$. Clearly, the procedure is symmetric and we can start with λ and then define ν. The relations (11.5) and (11.6) still remain valid.

A very special, but important, case is the Lebesgue measure $d\tau$ on $\mathbb{T}$. Using differential calculus techniques, the relation $e^{i\tau} = (i - t)/(i + t)$ immediately implies that

$$d\tau = \frac{2\, dt}{1 + t^2} \tag{11.7}$$

and thus (11.6) is written as

$$\int_{-\pi}^{\pi} \psi(e^{i\tau})\, d\tau = \int_{-\infty}^{\infty} \psi\left(\frac{i - t}{i + t} \right) \frac{2dt}{1 + t^2}. \tag{11.8}$$

11.6 Poisson representation of positive harmonic functions

In Section 3.5, we saw that positive harmonic functions in the unit disc are those elements of $h^1(\mathbb{D})$ which are generated by positive Borel measures. As the elementary example

$$U(x + iy) = y \tag{11.9}$$

shows, in the upper half plane, positive harmonic functions are not necessarily in $h^1(\mathbb{C}_+)$. Another observation is based on the contents of Section 10.2. If μ is a positive Borel measure on $\mathbb{R}$ such that

$$\int_{\mathbb{R}} \frac{d\mu(t)}{1 + t^2} < \infty,$$

then

$$U(z) = \int_{\mathbb{R}} \frac{y}{(t - x)^2 + y^2} \, d\mu(t) \tag{11.10}$$

is a well-defined positive harmonic function on $\mathbb{C}_+$. In this section, we show that any positive harmonic function in the upper half plane is a positive linear combination of (11.9) and (11.10).

Theorem 11.8 (Herglotz) *Let U be a positive harmonic function on $\mathbb{C}_+$. Then the limit*

$$\alpha = \lim_{y \to \infty} \frac{U(iy)}{y}$$

exists and $\alpha \in [0, \infty)$. Moreover, there exists a positive Borel measure μ on $\mathbb{R}$ satisfying

$$\int_{\mathbb{R}} \frac{d\mu(t)}{1 + t^2} < \infty$$

such that

$$U(x + iy) = \alpha y + \frac{1}{\pi} \int_{\mathbb{R}} \frac{y}{(x - t)^2 + y^2} \, d\mu(t), \qquad (x + iy \in \mathbb{C}_+).$$

Proof. Let

$$\mathcal{U}(w) = U\left(i \frac{1 - w}{1 + w} \right), \qquad (w \in \mathbb{D}).$$

Then $\mathcal{U}$ is a positive harmonic function on $\mathbb{D}$. Hence, by Theorem 3.8, there is a finite positive Borel measure ν on $\mathbb{T}$ such that

$$\mathcal{U}(re^{i\theta}) = \int_{\mathbb{T}} \frac{1 - r^2}{1 + r^2 - 2r\cos(\theta - \tau)} \, d\nu(e^{i\tau}), \qquad (re^{i\theta} \in \mathbb{D}).$$

Thus,

$$\mathcal{U}(re^{i\theta}) = \frac{1 - r^2}{1 + r^2 + 2r\cos\theta} \, \nu(\{-1\}) + \int_{\mathbb{T}\setminus\{-1\}} \frac{1 - r^2}{1 + r^2 - 2r\cos(\theta - \tau)} \, d\nu(e^{i\tau}).$$

To apply (11.6), we need two elementary identities which are easy to verify. If $z = x + iy \in \mathbb{C}_+$, $t \in \mathbb{R}$, $w = re^{i\theta} \in \mathbb{D}$ and $e^{i\tau} \in \mathbb{T}$, and they are related as explained in the preceding section, then we have

$$\frac{1 - r^2}{1 + r^2 + 2r\cos\theta} = y$$

and

$$\frac{1 - r^2}{1 + r^2 - 2r\cos(\theta - \tau)} = \frac{y}{(x - t)^2 + y^2}(1 + t^2).$$

Therefore, by (11.6),

$$U(x + iy) = \nu(\{-1\})\, y + \int_{\mathbb{R}} \frac{y}{(x - t)^2 + y^2}(1 + t^2)\, d\lambda(t),$$

where λ is defined by (11.4). Let $\alpha = \nu(\{-1\})$ and let

$$d\mu(t) = \pi\,(1 + t^2)\, d\lambda(t).$$

Then clearly $\alpha \in [0, \infty)$ and μ is a positive Borel measure on $\mathbb{R}$ such that

$$\frac{1}{\pi}\int_{\mathbb{R}} \frac{d\mu(t)}{1 + t^2} = \lambda(\mathbb{R}) = \nu(\mathbb{T} \setminus \{-1\}) < \infty$$

and

$$U(x + iy) = \alpha y + \frac{1}{\pi}\int_{\mathbb{R}} \frac{y}{(x - t)^2 + y^2}\, d\mu(t), \qquad (x + iy \in \mathbb{C}_+).$$

Finally, since

$$\frac{U(iy)}{y} = \alpha + \int_{\mathbb{R}} \frac{t^2 + 1}{t^2 + y^2}\, d\lambda(t),$$

and $\lambda(\mathbb{R}) < \infty$, by the dominated convergence theorem, we see that

$$\lim_{y \to \infty} \frac{U(iy)}{y} = \alpha.$$

$\square$

Let μ be a Borel measure satisfying

$$\int_{\mathbb{R}} \frac{d|\mu|(t)}{1 + t^2} < \infty,$$

and let $\alpha \in \mathbb{C}$. Let

$$U(x + iy) = \alpha y + \frac{1}{\pi}\int_{\mathbb{R}} \frac{y}{(x - t)^2 + y^2}\, d\mu(t), \qquad (x + iy \in \mathbb{C}_+).$$

Then U is a harmonic function on the upper half plane. As we saw in Theorem 11.8, a positive harmonic function on $\mathbb{C}_+$ is a special case obtained by a positive constant α and a positive measure μ. The family discussed in Corollary 10.8 is an even smaller subclass. Nevertheless, we expect the horizontal sections U_y to somehow converge to μ as $y \to 0$. Since μ satisfies a weaker condition in this case, we have to consider a more restrictive type of convergence. This is done by considering $\mathcal{C}_c(\mathbb{R})$, the class of continuous functions of compact support.

Theorem 11.9 *Let $\alpha \in \mathbb{C}$ and let μ be a Borel measure on $\mathbb{R}$ satisfying*

$$\int_{\mathbb{R}} \frac{d|\mu|(t)}{1+t^2} < \infty.$$

Let

$$U(x+iy) = \alpha y + \frac{1}{\pi} \int_{\mathbb{R}} \frac{y}{(x-t)^2 + y^2}\, d\mu(t), \qquad (x+iy \in \mathbb{C}_+).$$

Then

$$\lim_{y \to 0} \int_{\mathbb{R}} \varphi(t)\, U(t+iy)\, dt = \int_{\mathbb{R}} \varphi(t)\, d\mu(t)$$

for all $\varphi \in \mathcal{C}_c(\mathbb{R})$.

Proof. Since φ has compact support,

$$A = \int_{\mathbb{R}} \varphi(x)\, dx$$

is finite. Hence,

$$
\begin{aligned}
\int_{\mathbb{R}} \varphi(x)\, U(x+iy)\, dx &= \alpha A\, y + \int_{\mathbb{R}} \varphi(x) \left(\int_{\mathbb{R}} P_y(x-t)\, d\mu(t) \right) dx \\
&= \alpha A\, y + \int_{\mathbb{R}} (P_y * \varphi)(t)\, d\mu(t).
\end{aligned}
$$

By Corollary 10.4, $P_y * \varphi$ converges uniformly and boundedly to φ. Let supp $\varphi \subset [-M, M]$. Then, for $|x| \geq 2M$ and $0 < y < 1$,

$$
\begin{aligned}
|(P_y * \varphi)(x)| &\leq \frac{\|\varphi\|_\infty}{\pi} \int_{-M}^{M} \frac{y}{(x-t)^2 + y^2}\, dt \\
&\leq \frac{\|\varphi\|_\infty}{\pi} \int_{-M}^{M} \frac{dt}{(x-t)^2} \\
&\leq \frac{\|\varphi\|_\infty}{\pi} \frac{2M}{x^2 - M^2}.
\end{aligned}
$$

Hence, for all $x \in \mathbb{R}$ and all $0 < y < 1$,

$$|(P_y * \varphi)(x)| \leq \frac{C}{1+x^2},$$

where $C = C(\varphi)$ is an absolute constant. Therefore, by the dominated convergence theorem,

$$\int_{\mathbb{R}} (P_y * \varphi)(t)\, d\mu(t) \longrightarrow \int_{\mathbb{R}} \varphi(t)\, d\mu(t)$$

as $y \to 0$. $\square$

Exercises

Exercise 11.6.1   Let U be a real harmonic function in $h^1(\mathbb{C}_+)$. Show that U is the difference of two positive harmonic functions.

Exercise 11.6.2   Let U_1 and U_2 be two positive harmonic functions on $\mathbb{C}_+$. Let $U = U_1 - U_2$. Can we conclude that $U \in h^1(\mathbb{C}_+)$?
Remark: Compare with Exercises 3.5.2 and 11.6.1.

11.7   Vertical limits of $h^p(\mathbb{C}_+)$ functions $(1 \le p \le \infty)$

Using the conformal mapping given in Section 11.5, we are able to exploit Fatou's theorem (Theorem 3.12) and obtain a similar result about the vertical limits of harmonic functions defined in the upper half plane. The following result is rather general and applies for the elements of $h^p(\mathbb{C}_+)$, $1 \le p \le \infty$, as a special case.

Theorem 11.10 *Let μ be a Borel measure on $\mathbb{R}$ such that*

$$\int_{\mathbb{R}} \frac{d|\mu|(t)}{1+t^2} < \infty, \tag{11.11}$$

and let

$$U(x + iy) = \frac{1}{\pi} \int_{\mathbb{R}} \frac{y}{(x-t)^2 + y^2}\, d\mu(t), \qquad (x + iy \in \mathbb{C}_+).$$

Let $x_0 \in \mathbb{R}$. Suppose that

$$A = \lim_{s \to 0} \frac{\mu([x_0 - s, x_0 + s])}{2s}$$

exists and is finite. Then

$$\lim_{y \to 0} U(x_0 + iy) = A. \tag{11.12}$$

Proof. Without loss of generality, assume that $\mu \ge 0$, since otherwise, using Hahn's decomposition theorem, we can write $\mu = (\mu_1 - \mu_2) + i(\mu_3 - \mu_4)$ where each μ_k is a positive Borel measure satisfying (11.11).
    Without loss of generality, assume that $x = 0$. Let

$$\mathcal{U}(w) = U\left(i\,\frac{1-w}{1+w}\right), \qquad (w \in \mathbb{D}).$$

Then, on the one hand,

$$\lim_{y \to 0} U(iy) = \lim_{r \to 1} \mathcal{U}(r) \tag{11.13}$$

and, on the other hand, since

$$\frac{y}{(x-t)^2+y^2} = \frac{1-r^2}{1+r^2-2r\cos(\theta-\tau)}\cos^2(\tau/2),$$

we have

$$\mathcal{U}(re^{i\theta}) = \frac{1}{\pi}\int_{\mathbb{T}\setminus\{-1\}}\frac{1-r^2}{1+r^2-2r\cos(\theta-\tau)}\cos^2(\tau/2)\,d\nu(e^{i\tau}),$$

where ν corresponds to μ as explained in (11.4). Let

$$d\lambda(e^{i\tau}) = 2\cos^2(\tau/2)\,d\nu(e^{i\tau}).$$

Hence,

$$\mathcal{U}(re^{i\theta}) = \frac{1}{2\pi}\int_{\mathbb{T}}\frac{1-r^2}{1+r^2-2r\cos(\theta-\tau)}\,d\lambda(e^{i\tau})$$

and

$$\begin{aligned}
\frac{1}{2s}\int_{-s}^{s}d\lambda(e^{i\tau}) &= \frac{1}{s}\int_{-s}^{s}\cos^2(\tau/2)\,d\nu(e^{i\tau})\\
&= \frac{1}{s}\int_{-s}^{s}\left(\cos^2(\tau/2)-1\right)d\nu(e^{i\tau}) + \frac{1}{s}\int_{-s}^{s}d\nu(e^{i\tau})\\
&= \frac{1}{s}\int_{-s}^{s}\left(\cos^2(\tau/2)-1\right)d\nu(e^{i\tau}) + \frac{1}{s}\int_{-s'}^{s'}d\mu(t),
\end{aligned}$$

where

$$s' = i\,\frac{1-e^{is}}{1+e^{is}}.$$

Since $\cos^2(\tau/2)-1 = O(\tau^2)$ and $s'/s \to 1/2$, as $s \to 0$, we thus have

$$\lim_{s\to 0}\frac{1}{2s}\int_{-s}^{s}d\lambda(e^{i\tau}) = \lim_{s'\to 0}\frac{1}{2s'}\int_{-s'}^{s'}d\mu(t) = A.$$

Since, by Theorem 3.12,

$$\lim_{r\to 1}\mathcal{U}(r) = \lim_{s\to 0}\frac{1}{2s}\int_{-s}^{s}d\lambda(e^{i\tau}),$$

the relation (11.13) implies that $\lim_{y\to 0}U(iy) = A$.   □

For each Borel measure μ satisfying (11.11), we know that, for almost all $x \in \mathbb{R}$,

$$\lim_{s\to 0}\frac{\mu([x-s,x+s])}{2s} = \mu'(x)$$

exists and $\mu'(x)$ is finite. Hence, by Theorem 11.10,

$$\lim_{y\to 0}\frac{1}{\pi}\int_{\mathbb{R}}\frac{y}{(x-t)^2+y^2}\,d\mu(t) = \mu'(x)$$

for almost all $x \in \mathbb{R}$. A special case of this fact is mentioned below.

Corollary 11.11 *Let*

$$\int_{-\infty}^{\infty} \frac{|u(t)|}{1+t^2}\, dt < \infty \qquad (11.14)$$

and let

$$U(x+iy) = \frac{1}{\pi}\int_{-\infty}^{\infty} \frac{y}{(x-t)^2+y^2}\, u(t)\, dt, \qquad (x+iy \in \mathbb{C}_+).$$

Then

$$\lim_{y\to 0} U(x+iy) = u(x)$$

for almost all $x \in \mathbb{R}$.

Note that (11.14) is fulfilled if $u \in L^p(\mathbb{R})$, $1 \le p \le \infty$.

Exercises

Exercise 11.7.1   Let μ be a real signed Borel measure on $\mathbb{R}$ such that

$$\int_{\mathbb{R}} \frac{d|\mu|(t)}{1+t^2} < \infty,$$

and let

$$U(x+iy) = \frac{1}{\pi}\int_{\mathbb{R}} \frac{y}{(x-t)^2+y^2}\, d\mu(t), \qquad (x+iy \in \mathbb{C}_+).$$

Let $x_0 \in \mathbb{R}$ and suppose that

$$\mu'(x_0) = +\infty.$$

Show that

$$\lim_{y\to 0} U(x_0+iy) = +\infty.$$

Exercise 11.7.2   Let μ be a Borel measure on $\mathbb{R}$ such that

$$\int_{\mathbb{R}} \frac{d|\mu|(t)}{1+t^2} < \infty,$$

and let

$$U(x+iy) = \frac{1}{\pi}\int_{\mathbb{R}} \frac{y}{(x-t)^2+y^2}\, d\mu(t), \qquad (x+iy \in \mathbb{C}_+).$$

Show that, for almost all $x_0 \in \mathbb{R}$,

$$\lim_{\substack{z\to x_0 \\ z\in S_\alpha(x_0)}} U(z) = \mu'(x_0)$$

where $S_\alpha(x_0)$, $0 \leq \alpha < \pi/2$, is the Stoltz domain

$$S_\alpha(x_0) = \{\, x + iy \in \mathbb{C}_+ \; : \; |x - x_0| \leq (\tan \alpha)\, y \,\}.$$

(See Figure 11.1.)

Remark: We say that $\mu'(x_0)$ is the nontangential limit of U at the point x_0.

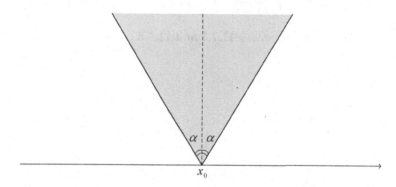

Fig. 11.1. The Stoltz domain $S_\alpha(x_0)$.

Exercise 11.7.3   Let μ be a positive Borel measure on $\mathbb{R}$ such that

$$\int_{\mathbb{R}} \frac{d\mu(t)}{1 + t^2} < \infty,$$

and let

$$U(x + iy) = \frac{1}{\pi} \int_{\mathbb{R}} \frac{y}{(x - t)^2 + y^2}\, d\mu(t), \qquad (x + iy \in \mathbb{C}_+).$$

Let $x_0 \in \mathbb{R}$ and suppose that

$$\mu'(x_0) = +\infty.$$

Show that, for any $\alpha \geq 0$,

$$\lim_{\substack{z \to x_0 \\ z \in S_\alpha(x_0)}} U(z) = +\infty.$$

Exercise 11.7.4   Construct a real signed Borel measure μ such that

$$\int_{\mathbb{R}} \frac{d|\mu|(t)}{1 + t^2} < \infty$$

and
$$\mu'(0) = +\infty.$$

But
$$\lim_{t \to 0} U(t + it) \neq +\infty,$$

where
$$U(x + iy) = \frac{1}{\pi} \int_{\mathbb{R}} \frac{y}{(x - t)^2 + y^2} \, d\mu(t), \qquad (x + iy \in \mathbb{C}_+).$$

Remark: Compare with Exercises 11.7.1 and 11.7.3.

Chapter 12

The Plancherel transform

12.1 The inversion formula

The representation formulas given in Theorem 10.1 and Corollary 10.2 represent a harmonic function in $h^1(\mathbb{C}_+)$ in terms of the Fourier transform of a boundary measure or function on $\mathbb{R}$. Our goal is to generalize these results for other $h^p(\mathbb{C}_+)$ classes. Of course, the first and most important step would be to generalize the definition of Fourier transform for other $L^p(\mathbb{R})$ spaces. Then we will be able to obtain further representations of elements of $h^p(\mathbb{C}_+)$.

The case $p = 1$ of Corollary 10.12 is of particular interest and helps us to say a bit more about the Fourier transform of functions in $L^1(\mathbb{R})$.

Theorem 12.1 (Inversion formula) *Let $f \in L^1(\mathbb{R})$ and suppose that $\hat{f} \in L^1(\mathbb{R})$. Then, for almost all $t \in \mathbb{R}$,*

$$f(t) = \int_{-\infty}^{\infty} \hat{f}(\tau) \, e^{i\, 2\pi t\tau} \, d\tau.$$

Proof. Let

$$F_y(x) = \frac{1}{\pi} \int_{-\infty}^{\infty} \frac{y}{(x-t)^2 + y^2} \, f(t) \, dt, \qquad (x + iy \in \mathbb{C}_+).$$

By Corollary 10.12, for each $y > 0$, $F_y \in L^1(\mathbb{R})$ and moreover

$$\lim_{y \to 0} \|F_y - f\|_1 = 0.$$

A convergent sequence in $L^1(\mathbb{R})$ has a subsequence converging at almost all points of $\mathbb{R}$. Hence, there is $y_n > 0$, $y_n \to 0$, such that

$$\lim_{n \to \infty} F_{y_n}(x) = f(x)$$

for almost all $x \in \mathbb{R}$. On the other hand, by Corollary 10.2, the identity

$$F_y(x) = \int_{-\infty}^{\infty} e^{-2\pi y|t|} \, \hat{f}(t) \, e^{i\, 2\pi x t} \, dt$$

holds for all $x \in \mathbb{R}$ and $y > 0$. Now, the assumption $\hat{f} \in L^1(\mathbb{R})$ together with the dominated convergence theorem imply that

$$\lim_{y \to 0} F_y(x) = \int_{-\infty}^{\infty} \hat{f}(t)\, e^{i\, 2\pi x t}\, dt$$

for all $x \in \mathbb{R}$. Therefore, we immediately obtain the inversion formula. $\qquad\square$

Writing the inversion formula as

$$f(t) = \int_{-\infty}^{\infty} \hat{f}(\tau)\, e^{-i\, 2\pi(-t)\tau}\, d\tau,$$

the last integral can be interpreted as the Fourier transform of $\hat{f}$, and we obtain

$$f(t) = \hat{\hat{f}}(-t), \qquad (t \in \mathbb{R}), \tag{12.1}$$

provided that $f, \hat{f} \in L^1(\mathbb{R})$.

Suppose that $f \in L^1(\mathbb{R})$, fix $y > 0$, and consider

$$F_y(x) = \int_{-\infty}^{\infty} e^{-2\pi y|t|}\, \hat{f}(t)\, e^{i\, 2\pi x t}\, dt.$$

By Corollary 10.12, $F_y \in L^1(\mathbb{R})$. Hence, according to (12.1), we have

$$\widehat{F_y}(t) = e^{-2\pi y|t|}\, \hat{f}(t).$$

Clearly $\widehat{F_y} \in C_0(\mathbb{R})$, for all $y > 0$. On the other hand, $\|F_y - f\|_1 \to 0$ and thus, by Lemma 9.1,

$$\lim_{y \to 0} \|\widehat{F_y} - \hat{f}\|_\infty = 0.$$

Therefore,

$$\hat{f} \in C_0(\mathbb{R}).$$

This is the Riemann–Lebesgue lemma on the real line. The Riemann–Lebesgue lemma combined with the uniqueness theorem (Corollary 10.10) say that the mapping

$$L^1(\mathbb{R}) \quad \longrightarrow \quad C_0(\mathbb{R})$$
$$f \quad \longmapsto \quad \hat{f}$$

is well-defined and one-to-one.

The following result is a simple consequence of Corollaries 10.2, 10.6 and (11.11). However, this simple observation has an important application in extending the definition of Fourier transform to $L^p(\mathbb{R})$, $1 < p \le 2$, spaces.

Corollary 12.2 *Let $f \in L^1(\mathbb{R})$. Suppose that $\hat{f} \ge 0$ and that f is continuous at 0. Then $\hat{f} \in L^1(\mathbb{R})$ and*

$$f(t) = \int_{-\infty}^{\infty} \hat{f}(\tau)\, e^{i\, 2\pi t\tau}\, d\tau$$

for almost all $t \in \mathbb{R}$. In particular,

$$f(0) = \int_{-\infty}^{\infty} \hat{f}(\tau) \, d\tau.$$

Proof. By Corollary 10.2,

$$\frac{1}{\pi} \int_{-\infty}^{\infty} \frac{y}{(x-t)^2 + y^2} f(t) \, dt = \int_{-\infty}^{\infty} e^{-2\pi y \, |t|} \, \hat{f}(t) \, e^{i \, 2\pi x t} \, dt \qquad (12.2)$$

for all $y > 0$ and all $x \in \mathbb{R}$. As a special case, on the imaginary axis we have

$$\frac{1}{\pi} \int_{-\infty}^{\infty} \frac{y}{t^2 + y^2} f(t) \, dt = \int_{-\infty}^{\infty} e^{-2\pi y \, |t|} \, \hat{f}(t) \, dt$$

for all $y > 0$. Let $y \to 0$. Then the monotone convergence theorem ensures that the right side tends to

$$\int_{-\infty}^{\infty} \hat{f}(t) \, dt,$$

and Corollary 10.6 says that the left side converges to $f(0)$. Hence

$$f(0) = \int_{-\infty}^{\infty} \hat{f}(t) \, dt,$$

which incidentally also implies that $\hat{f} \in L^1(\mathbb{R})$. Knowing that $\hat{f} \in L^1(\mathbb{R})$, the inversion formula (Theorem 12.1) applies and ensures that the first identity holds for almost all $t \in \mathbb{R}$. $\qquad\square$

Exercises

Exercise 12.1.1 Let $f \in L^1(\mathbb{R})$, and let

$$f_\lambda(t) = \int_{-\lambda}^{\lambda} \left(1 - |\tau|/\lambda\right) \hat{f}(\tau) \, e^{i \, 2\pi t \tau} \, d\tau, \qquad (t \in \mathbb{R}).$$

Show that, for each $\lambda > 0$,

$$f_\lambda \in L^1(\mathbb{R})$$

with

$$\|f_\lambda\|_1 \le \|f\|_1,$$

$$\hat{f_\lambda}(t) = \begin{cases} \left(1 - |t|/\lambda\right) \hat{f}(t) & \text{if} \quad |t| \le \lambda, \\[2mm] 0 & \text{if} \quad |t| \ge \lambda \end{cases}$$

and

$$\lim_{\lambda \to \infty} \|f_\lambda - f\|_1 = 0.$$

Hint: Use Exercises 9.3.2 and 10.6.1.

Exercise 12.1.2   Let $f \in L^1(\mathbb{R})$, and let

$$f_\varepsilon(t) = \int_{-\infty}^{\infty} e^{-\pi\varepsilon\tau^2}\, \hat{f}(\tau)\, e^{i\,2\pi t\tau}\, d\tau, \qquad (t \in \mathbb{R}).$$

Show that, for each $\varepsilon > 0$,

$$f_\varepsilon \in L^1(\mathbb{R})$$

with

$$\|f_\varepsilon\|_1 \le \|f\|_1,$$
$$\hat{f}_\varepsilon(t) = e^{-\pi\varepsilon t^2}\, \hat{f}(t)$$

and

$$\lim_{\varepsilon \to 0} \|f_\varepsilon - f\|_1 = 0.$$

Hint: Use Exercises 9.3.1 and 10.6.2.

Exercise 12.1.3   We saw that the map

$$L^1(\mathbb{R}) \longrightarrow C_0(\mathbb{R})$$
$$f \longmapsto \hat{f}$$

is injective. However, show that it is not surjective.

12.2   The Fourier–Plancherel transform

Using Corollary 12.2, we show that the Fourier transform maps $L^1(\mathbb{R}) \cap L^2(\mathbb{R})$ into $L^2(\mathbb{R})$. This is the first step in generalizing the definition of Fourier transform.

Theorem 12.3 (Plancherel) *Let $f \in L^1(\mathbb{R}) \cap L^2(\mathbb{R})$. Then $\hat{f} \in L^2(\mathbb{R})$ and moreover*

$$\|\hat{f}\|_2 = \|f\|_2.$$

Proof. Let $g(x) = \overline{f(-x)}$ and let $h = f * g$. By Young's inequality (Lemma 9.5),

$$h \in L^1(\mathbb{R}) \cap C(\mathbb{R})$$

and, by Lemma 9.3,

$$\hat{h} = \hat{f}\,\hat{g} = |\hat{f}|^2 \ge 0.$$

Hence, by Corollary 12.2, $\hat{h} \in L^1(\mathbb{R})$, which is equivalent to $\hat{f} \in L^2(\mathbb{R})$, and moreover

$$h(0) = \int_{-\infty}^{\infty} \hat{h}(t)\, dt.$$

Let us calculate both sides according to their definitions. On the one hand, we have

$$\int_{-\infty}^{\infty} \hat{h}(t) \, dt = \int_{-\infty}^{\infty} |\hat{f}(t)|^2 \, dt = \|\hat{f}\|_2^2$$

and, on the other hand,

$$h(0) = \int_{-\infty}^{\infty} f(t) \, g(0-t) \, dt = \int_{-\infty}^{\infty} f(t) \, \overline{f(t)} \, dt = \|f\|_2^2.$$

Comparing the last three identities implies $\|\hat{f}\|_2 = \|f\|_2$. □

Theorem 12.3 is a valuable result. Since $C_c(\mathbb{R}) \subset L^1(\mathbb{R}) \cap L^2(\mathbb{R})$, the latter is also dense in $L^2(\mathbb{R})$. This fact enables us to extend the definition of Fourier transform to $L^2(\mathbb{R})$. Indeed, let $f \in L^2(\mathbb{R})$. Pick any sequence $f_n \in L^1(\mathbb{R}) \cap L^2(\mathbb{R})$, $n \geq 1$, such that

$$\lim_{n\to\infty} \|f_n - f\|_2 = 0.$$

Then $(f_n)_{n\geq 1}$ is a Cauchy sequence in $L^2(\mathbb{R})$ and, by Theorem 12.3, we have

$$\|\hat{f}_n - \hat{f}_m\|_2 = \|\widehat{f_n - f_m}\|_2 = \|f_n - f_m\|_2,$$

and thus $(\hat{f}_n)_{n\geq 1}$ is also a Cauchy sequence in $L^2(\mathbb{R})$. Since $L^2(\mathbb{R})$ is complete,

$$\lim_{n\to\infty} \hat{f}_n$$

exists in $L^2(\mathbb{R})$. If $(g_n)_{n\geq 1}$ is another sequence in $L^1(\mathbb{R}) \cap L^2(\mathbb{R})$ satisfying similar properties, then $\lim_{n\to\infty} \hat{g}_n$ also exists in $L^2(\mathbb{R})$. However, again by Theorem 12.3, we have

$$\begin{aligned} \|\hat{f}_n - \hat{g}_n\|_2 &= \|\widehat{f_n - g_n}\|_2 = \|f_n - g_n\|_2 \\ &\leq \|f_n - f\|_2 + \|g_n - f\|_2 \longrightarrow 0, \end{aligned}$$

and thus

$$\lim_{n\to\infty} \hat{g}_n = \lim_{n\to\infty} \hat{f}_n.$$

Therefore, as an element of $L^2(\mathbb{R})$, the limit of $\hat{f}_n$ is independent of the choice of sequence as long as the sequence is in $L^1(\mathbb{R}) \cap L^2(\mathbb{R})$ and converges to f in $L^2(\mathbb{R})$. Hence, we define the *Fourier–Plancherel transform* of $f \in L^2(\mathbb{R})$ by

$$\mathcal{F}f = \lim_{n\to\infty} \hat{f}_n,$$

where f_n is any sequence in $L^1(\mathbb{R}) \cap L^2(\mathbb{R})$ satisfying

$$\lim_{n\to\infty} \|f_n - f\|_2 = 0.$$

In particular, if $f \in L^1(\mathbb{R}) \cap L^2(\mathbb{R})$, we can take $f_n = f$, and thus we have

$$\mathcal{F}f = \hat{f}. \tag{12.3}$$

In other words, the two definitions coincide on $L^1(\mathbb{R}) \cap L^2(\mathbb{R})$. Therefore, without any ambiguity we can also use the notation $\hat{f}$ for the Fourier–Plancherel transform of functions in $L^2(\mathbb{R})$. Finally, by Theorem 12.3,

$$\|\hat{f}\|_2 = \lim_{n\to\infty} \|\hat{f}_n\|_2 = \lim_{n\to\infty} \|f_n\|_2 = \|f\|_2 \qquad (12.4)$$

for all $f \in L^2(\mathbb{R})$. Thus, the Fourier–Plancherel transformation

$$\begin{array}{rcl} \mathcal{F} : L^2(\mathbb{R}) & \longrightarrow & L^2(\mathbb{R}) \\ f & \longmapsto & \hat{f} \end{array}$$

is an operator on $L^2(\mathbb{R})$ which preserves the norm.

For an arbitrary function $f \in L^2(\mathbb{R})$ we usually take

$$f_n(t) = \begin{cases} f(t) & \text{if} \quad |t| \le n, \\[2mm] 0 & \text{if} \quad |t| > n. \end{cases}$$

Hence, we have

$$\hat{f}(t) = \operatorname*{l.i.m.}_{n\to\infty} \int_{-n}^{n} f(\tau)\, e^{-i\,2\pi t\tau}\, d\tau,$$

where l.i.m. stands for *limit in mean* and implies that the limit is taken in $L^2(\mathbb{R})$.

Let us discuss a relevant example. The conjugate Poisson kernel is in $L^2(\mathbb{R})$. We evaluate its Fourier–Plancherel transform. Fix $x + iy \in \mathbb{C}_+$, and let

$$f(t) = \frac{1}{\pi} \frac{x-t}{(x-t)^2 + y^2}, \qquad (t \in \mathbb{R}).$$

To obtain the Fourier–Plancherel transform of f, we first calculate the Fourier transform of

$$f_n(t) = \begin{cases} \dfrac{1}{\pi} \dfrac{x-\tau}{(x-\tau)^2 + y^2} & \text{if} \quad |t-x| \le n, \\[4mm] 0 & \text{if} \quad |t-x| > n. \end{cases}$$

Hence,

$$\begin{aligned} \hat{f}_n(t) &= \frac{1}{\pi} \int_{x-n}^{x+n} \frac{x-\tau}{(x-\tau)^2 + y^2}\, e^{-i\,2\pi t\tau}\, d\tau \\[2mm] &= \frac{1}{\pi} \int_{-n}^{n} \frac{\tau}{\tau^2 + y^2}\, e^{-i\,2\pi t(x-\tau)}\, d\tau \\[2mm] &= \frac{e^{-i\,2\pi x t}}{\pi} \int_{-n}^{n} \frac{\tau}{\tau^2 + y^2}\, e^{i\,2\pi t\tau}\, d\tau. \end{aligned}$$

Suppose that $t > 0$ and $n > y$. Hence, by Cauchy's integral formula,

$$
\begin{aligned}
\hat{f}_n(t) &= \frac{e^{-i\,2\pi xt}}{\pi} \left(\int_{\Gamma_n} - \int_{C_n} \right) \frac{\zeta}{\zeta^2 + y^2}\, e^{i\,2\pi t\zeta}\, d\zeta \\
&= i\,e^{-2\pi yt - i\,2\pi xt} - \frac{e^{-i\,2\pi xt}}{\pi} \int_{C_n} \frac{\zeta}{\zeta^2 + y^2}\, e^{i\,2\pi t\zeta}\, d\zeta.
\end{aligned}
$$

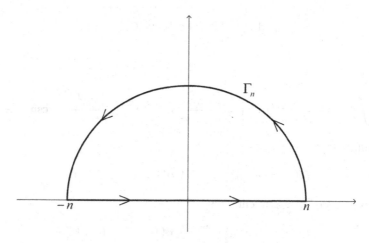

Fig. 12.1. The curve Γ_n.

The curve C_n is the semicircle in Γ_n. (See Figures 12.1 and 12.2.) To estimate the last integral, note that

$$
\left| \frac{\zeta}{\zeta^2 + y^2}\, e^{i\,2\pi t\zeta} \right| \le \frac{n}{n^2 - y^2}\, e^{-2\pi t\,\Im\zeta},
$$

for all $\zeta \in C_n$, and write

$$
\int_{C_n} = \int_{C_n \cap \{\Im\zeta > \sqrt{n}\}} + \int_{C_n \cap \{\Im\zeta < \sqrt{n}\}}.
$$

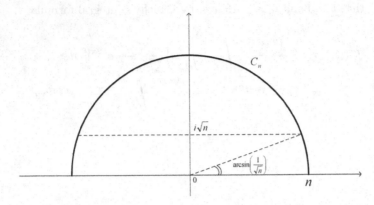

Fig. 12.2. The curve C_n.

Therefore,

$$\left| \int_{C_n} \frac{\zeta}{\zeta^2 + y^2} \, e^{i\,2\pi t\zeta} \, d\zeta \right| \le \frac{\pi n^2}{n^2 - y^2} \, e^{-2\pi t \sqrt{n}} + \frac{2n^2}{n^2 - y^2} \, \arcsin \frac{1}{\sqrt{n}},$$

which implies

$$\hat{f}_n(t) = i \, e^{-2\pi y t - i\,2\pi x t} + o(1), \qquad (t > 0),$$

as $n \to \infty$. A similar argument shows that

$$\hat{f}_n(t) = -i \, e^{2\pi y t - i\,2\pi x t} + o(1), \qquad (t < 0).$$

We know that $\hat{f}_n$ converges to $\hat{f}$ in $L^2(\mathbb{R})$. Hence a subsequence of $\hat{f}_n$ converges almost everywhere to $\hat{f}$. But, as the last two identities show,

$$\hat{f}_n(t) \longrightarrow i \operatorname{sgn}(t) \, e^{-2\pi y|t| - i\,2\pi x t}$$

at all $t \in \mathbb{R} \setminus \{0\}$. Hence

$$\hat{f}(t) = i \operatorname{sgn}(t) \, e^{-2\pi y|t| - i\,2\pi x t}, \qquad (t \in \mathbb{R}).$$

In particular, the Fourier–Plancherel transform of

$$Q_y(t) = \frac{1}{\pi} \, \frac{t}{t^2 + y^2}$$

is

$$\widehat{Q}_y(t) = -i \operatorname{sgn}(t) \, e^{-2\pi y|t|}.$$

(See Figure 12.3.)

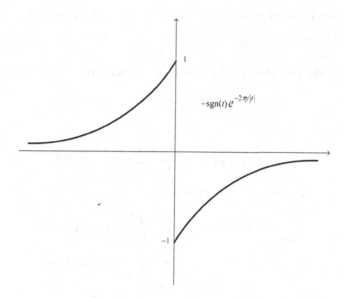

Fig. 12.3. The spectrum of $\frac{1}{i} Q_y$.

Exercises

Exercise 12.2.1 Let $f \in L^1(\mathbb{R})$ and suppose that $\hat{f} \in L^2(\mathbb{R})$. Show that we also have $f \in L^2(\mathbb{R})$.

Exercise 12.2.2 Let $f \in L^1(\mathbb{R})$ and $g \in L^2(\mathbb{R})$. Show that

$$\widehat{f * g} = \hat{f}\,\hat{g}.$$

12.3 The multiplication formula on $L^p(\mathbb{R})$ $(1 \leq p \leq 2)$

The identity $\|\hat{f}\|_2 = \|f\|_2$ implies that the Fourier–Plancherel transform is injective on $L^2(\mathbb{R})$. To show that it is also surjective, and thus a unitary operator on $L^2(\mathbb{R})$, we need the following generalization of the multiplication formula.

Lemma 12.4 (Multiplication formula) *Let $f, g \in L^2(\mathbb{R})$. Then*

$$\int_{-\infty}^{\infty} f(t)\,\hat{g}(t)\,dt = \int_{-\infty}^{\infty} \hat{f}(t)\,g(t)\,dt.$$

Proof. Pick two sequences $f_n, g_n \in L^1(\mathbb{R}) \cap L^2(\mathbb{R})$, $n \geq 1$, such that

$$\lim_{n \to \infty} \|f_n - f\|_2 = \lim_{n \to \infty} \|g_n - g\|_2 = 0.$$

According to the definition of the Fourier–Plancherel transform, we also have

$$\lim_{n \to \infty} \|\hat{f}_n - \hat{f}\|_2 = \lim_{n \to \infty} \|\hat{g}_n - \hat{g}\|_2 = 0.$$

Thus, by Hölder's inequality,

$$\lim_{n \to \infty} \|f_n \hat{g}_n - f\hat{g}\|_1 = \lim_{n \to \infty} \|\hat{f}_n g_n - \hat{f}g\|_1 = 0.$$

But $f_n, g_n \in L^1(\mathbb{R})$. Hence, by Lemma 9.2,

$$
\begin{aligned}
\int_{-\infty}^{\infty} f(t)\, \hat{g}(t)\, dt &= \lim_{n \to \infty} \int_{-\infty}^{\infty} f_n(t)\, \hat{g}_n(t)\, dt \\
&= \lim_{n \to \infty} \int_{-\infty}^{\infty} \hat{f}_n(t)\, g_n(t)\, dt \\
&= \int_{-\infty}^{\infty} \hat{f}(t)\, g(t)\, dt.
\end{aligned}
$$

$\square$

We are now able to show that the Fourier–Plancherel transform $\mathcal{F}$ is surjective on $L^2(\mathbb{R})$. As a matter of fact, if g is orthogonal to the range of $\mathcal{F}$, i.e.

$$\int_{-\infty}^{\infty} \hat{f}(t)\, \overline{g(t)}\, dt = 0$$

for all $f \in L^2(\mathbb{R})$, then, by the multiplication formula, we have

$$\int_{-\infty}^{\infty} f(t)\, \hat{\bar{g}}(t)\, dt = 0,$$

which immediately implies $\hat{\bar{g}} = 0$. Hence, by (12.4), $\bar{g} = 0$. Thus $\mathcal{F}$ is surjective.

Using (12.1), we can also provide an inversion formula for the Fourier–Plancherel transform. Let T denote the unitary operator

$$
\begin{aligned}
T : L^2(\mathbb{R}) &\longrightarrow L^2(\mathbb{R}) \\
f(t) &\longmapsto f(-t).
\end{aligned}
$$

Hence $T \circ \mathcal{F} \circ \mathcal{F}$ is also a unitary operator on $L^2(\mathbb{R})$, and moreover, by (12.1), the identity

$$T \circ \mathcal{F} \circ \mathcal{F} = id$$

holds at least on a dense subset of $L^2(\mathbb{R})$, e.g. on $C_c^\infty(\mathbb{R})$. Note that if $f \in C_c^\infty(\mathbb{R})$, then certainly $\hat{f} \in L^1(\mathbb{R})$. Therefore, the identity holds on $L^2(\mathbb{R})$ and thus

$$\mathcal{F}^{-1} = T \circ \mathcal{F}. \tag{12.5}$$

Exercises

Exercise 12.3.1   Let $f \in C_c^\infty(\mathbb{R})$. Show that

$$\hat{f} \in C^\infty(\mathbb{R}) \cap L^p(\mathbb{R})$$

for all $0 < p \leq \infty$.
Hint: Use Exercise 9.2.9.

Exercise 12.3.2   Let $f, g \in L^2(\mathbb{R})$. Show that

$$\int_{-\infty}^\infty \hat{f}(t)\, \hat{g}(t)\, dt = \int_{-\infty}^\infty f(t)\, g(-t)\, dt.$$

Hint: Use Lemma 12.4 and (12.5).

12.4   The Fourier transform on $L^p(\mathbb{R})$ $(1 \leq p \leq 2)$

Let

$$f \in L^1(\mathbb{R}) + L^2(\mathbb{R}).$$

Then for each $f = f_1 + f_2$, where $f_1 \in L^1(\mathbb{R})$ and $f_2 \in L^2(\mathbb{R})$, define its Fourier transform by

$$\hat{f} = \hat{f}_1 + \mathcal{F}f_2.$$

If f has another representation, say $f = g_1 + g_2$, where $g_1 \in L^1(\mathbb{R})$ and $g_2 \in L^2(\mathbb{R})$, then

$$f_1 - g_1 = g_2 - f_2 \in L^1(\mathbb{R}) \cap L^2(\mathbb{R}),$$

and thus, by (12.3),

$$\widehat{f_1 - g_1} = \mathcal{F}(g_2 - f_2).$$

Therefore,

$$\hat{f}_1 + \mathcal{F}f_2 = \hat{g}_1 + \mathcal{F}g_2,$$

which implies that $\hat{f}$ is well-defined. Since, for all p with $1 \leq p \leq 2$,

$$L^p(\mathbb{R}) \subset L^1(\mathbb{R}) + L^2(\mathbb{R}),$$

the Fourier transforms of elements of $L^p(\mathbb{R})$, $1 \leq p \leq 2$, are well-defined now.

Theorem 12.5 (Hausdorff–Young theorem) *Let $f \in L^p(\mathbb{R})$, $1 \leq p \leq 2$. Then $\hat{f} \in L^q(\mathbb{R})$, where q is the conjugate exponent of p, and*

$$\|\hat{f}\|_q \leq \|f\|_p.$$

Proof. By Lemma 9.1,

$$\mathcal{F}: L^1(\mathbb{R}) \longrightarrow L^\infty(\mathbb{R})$$
$$f \longmapsto \hat{f}$$

is of type $(1, \infty)$ with

$$\|\mathcal{F}\|_{(1,\infty)} \leq 1$$

and, by Plancherel's theorem (see (12.4)),

$$\mathcal{F}: L^2(\mathbb{R}) \longrightarrow L^2(\mathbb{R})$$
$$f \longmapsto \hat{f}$$

is of type $(2, 2)$ with

$$\|\mathcal{F}\|_{(2,2)} \leq 1.$$

The rest is exactly as in the proof of Theorem 8.3. $\square$

Exercises

Exercise 12.4.1   Let $f \in L^1(\mathbb{R})$ and $g \in L^p(\mathbb{R})$, $1 \leq p \leq 2$. Show that

$$\widehat{f * g} = \hat{f}\,\hat{g}.$$

Hint: Use Exercise 12.2.2.

Exercise 12.4.2   Let $1 \leq p \leq 2$, and let q be the conjugate exponent of p. Let $g \in L^p(\mathbb{R})$. Show that there is an $f \in L^q(\mathbb{R})$ such that

$$\hat{f} = g$$

and that

$$\|f\|_q \leq \|\hat{f}\|_p.$$

Hint 1: Use duality and Theorem 12.5.
Hint 2: Give a direct proof using the Riesz–Thorin interpolation theorem.

12.5   An application of the multiplication formula on $L^p(\mathbb{R})$ $(1 \leq p \leq 2)$

The contents of this chapter are designed to obtain the following result, which reveals the spectral properties of elements in $h^p(\mathbb{C}_+)$, $1 \leq p \leq 2$.

Theorem 12.6 *Let* $u \in L^p(\mathbb{R})$, $1 \leq p \leq 2$. *Then we have*

$$\frac{1}{\pi} \int_{-\infty}^{\infty} \frac{y}{(x-t)^2 + y^2}\, u(t)\, dt \;=\; \int_{-\infty}^{\infty} e^{-2\pi y|t|}\, \hat{u}(t)\, e^{i\,2\pi xt}\, dt,$$

$$\frac{1}{\pi} \int_{-\infty}^{\infty} \frac{x-t}{(x-t)^2 + y^2}\, u(t)\, dt \;=\; \int_{-\infty}^{\infty} -i\, \mathrm{sgn}(t)\, e^{-2\pi y\,|t|}\, \hat{u}(t)\, e^{i\,2\pi xt}\, dt,$$

$$\frac{1}{2\pi i} \int_{-\infty}^{\infty} \frac{u(t)}{t-z}\, dt \;=\; \int_{0}^{\infty} \hat{u}(t)\, e^{i\,2\pi zt}\, dt,$$

$$\frac{1}{2\pi i} \int_{-\infty}^{\infty} \frac{u(t)}{t-\bar{z}}\, dt \;=\; -\int_{-\infty}^{0} \hat{u}(t)\, e^{i\,2\pi \bar{z}t}\, dt,$$

for all $z = x + iy \in \mathbb{C}_+$. *The integrals are absolutely convergent on compact subsets of* $\mathbb{C}_+$. *Moreover, each line represents a harmonic function on* $\mathbb{C}_+$. *The third line gives an analytic function.*

Proof. The case $p = 1$ was considered in Corollary 10.2. We start by proving the first identity for $p = 2$. If $u \in L^2(\mathbb{R})$, pick a sequence $u_n \in L^1(\mathbb{R}) \cap L^2(\mathbb{R})$, $n \geq 1$, such that

$$\lim_{n \to \infty} \|u_n - u\|_2 = 0.$$

Then, again by Corollary 10.2,

$$\frac{1}{\pi} \int_{-\infty}^{\infty} \frac{y}{(x-t)^2 + y^2}\, u_n(t)\, dt = \int_{-\infty}^{\infty} e^{-2\pi y\,|t|}\, \hat{u}_n(t)\, e^{i\,2\pi xt}\, dt.$$

Let $n \to \infty$. The Cauchy–Schwarz inequality ensures that the left side converges to

$$\frac{1}{\pi} \int_{-\infty}^{\infty} \frac{y}{(x-t)^2 + y^2}\, u(t)\, dt,$$

and the right side to

$$\int_{-\infty}^{\infty} e^{-2\pi y\,|t|}\, \hat{u}(t)\, e^{i\,2\pi xt}\, dt,$$

as $n \to \infty$. Hence, we obtain the required equality. Other identities are proved similarly if $p = 2$.

If $u \in L^p(\mathbb{R}) \subset L^1(\mathbb{R}) + L^2(\mathbb{R})$, write $u = u_1 + u_2$, where $u_1 \in L^1(\mathbb{R})$ and $u_2 \in L^2(\mathbb{R})$. Each identity holds for u_1 and u_2. Hence, taking their linear combination, each identity also holds for u.

Since

$$\hat{u} = \hat{u}_1 + \hat{u}_2 \subset C_0(\mathbb{R}) + L^2(\mathbb{R}),$$

the presence of $e^{-2\pi y\,|t|}$ ensures that the right-side integral converges absolutely and uniformly on compact subsets of $\mathbb{C}_+$. Finally, the same method used in Theorem 10.1 shows that each row represents a harmonic function. $\square$

Corollary 12.7 (Uniqueness theorem) *Let* $f \in L^p(\mathbb{R})$, $1 \leq p \leq 2$. *Suppose that*

$$\hat{f}(t) = 0$$

for almost all $t \in \mathbb{R}$. Then

$$f(t) = 0$$

for almost all $t \in \mathbb{R}$.

Proof. By Theorem 12.6, we have

$$\frac{1}{\pi} \int_{-\infty}^{\infty} \frac{y}{(x-t)^2 + y^2}\, f(t)\, dt = \int_{-\infty}^{\infty} e^{-2\pi y |t|}\, \hat{f}(t)\, e^{i\, 2\pi x t}\, dt = 0$$

for all $x + iy \in \mathbb{C}_+$. Hence, by Corollary 11.11, $f = 0$. $\square$

Exercises

Exercise 12.5.1   [Multiplication formula] Let $f, g \in L^p(\mathbb{R})$, $1 \le p \le 2$. Show that

$$\int_{-\infty}^{\infty} f(t)\, \hat{g}(t)\, dt = \int_{-\infty}^{\infty} \hat{f}(t)\, g(t)\, dt.$$

Exercise 12.5.2   Let $1 < p < 2$ and let q be the conjugate exponent of p. By Theorem 12.5 and Corollary 12.7, the map

$$
\begin{array}{ccc}
L^p(\mathbb{R}) & \longrightarrow & L^q(\mathbb{R}) \\
f & \longmapsto & \hat{f}
\end{array}
$$

is well-defined and injective. However, show that it is not surjective.

12.6   A complete characterization of $h^p(\mathbb{C}_+)$ spaces

Similar to what we did in Section 4.7, we summarize the results of preceding sections to demonstrate a complete characterization of $h^p(\mathbb{C}_+)$ spaces. In the following, U is harmonic in the upper half plane $\mathbb{C}_+$.

Case 1: $p = 1$.

$U \in h^1(\mathbb{C}_+)$ if and only if there exists $\mu \in \mathcal{M}(\mathbb{R})$ such that

$$U(x + iy) = \frac{1}{\pi} \int_{\mathbb{R}} \frac{y}{(x-t)^2 + y^2}\, d\mu(t), \qquad (x + iy \in \mathbb{C}_+).$$

The measure μ is unique and

$$U(x + iy) = \int_{-\infty}^{\infty} \hat{\mu}(t)\, e^{-2\pi y |t| + i 2\pi x t}\, dt, \qquad (x + iy \in \mathbb{C}_+),$$

where $\hat{\mu} \in \mathcal{C}(\mathbb{R}) \cap L^\infty(\mathbb{R})$ and

$$\|\hat{\mu}\|_\infty \leq \|\mu\|.$$

The measures $d\mu_y(t) = U(t + iy)\,dt$ converge to $d\mu(t)$ in the weak* topology, as $y \to 0$, i.e.

$$\lim_{y \to 0} \int_{\mathbb{R}} \varphi(t)\,d\mu_y(t) = \int_{\mathbb{T}} \varphi(t)\,d\mu(t)$$

for all $\varphi \in \mathcal{C}_0(\mathbb{R})$, and we have

$$\|U\|_1 = \|\mu\| = |\mu|(\mathbb{R}).$$

Case 2: $1 < p \leq 2$.

$U \in h^p(\mathbb{C}_+)$ if and only if there exists $u \in L^p(\mathbb{R})$ such that

$$U(x + iy) = \frac{1}{\pi} \int_{\mathbb{R}} \frac{y}{(x - t)^2 + y^2}\, u(t)\,dt, \qquad (x + iy \in \mathbb{C}_+).$$

The function u is unique and

$$U(x + iy) = \int_{-\infty}^{\infty} \hat{u}(t)\, e^{-2\pi y|t| + i2\pi xt}\,dt, \qquad (x + iy \in \mathbb{C}_+),$$

where $\hat{u} \in L^q(\mathbb{R})$, with $1/p + 1/q = 1$, and

$$\|\hat{u}\|_q \leq \|u\|_p.$$

Moreover,

$$\lim_{y \to 0} \|U_y - u\|_p = 0,$$

and thus

$$\|U\|_p = \|u\|_p.$$

Case 3: $2 < p < \infty$.

$U \in h^p(\mathbb{C}_+)$ if and only if there exists $u \in L^p(\mathbb{R})$ such that

$$U(x + iy) = \frac{1}{\pi} \int_{\mathbb{R}} \frac{y}{(x - t)^2 + y^2}\, u(t)\,dt, \qquad (x + iy \in \mathbb{C}_+).$$

The function u is unique,

$$\lim_{y \to 0} \|U_y - u\|_p = 0,$$

and thus

$$\|U\|_p = \|u\|_p.$$

Case 4: $p = \infty$.

$U \in h^\infty(\mathbb{C}_+)$ if and only if there exists $u \in L^\infty(\mathbb{R})$ such that

$$U(x + iy) = \frac{1}{\pi} \int_\mathbb{R} \frac{y}{(x-t)^2 + y^2} \, u(t) \, dt, \qquad (x + iy \in \mathbb{C}_+).$$

The function u is unique and

$$\|U\|_\infty = \|u\|_\infty.$$

Moreover, U_y converges to u in the weak* topology, as $y \to 0$, i.e.

$$\lim_{y \to 0} \int_\mathbb{R} \varphi(t) \, U(t + iy) \, dt = \int_\mathbb{R} \varphi(t) \, u(t) \, dt$$

for all $\varphi \in L^1(\mathbb{R})$.

The first case shows that $h^1(\mathbb{C}_+)$ and $\mathcal{M}(\mathbb{R})$ are isometrically isomorphic Banach spaces. Similarly, by the other three cases, $h^p(\mathbb{C}_+)$ and $L^p(\mathbb{R})$, $1 < p \le \infty$, are isometrically isomorphic.

Exercises

Exercise 12.6.1   We saw that $h^1(\mathbb{C}_+)$ and $\mathcal{M}(\mathbb{R})$ are isometrically isomorphic Banach spaces. Consider $L^1(\mathbb{R})$ as a subspace of $\mathcal{M}(\mathbb{R})$. What is the image of $L^1(\mathbb{R})$ in $h^1(\mathbb{C}_+)$ under this correspondence? What is the image of positive measures? Do they give all the positive harmonic functions in the upper half plane?

Exercise 12.6.2   We know that $h^\infty(\mathbb{C}_+)$ and $L^\infty(\mathbb{R})$ are isometrically isomorphic Banach spaces. Consider $C_0(\mathbb{R})$ as a subspace of $L^\infty(\mathbb{R})$. What is the image of $C_0(\mathbb{R})$ in $h^\infty(\mathbb{C}_+)$ under this correspondence? What is the image of $C_c(\mathbb{R})$?

Chapter 13

Analytic functions in the upper half plane

13.1 Representation of $H^p(\mathbb{C}_+)$ functions $(1 < p \leq \infty)$

Since $H^p(\mathbb{C}_+) \subset h^p(\mathbb{C}_+)$, the representation theorems that we have obtained for harmonic functions apply to analytic functions too. Some of these results have been gathered in Section 12.6. In this chapter we study the analytic Hardy spaces $H^p(\mathbb{C}_+)$ in more detail.

Lemma 13.1 *Let $\beta > 0$. If $F \in H^p(\mathbb{C}_+)$, $1 \leq p < \infty$, then*

$$\int_{-\infty}^{\infty} \frac{F(t+i\beta)}{t-\bar{z}} \, dt = 0, \qquad (z \in \mathbb{C}_+).$$

If $F \in H^\infty(\mathbb{C}_+)$, then

$$\int_{-\infty}^{\infty} \frac{F(t+i\beta)}{(t-\bar{z})(t+i)} \, dt = 0, \qquad (z \in \mathbb{C}_+).$$

Proof. Fix $z \in \mathbb{C}_+$ and $\beta > 0$. Let $R > \beta + |z|$ and let $\Gamma_{R,\beta}$ be the curve shown in Figure 13.1. Therefore, $i\beta + \bar{z}$ is not in the interior of $\Gamma_{R,\beta}$, and thus, by Cauchy's integral formula,

$$\int_{\Gamma_{R,\beta}} \frac{F(\zeta)}{\zeta - i\beta - \bar{z}} \, d\zeta = 0.$$

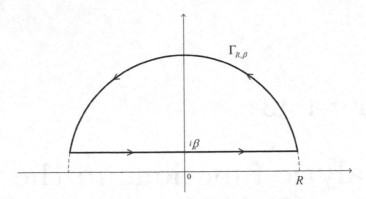

Fig. 13.1. The curve $\Gamma_{R,\beta}$.

Hence,

$$\int_{-R\cos\theta_R}^{R\cos\theta_R} \frac{F(t+i\beta)}{t-\bar{z}}\, dt = -\int_{\theta_R}^{\pi-\theta_R} \frac{F(Re^{i\theta})}{Re^{i\theta}-i\beta-\bar{z}}\, iRe^{i\theta}\, d\theta,$$

where, according to triangle $\Delta_{R,\beta}$, we have $\theta_R = \arcsin(\beta/R)$. (See Figure 13.2.)

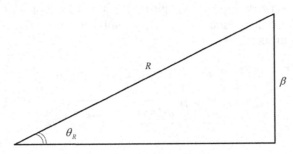

Fig. 13.2. The triangle $\Delta_{R,\beta}$.

By Lemma 11.3, we now have

$$
\left| \int_{\theta_R}^{\pi-\theta_R} \frac{F(Re^{i\theta})}{Re^{i\theta}-i\beta-\bar{z}}\, iRe^{i\theta}\, d\theta \right|
\leq \frac{R\,(2/\pi)^{1/p}\,\|F\|_p}{R-|i\beta+\bar{z}|} \int_{\theta_R}^{\pi-\theta_R} \frac{1}{(R\sin\theta)^{1/p}}\, d\theta
$$

$$
\leq \frac{2R\,(2/\pi)^{1/p}\,\|F\|_p}{(R-|i\beta+\bar{z}|)\,R^{1/p}} \int_{\theta_R}^{\pi/2} \frac{1}{(2\theta/\pi)^{1/p}}\, d\theta
$$

$$
\leq \frac{2R\,(2/\pi)^{1/p}\,\|F\|_p}{(R-|i\beta+\bar{z}|)\,R^{1/p}} \int_{\theta_R}^{\pi/2} \frac{1}{2\theta/\pi}\, d\theta
$$

$$
= \frac{\pi R\,(2/\pi)^{1/p}\,\|F\|_p}{(R-|i\beta+\bar{z}|)\,R^{1/p}} \log(\pi/2\theta_R).
$$

But $\theta_R \geq \sin \theta_R = \beta/R$, and thus

$$\left| \int_{\theta_R}^{\pi-\theta_R} \frac{F(Re^{i\theta})}{Re^{i\theta} - i\beta - \bar{z}} iRe^{i\theta} d\theta \right| \leq \frac{\pi R (2/\pi)^{1/p} \|F\|_p}{(R - |i\beta + \bar{z}|) R^{1/p}} \log(\pi R/2\beta).$$

Hence,

$$\int_{-R\cos\theta_R}^{R\cos\theta_R} \frac{F(t + i\beta)}{t - \bar{z}} dt = O(\log R/R^{1/p}).$$

Now let $R \to \infty$.

The proof for the case $p = \infty$ is even simpler. The reason we add the extra factor $(t + i)$ in the denominator is that, for an arbitrary bounded analytic function F,

$$\frac{F(t + i\beta)}{t - \bar{z}}$$

is not necessarily integrable over $\mathbb{R}$. □

No wonder we want to let $\beta \to 0$ in the preceding lemma to obtain a representation theorem based on the boundary values of F on the real line $\mathbb{R}$. Let us see what happens. If $F \in H^p(\mathbb{C}_+)$, $1 < p < \infty$, then, by Theorem 11.6, there is a unique $f \in L^p(\mathbb{R})$ such that

$$F(x + iy) = \frac{1}{\pi} \int_{-\infty}^{\infty} \frac{y}{(x - t)^2 + y^2} f(t)\, dt, \qquad (x + iy \in \mathbb{C}_+). \qquad (13.1)$$

Moreover, by Corollary 10.12, we have

$$\lim_{\beta \to 0} \|F_\beta - f\|_p = 0.$$

Hence, by Lemma 13.1,

$$\int_{-\infty}^{\infty} \frac{f(t)}{t - \bar{z}} dt = \lim_{\beta \to 0} \int_{-\infty}^{\infty} \frac{F_\beta(t)}{t - \bar{z}} dt = 0 \qquad (13.2)$$

for all $z \in \mathbb{C}_+$. Since

$$\frac{y}{(x - t)^2 + y^2} = \frac{i/2}{t - \bar{z}} - \frac{i/2}{t - z} \qquad (13.3)$$

and

$$\frac{(x - t)}{(x - t)^2 + y^2} = \frac{-1/2}{t - z} + \frac{-1/2}{t - \bar{z}},$$

the representation (13.1) along with the property (13.2) imply

$$F(z) = \frac{1}{2\pi i} \int_{-\infty}^{\infty} \frac{f(t)}{t - z} dt$$

$$= \frac{i}{\pi} \int_{-\infty}^{\infty} \frac{x - t}{(x - t)^2 + y^2} f(t)\, dt,$$

for all $z = x + iy \in \mathbb{C}_+$.

On the other hand, if $f \in L^p(\mathbb{R})$, $1 < p < \infty$, satisfying

$$\int_{-\infty}^{\infty} \frac{f(t)}{t - \bar{z}}\, dt = 0, \qquad (z \in \mathbb{C}_+),$$

is given, we can define

$$F(x + iy) = \frac{1}{\pi} \int_{-\infty}^{\infty} \frac{y}{(x - t)^2 + y^2} f(t)\, dt, \qquad (x + iy \in \mathbb{C}_+).$$

Then, on the one hand, by Corollary 10.12, $F \in h^p(\mathbb{C}_+)$, and on the other hand, by (13.3) and our assumption on f, we obtain

$$F(z) = \frac{1}{2\pi i} \int_{-\infty}^{\infty} \frac{f(t)}{t - z}\, dt, \qquad (z \in \mathbb{C}_+).$$

Thus F indeed represents an element of $H^p(\mathbb{C}_+)$.

Finally, since F_y converges to f in $L^p(\mathbb{R})$, we have

$$\|F\|_p = \|f\|_p.$$

The preceding discussion shows that $H^p(\mathbb{C}_+)$, $1 < p < \infty$, is isomorphically isometric to

$$H^p(\mathbb{R}) = \left\{ f \in L^p(\mathbb{R}) : \int_{-\infty}^{\infty} \frac{f(t)}{t - \bar{z}}\, dt = 0,\ \text{for all } z \in \mathbb{C}_+ \right\}, \qquad (13.4)$$

which is a closed subspace of $L^p(\mathbb{R})$.

Theorem 13.2 *Let F be analytic in the upper half plane $\mathbb{C}_+$. Then $F \in H^p(\mathbb{C}_+)$, $1 < p < \infty$, if and only if there is $f \in H^p(\mathbb{R})$ such that*

$$F(x + iy) = \frac{1}{\pi} \int_{-\infty}^{\infty} \frac{y}{(x - t)^2 + y^2} f(t)\, dt, \qquad (x + iy \in \mathbb{C}_+).$$

If so, f is unique and we also have

$$\begin{aligned} F(z) &= \frac{i}{\pi} \int_{-\infty}^{\infty} \frac{x - t}{(x - t)^2 + y^2} f(t)\, dt \\ &= \frac{1}{2\pi i} \int_{-\infty}^{\infty} \frac{f(t)}{t - z}\, dt, \qquad (z = x + iy \in \mathbb{C}_+). \end{aligned}$$

Moreover,

$$\lim_{y \to 0} \|F_y - f\|_p = 0$$

and

$$\|F\|_p = \|f\|_p.$$

A slight modification of the argument provided above for $H^p(\mathbb{C}_+)$, $1 < p < \infty$, gives a representation for $H^\infty(\mathbb{C}_+)$. If $F \in H^\infty(\mathbb{C}_+)$, then, by Theorem 11.6, there is a unique $f \in L^\infty(\mathbb{R})$ such that

$$F(x + iy) = \frac{1}{\pi} \int_{-\infty}^{\infty} \frac{y}{(x-t)^2 + y^2} f(t)\, dt, \qquad (x + iy \in \mathbb{C}_+), \qquad (13.5)$$

and

$$\lim_{\beta \to 0} \int_{-\infty}^{\infty} \varphi(t)\, F(t + i\beta)\, dt = \int_{-\infty}^{\infty} \varphi(t)\, f(t)\, dt$$

for all $\varphi \in L^1(\mathbb{R})$. In particular, with

$$\varphi(t) = \frac{1}{(t - \bar{z})(t + i)}, \qquad (t \in \mathbb{R},\ z \in \mathbb{C}_+),$$

we obtain

$$\int_{-\infty}^{\infty} \frac{f(t)}{(t - \bar{z})(t + i)}\, dt = 0, \qquad (z \in \mathbb{C}_+). \qquad (13.6)$$

Since

$$\frac{y}{(x-t)^2 + y^2} = \frac{(z + i)/2i}{(t - z)(t + i)} - \frac{(\bar{z} + i)/2i}{(t - \bar{z})(t + i)}, \qquad (13.7)$$

the representation (13.5) along with the properties (13.6) and (13.7) imply

$$F(z) = \frac{1}{2\pi i} \int_{-\infty}^{\infty} \left(\frac{1}{t - z} - \frac{1}{t + i} \right) f(t)\, dt, \qquad (z \in \mathbb{C}_+).$$

On the other hand, if $f \in L^\infty(\mathbb{R})$ satisfying

$$\int_{-\infty}^{\infty} \frac{f(t)}{(t - \bar{z})(t + i)}\, dt = 0, \qquad (z \in \mathbb{C}_+),$$

is given, we can define

$$F(x + iy) = \frac{1}{\pi} \int_{-\infty}^{\infty} \frac{y}{(x-t)^2 + y^2} f(t)\, dt, \qquad (x + iy \in \mathbb{C}_+).$$

Then, on the one hand, by Corollary 10.14, $F \in h^\infty(\mathbb{C}_+)$, and on the other hand, by (13.7) and our assumption on f, we obtain

$$F(z) = \frac{1}{2\pi i} \int_{-\infty}^{\infty} \left(\frac{1}{t - z} - \frac{1}{t + i} \right) f(t)\, dt, \qquad (z \in \mathbb{C}_+).$$

Thus F represents an element of $H^\infty(\mathbb{C}_+)$. Finally, by Corollary 10.14, we have

$$\|F\|_\infty = \|f\|_\infty.$$

The preceding discussion shows that $H^\infty(\mathbb{C}_+)$ is isomorphically isometric to

$$H^\infty(\mathbb{R}) = \left\{ f \in L^\infty(\mathbb{R}) : \int_{-\infty}^{\infty} \frac{f(t)}{(t - \bar{z})(t + i)}\, dt = 0,\ \text{for all } z \in \mathbb{C}_+ \right\},$$

which is a closed subspace of $L^\infty(\mathbb{R})$.

Theorem 13.3 *Let F be analytic in the upper half plane $\mathbb{C}_+$. Then $F \in H^\infty(\mathbb{C}_+)$ if and only if there is an $f \in H^\infty(\mathbb{R})$ such that*

$$F(x+iy) = \frac{1}{\pi} \int_{-\infty}^{\infty} \frac{y}{(x-t)^2 + y^2} f(t)\, dt, \qquad (x+iy \in \mathbb{C}_+).$$

If so, f is unique and we also have

$$
\begin{aligned}
F(z) &= \frac{i}{\pi} \int_{-\infty}^{\infty} \left(\frac{x-t}{(x-t)^2 + y^2} + \frac{1}{t+i} \right) f(t)\, dt \\
&= \frac{1}{2\pi i} \int_{-\infty}^{\infty} \left(\frac{1}{t-z} - \frac{1}{t+i} \right) f(t)\, dt, \qquad (z = x+iy \in \mathbb{C}_+).
\end{aligned}
$$

Moreover,

$$\|F\|_\infty = \|f\|_\infty.$$

Exercises

Exercise 13.1.1   Show that

$$H^\infty(\mathbb{R}) = \left\{ f \in L^\infty(\mathbb{R}) : \int_{-\infty}^{\infty} \frac{f(t)}{(t-\bar{z})(t-\bar{w})}\, dt = 0, \text{ for all } z, w \in \mathbb{C}_+ \right\}.$$

Exercise 13.1.2   Let $f \in L^\infty(\mathbb{R})$ be such that

$$\int_{-\infty}^{\infty} \frac{f(t)}{(t-\bar{z})^2}\, dt = 0$$

for all $z \in \mathbb{C}_+$. Can we conclude that $f \in H^\infty(\mathbb{R})$?

13.2   Analytic measures on $\mathbb{R}$

One of our main goals in this chapter is to show that Theorem 13.2 remains valid even if $p = 1$. To prove this result, we need to study a special class of Borel measures. A measure $\mu \in \mathcal{M}(\mathbb{R})$ is called *analytic* if

$$\hat{\mu}(t) = 0 \tag{13.8}$$

for all $t \le 0$. We use Riesz's theorem (Theorem 5.10) to characterize these measures.

Lemma 13.4 *Let $\mu \in \mathcal{M}(\mathbb{R})$. Then the following are equivalent:*

(a) μ is analytic;

(b) *for all* $z \in \mathbb{C}_+$,

$$\int_{\mathbb{R}} \frac{d\mu(t)}{t - \bar{z}} = 0; \tag{13.9}$$

(c) $d\mu(t) = u(t)\,dt$ *where* $u \in L^1(\mathbb{R})$ *and* $\hat{u}(t) = 0$ *for all* $t \leq 0$.

Proof. (a) $\Longrightarrow$ (b) : According to Theorem 10.1, for any $\mu \in \mathcal{M}(\mathbb{R})$, we have

$$\frac{1}{2\pi i} \int_{\mathbb{R}} \frac{d\mu(t)}{t - \bar{z}} = -\int_{-\infty}^{0} \hat{\mu}(t)\, e^{i\,2\pi \bar{z}\,t}\,dt, \qquad (z \in \mathbb{C}_+).$$

Hence, (a) clearly implies (b).

(b) $\Longrightarrow$ (c) : Taking the successive derivative of (13.9) with respect to $\bar{z}$ gives

$$\int_{\mathbb{R}} \frac{d\mu(t)}{(t - \bar{z})^n} = 0$$

for all $n \geq 1$ and $z \in \mathbb{C}_+$. Since

$$\frac{(t + \bar{z})^{n-1}}{(t - \bar{z})^n} = \sum_{k=0}^{n-1} \binom{n-1}{k} \frac{(2t)^k\,(-1)^{n-1-k}}{(t - \bar{z})^{k+1}},$$

we thus have

$$\int_{\mathbb{R}} \frac{(t + \bar{z})^{n-1}}{(t - \bar{z})^n}\,d\mu(t) = 0, \qquad (n \geq 1).$$

In particular, for $z = i$, we obtain

$$\int_{\mathbb{R}} \left(\frac{t - i}{t + i}\right)^n \frac{d\mu(t)}{i - t} = 0, \qquad (n \geq 1).$$

Now we make the change of variable $e^{i\tau} = \frac{i-t}{i+t}$ to write the preceding integral over $\mathbb{T}$. Therefore, by (11.6), we have

$$\int_{\mathbb{T}} e^{in\tau}\,d\nu(e^{i\tau}) = 0, \qquad (n \geq 1),$$

where $d\nu(e^{i\tau})$ is the pullback of $d\mu(t)/(i - t)$ and we assume that $\nu(\{-1\}) = 0$. The celebrated theorem of F. and M. Riesz (Theorem 5.10) implies that $d\nu(e^{i\tau})$ is absolutely continuous with respect to the Lebesgue measure $d\tau$. Hence, by (11.4) and (11.7), $d\mu(t)$ is absolutely continuous with respect to the Lebesgue measure dt, i.e.

$$d\mu(t) = u(t)\,dt,$$

where $u \in L^1(\mathbb{R})$. Moreover, based on the assumption, u satisfies

$$\int_{-\infty}^{\infty} \frac{u(t)}{t - \bar{z}}\,dt = 0, \qquad (z \in \mathbb{C}_+).$$

Thus, by Corollary 10.2, we have

$$-\int_{-\infty}^{0} \hat{u}(t)\, e^{i\,2\pi \bar{z} t}\, dt = \int_{-\infty}^{\infty} \frac{u(t)}{t - \bar{z}}\, dt = 0, \qquad (z \in \mathbb{C}_+).$$

In particular, with $z = x + i$, $x \in \mathbb{R}$, we obtain

$$\int_{-\infty}^{\infty} \left(\hat{u}(t)\, e^{-2\pi t}\, \chi_{(-\infty,0)}(t) \right) e^{i\,2\pi x t}\, dt = 0$$

for all $x \in \mathbb{R}$. Hence, by the uniqueness theorem (Corollary 12.7),

$$\hat{u}(t) = 0$$

for almost all $t < 0$. Since $\hat{u}$ is a continuous function, the last identity holds for all $t \leq 0$.

(c) $\Longrightarrow$ (a) : This part is obvious. $\square$

13.3   Representation of $H^1(\mathbb{C}_+)$ functions

The definition (13.4) is valid even if $p = 1$. Now, using Lemma 13.4, we show that $H^1(\mathbb{R})$ also deserves attention. If $F \in H^1(\mathbb{C}_+)$, then, by Theorem 11.7, there is a unique $\mu \in \mathcal{M}(\mathbb{R})$ such that

$$\lim_{\beta \to 0} \int_{-\infty}^{\infty} \varphi(t)\, F(t + i\beta)\, dt = \int_{\mathbb{R}} \varphi(t)\, d\mu(t)$$

for all $\varphi \in \mathcal{C}_0(\mathbb{R})$. In particular, if $\varphi(t) = 1/(t - \bar{z})$, then, by Lemma 13.1,

$$\int_{-\infty}^{\infty} \frac{d\mu(t)}{t - \bar{z}} = 0, \qquad (z \in \mathbb{C}_+).$$

Hence, μ is an analytic measure and, by Lemma 13.4, there is an $f \in L^1(\mathbb{R})$ such that

$$d\mu(t) = f(t)\, dt$$

and

$$\int_{-\infty}^{\infty} \frac{f(t)}{t - \bar{z}}\, dt = 0, \qquad (z \in \mathbb{C}_+).$$

The big step is now taken and the rest of the discussion is exactly like the one given for the case $1 < p < \infty$ in Section 13.1.

Theorem 13.5 *Let F be analytic in the upper half plane $\mathbb{C}_+$. Then $F \in H^1(\mathbb{C}_+)$ if and only if there is an $f \in H^1(\mathbb{R})$ such that*

$$F(x + iy) = \frac{1}{\pi} \int_{-\infty}^{\infty} \frac{y}{(x - t)^2 + y^2}\, f(t)\, dt, \qquad (x + iy \in \mathbb{C}_+).$$

If so, f is unique and we also have

$$
\begin{aligned}
F(z) &= \frac{i}{\pi} \int_{-\infty}^{\infty} \frac{x-t}{(x-t)^2 + y^2} f(t)\, dt \\
&= \frac{1}{2\pi i} \int_{-\infty}^{\infty} \frac{f(t)}{t-z}\, dt, \qquad (z = x + iy \in \mathbb{C}_+).
\end{aligned}
$$

Moreover,

$$
\lim_{y \to 0} \| F_y - f \|_1 = 0,
$$

and

$$
\| F \|_1 = \| f \|_1.
$$

13.4 Spectral analysis of $H^p(\mathbb{R})$ $(1 \le p \le 2)$

In the preceding sections we defined $H^p(\mathbb{R})$, $1 \le p \le \infty$, as a closed subspace of $L^p(\mathbb{R})$, and showed that it is isomorphically isometric to $H^p(\mathbb{C}_+)$. If $1 \le p \le 2$, the Fourier transforms of elements of $H^p(\mathbb{R})$ are well-defined. In this section we study the Fourier transform of these elements.

Theorem 13.6 *Let $f \in H^p(\mathbb{R})$, $1 \le p \le 2$, and let*

$$
F(x + iy) = \frac{1}{\pi} \int_{-\infty}^{\infty} \frac{y}{(x-t)^2 + y^2} f(t)\, dt, \qquad (x + iy \in \mathbb{C}_+).
$$

Then $\hat{f}(t) = 0$, for almost all $t \le 0$, and moreover

$$
F(z) = \int_{0}^{\infty} \hat{f}(t)\, e^{i 2\pi z t}\, dt, \qquad (z \in \mathbb{C}_+).
$$

Proof. By Theorem 12.6, we have

$$
\frac{1}{2\pi i} \int_{-\infty}^{\infty} \frac{f(t)}{t - \bar{z}}\, dt = -\int_{-\infty}^{0} \hat{f}(t)\, e^{i 2\pi \bar{z} t}\, dt,
$$

for all $z = x + iy \in \mathbb{C}_+$ and all $f \in L^p(\mathbb{R})$, $1 \le p \le 2$. If $f \in H^p(\mathbb{R})$, then, by definition,

$$
\frac{1}{2\pi i} \int_{-\infty}^{\infty} \frac{f(t)}{t - \bar{z}}\, dt = 0, \qquad (z \in \mathbb{C}_+).
$$

Hence,

$$
\int_{-\infty}^{0} \hat{f}(t)\, e^{i 2\pi \bar{z} t}\, dt = 0, \qquad (z \in \mathbb{C}_+).
$$

Thus, by the uniqueness theorem (Corollary 12.7), we necessarily have $\hat{f}(t) = 0$ for almost all $t \le 0$.

Finally, by Theorem 12.6, we have

$$
\begin{aligned}
F(z) &= \frac{1}{\pi} \int_{-\infty}^{\infty} \frac{y}{(x-t)^2 + y^2}\, f(t)\, dt \\
&= \int_{-\infty}^{\infty} e^{-2\pi y |t|}\, \hat{f}(t)\, e^{i\,2\pi x t}\, dt \\
&= \int_{0}^{\infty} e^{-2\pi y t}\, \hat{f}(t)\, e^{i\,2\pi x t}\, dt \\
&= \int_{0}^{\infty} \hat{f}(t)\, e^{i 2\pi z t}\, dt, \qquad (z = x + iy \in \mathbb{C}_+).
\end{aligned}
$$

$\square$

The case $p = 1$ of Theorem 13.6 is more interesting, since in this case $\hat{f}$ is continuous on $\mathbb{R}$ and we thus have

$$
\hat{f}(t) = 0
$$

for *all* $t \leq 0$. In particular, for $t = 0$, we obtain the following result.

Corollary 13.7 *Let $f \in H^1(\mathbb{R})$. Then*

$$
\int_{-\infty}^{\infty} f(\tau)\, d\tau = 0.
$$

Theorem 13.6 has a converse which enables us to give another characterization of $H^p(\mathbb{R})$, $1 \leq p \leq 2$. Suppose that $f \in L^p(\mathbb{R})$, $1 \leq p \leq 2$, satisfying $\hat{f}(t) = 0$, for almost all $t \leq 0$, is given. Hence, by Theorem 12.6, we immediately have

$$
\frac{1}{2\pi i} \int_{-\infty}^{\infty} \frac{f(t)}{t - z}\, dt = -\int_{-\infty}^{0} \hat{f}(t)\, e^{i 2\pi \bar{z} t}\, dt = 0
$$

for all $z = x + iy \in \mathbb{C}_+$. Therefore, $f \in H^p(\mathbb{R})$. Based on this observation and Theorem 13.6, we can say

$$
H^p(\mathbb{R}) = \{\, f \in L^p(\mathbb{R}) \,:\, \hat{f}(t) = 0 \text{ for almost all } t \leq 0 \,\}
$$

for $1 \leq p \leq 2$. Hence, by the Hausdorff–Young theorem (Theorem 12.5), the Hardy space $H^p(\mathbb{R})$, $1 \leq p \leq 2$, is mapped into $L^q(\mathbb{R}^+)$, where q is the conjugate exponent of p, under the Fourier transform. According to the uniqueness theorem (Corollary 12.7), the map is injective. Hence, we naturally ask if it is surjective too. Using Plancherel's theorem we prove this assertion for $p = 2$. In other cases the map is not surjective.

Theorem 13.8 *Let $\varphi \in L^2(\mathbb{R}^+)$. Then there is an $f \in H^2(\mathbb{R})$ such that*

$$
\hat{f} = \varphi.
$$

Proof. Let

$$f(t) = \hat{\varphi}(-t).$$

By Plancherel's theorem we certainly have $f \in L^2(\mathbb{R})$. Moreover, by (12.5),

$$\hat{f}(t) = \hat{\hat{\varphi}}(-t) = \varphi(t).$$

Hence $\hat{f} \in L^2(\mathbb{R}^+)$, which is equivalent to $f \in H^2(\mathbb{R})$. $\qquad\square$

Exercises

Exercise 13.4.1 By Theorem 13.6 and the uniqueness theorem (Corollary 12.7), the map

$$\begin{array}{ccc} H^1(\mathbb{R}) & \longrightarrow & \mathcal{C}_0(\mathbb{R}^+) \\ f & \longmapsto & \hat{f} \end{array}$$

is injective. However, show that it is not surjective.
Hint: Use Exercise 12.1.3.

Exercise 13.4.2 Let $1 < p < 2$ and let q be the conjugate exponent of p. By Theorem 13.6 and the uniqueness theorem (Corollary 12.7), the map

$$\begin{array}{ccc} H^p(\mathbb{R}) & \longrightarrow & L^q(\mathbb{R}^+) \\ f & \longmapsto & \hat{f} \end{array}$$

is injective. However, show that it is not surjective.
Hint: See Exercise 12.5.2.

13.5 A contraction from $H^p(\mathbb{C}_+)$ into $H^p(\mathbb{D})$

The following result gives a contraction from $H^p(\mathbb{C}_+)$ into $H^p(\mathbb{D})$. This is a valuable tool which enables us to easily establish some properties of Hardy spaces of the upper half plane from the known results on the unit disc.

Lemma 13.9 *Let $F \in H^p(\mathbb{C}_+)$, $0 < p \leq \infty$, and let*

$$G(w) = \pi^{1/p} F\left(i\frac{1-w}{1+w}\right), \qquad (w \in \mathbb{D}).$$

Then $G \in H^p(\mathbb{D})$ and

$$\|G\|_{H^p(\mathbb{D})} \leq \|F\|_{H^p(\mathbb{C}_+)}.$$

Proof. The result is obvious if $p = \infty$. Hence, suppose that $0 < p < \infty$. Fix $\beta > 0$. By Lemma 11.5, $|F(z + i\beta)|^p$, as a function of z, is a bounded subharmonic function on $\mathbb{C}_+$ which is also continuous on $\overline{\mathbb{C}_+}$. Thus, by Corollary 10.15,

$$|F(x + iy + i\beta)|^p \le \frac{1}{\pi} \int_{-\infty}^{\infty} \frac{y}{(x - t)^2 + y^2} \, |F(t + i\beta)|^p \, dt, \qquad (x + iy \in \mathbb{C}_+).$$

By assumption, $(|F_\beta|^p)_{\beta > 0}$ is a uniformly bounded family in $L^1(\mathbb{R})$. Hence, there is a measure $\lambda \in \mathcal{M}(\mathbb{R})$ and a subsequence $(|F_{\beta_n}|^p)_{n \ge 1}$ such that $|F(t + i\beta_n)|^p \, dt$ converges to $d\lambda(t)$ in the weak* topology. Hence,

$$\lambda(\mathbb{R}) = \|\lambda\| \le \liminf_{n \to \infty} \|F_{\beta_n}\|_p^p = \|F\|_p^p. \tag{13.10}$$

Moreover, by letting $\beta_n \to 0$, we obtain

$$|F(x + iy)|^p \le \frac{1}{\pi} \int_{\mathbb{R}} \frac{y}{(x - t)^2 + y^2} \, d\lambda(t), \qquad (x + iy \in \mathbb{C}_+).$$

Now we apply the change of variable $z = i\frac{1-w}{1+w}$, as discussed in Section 11.5, to get

$$|G(re^{i\theta})|^p \le \int_{\mathbb{T}} \frac{1 - r^2}{1 + r^2 - 2r\cos(\theta - \tau)} \, \cos^2(\tau/2) \, d\nu(\tau), \qquad (re^{i\theta} \in \mathbb{D}),$$

where ν is the pullback of λ with the extra assumption $\nu(\{-1\}) = 0$. The identity

$$\frac{y}{(x - t)^2 + y^2} = \frac{1 - r^2}{1 + r^2 - 2r\cos(\theta - \tau)} \, \cos^2(\tau/2)$$

was also used implicitly in the preceding calculation. Therefore, by Fubini's theorem and (13.10),

$$\frac{1}{2\pi} \int_{-\pi}^{\pi} |G(re^{i\theta})|^p \, d\theta \le \int_{\mathbb{T}} \cos^2(\tau/2) \, d\nu(\tau) \le \int_{\mathbb{T}} d\nu(\tau) = \nu(\mathbb{T}) = \lambda(\mathbb{R}) \le \|F\|_p^p.$$

$\square$

Lemma 13.9 says that the mapping

$$H^p(\mathbb{C}_+) \quad \longrightarrow \quad H^p(\mathbb{D})$$

$$F(z) \quad \longmapsto \quad \pi^{1/p} F\left(i\frac{1-w}{1+w} \right)$$

is a contraction. Even though this map is one-to-one, it is not onto. More precisely, if we start with $G \in H^p(\mathbb{D})$, $0 < p < \infty$, and define

$$F(z) = G\left(\frac{i - z}{i + z} \right), \qquad (z \in \mathbb{C}_+),$$

then F is not necessarily in $H^p(\mathbb{C}_+)$. A simple counterexample is $G(w) = w \in H^1(\mathbb{D})$, for which $F(z) = \frac{i-z}{i+z} \notin H^1(\mathbb{C}_+)$.

Theorem 13.10 *Let $F \in H^p(\mathbb{C}_+)$, $0 < p \le \infty$, $F \not\equiv 0$. Then*

(a)

$$f(x) = \lim_{y \to 0} F(x + iy)$$

exists for almost all $x \in \mathbb{R}$.

(b)

$$\int_{-\infty}^{\infty} \Big| \log |f(t)| \Big| \frac{dt}{1 + t^2} < \infty,$$

(c)

$$\lim_{y \to 0} \int_{-\infty}^{\infty} \Big| \log^+ |F(t + iy)| - \log^+ |f(t)| \Big| \frac{dt}{1 + t^2} = 0,$$

(d)

$$\log |F(x + iy)| \le \frac{1}{\pi} \int_{-\infty}^{\infty} \frac{y}{(x - t)^2 + y^2} \log |f(t)| \, dt, \qquad (x + iy \in \mathbb{C}_+).$$

Proof. Define

$$G(w) = F\left(i \frac{1 - w}{1 + w} \right), \qquad (w \in \mathbb{D}).$$

By Lemma 13.9, $G \in H^p(\mathbb{D})$. Hence, by Fatou's theorem (Theorem 3.12), G has boundary values almost everywhere on $\mathbb{T}$. Moreover, the results gathered in Theorem 7.9 hold for G. Applying again the change of variable $z = i \frac{1 - w}{1 + w}$, which was discussed in Section 11.5, implies all the parts. Note that $d\tau = 2 dt / (1 + t^2)$. □

The following result was partially obtained in the proof of Lemma 13.9. Knowing that F has boundary values almost everywhere on the real line, we obtain a candidate for the measure λ appearing there.

Corollary 13.11 *Let $F \in H^p(\mathbb{C}_+)$, $0 < p < \infty$, and let*

$$f(x) = \lim_{y \to 0} F(x + iy)$$

wherever the limit exists. Then $f \in L^p(\mathbb{R})$,

$$\|f\|_p = \|F\|_p$$

and

$$|F(x + iy)|^p \le \frac{1}{\pi} \int_{-\infty}^{\infty} \frac{y}{(x - t)^2 + y^2} |f(t)|^p \, dt, \qquad (x + iy \in \mathbb{C}_+).$$

Proof. The first assertion is a direct consequence of Fatou's lemma, which also implies $\|f\|_p \leq \|F\|_p$. By Theorem 13.10,

$$\log |F(x + iy)| \leq \frac{1}{\pi} \int_{-\infty}^{\infty} \frac{y}{(x-t)^2 + y^2} \log |f(t)|\, dt, \qquad (x + iy \in \mathbb{C}_+).$$

Hence

$$|F(x + iy)|^p \leq \exp \left\{ \frac{1}{\pi} \int_{-\infty}^{\infty} \frac{y}{(x-t)^2 + y^2} \log |f(t)|^p\, dt \right\}, \qquad (x + iy \in \mathbb{C}_+).$$

Now apply Jensen's inequality to obtain the last result. Moreover, by Fubini's theorem, we have

$$\int_{-\infty}^{\infty} |F(x + iy)|^p dx \leq \int_{-\infty}^{\infty} |f(t)|^p\, dt, \qquad (y > 0).$$

Hence, $\|F\|_p \leq \|f\|_p$. $\square$

Exercises

Exercise 13.5.1   Let $\varphi, \varphi_n \in L^p(\mathbb{R})$, $0 < p < \infty$, $n \geq 1$. Suppose that

$$\varphi_n(x) \longrightarrow \varphi(x)$$

for almost all $x \in \mathbb{R}$, and that

$$\lim_{n \to \infty} \|\varphi_n\|_p = \|\varphi\|_p.$$

Show that

$$\lim_{n \to \infty} \|\varphi_n - \varphi\|_p = 0.$$

Hint: Apply Egorov's theorem [5, page 21].

Exercise 13.5.2   Let $F \in H^p(\mathbb{C}_+)$, $0 < p < \infty$, and let

$$f(x) = \lim_{y \to 0} F(x + iy)$$

wherever the limit exists. Show that

$$\lim_{y \to 0} \|F_y - f\|_p = 0.$$

Hint: Use Corollary 13.11 and Exercise 13.5.1.

13.6 Blaschke products for the upper half plane

In Section 7.1, we called

$$b_{w_0}(w) = \frac{w_0 - w}{1 - \bar{w}_0 \, w}, \qquad (w \in \mathbb{D}),$$

a Blaschke factor for the open unit disc. If we apply the change of variables given in Section 11.5, we obtain

$$\frac{|w_0|}{w_0} \frac{w_0 - w}{1 - \bar{w}_0 \, w} = \frac{\left|\frac{i - z_0}{i - \bar{z}_0}\right|}{\frac{i - z_0}{i - \bar{z}_0}} \frac{z - z_0}{z - \bar{z}_0}, \tag{13.11}$$

where $z \in \mathbb{C}_+$ and $w \in \mathbb{D}$ are related by

$$w = \frac{i - z}{i + z}, \qquad \text{or equivalently} \qquad z = i \frac{1 - w}{1 + w}.$$

The points $z_0 \in \mathbb{C}_+$ and $w_0 \in \mathbb{D}$ are related by the same equations. As usual, we assume that

$$\frac{\left|\frac{i - z_0}{i - \bar{z}_0}\right|}{\frac{i - z_0}{i - \bar{z}_0}} = 1$$

if $z_0 = i$. Moreover, we also have

$$1 - |w_0|^2 = \frac{4 \Im z_0}{|i + z_0|^2}. \tag{13.12}$$

Therefore, for each $z_0 \in \mathbb{C}_+$, we call

$$b_{z_0}(z) = \frac{z - z_0}{z - \bar{z}_0}$$

a Blaschke factor for the upper half plane. Either by direct verification or by (13.11) and our knowledge of the Blaschke factors for the open unit disc, it is easy to verify that

$$|b_{z_0}(z)| < 1, \qquad (z \in \mathbb{C}_+),$$

and that

$$|b_{z_0}(t)| = 1, \qquad (t \in \mathbb{R}).$$

In the light of (13.12), a sequence $(z_n)_{n \geq 1} \subset \mathbb{C}_+$ is called a Blaschke sequence for the upper half plane if

$$\sum_{n=1}^{\infty} \frac{\Im z_n}{|i + z_n|^2} < \infty. \tag{13.13}$$

Hence, Theorem 7.1 is rewritten as follows for $\mathbb{C}_+$.

Theorem 13.12 *Let $\{z_n\}_{n\geqslant 1}$ be a sequence in the upper half plane with no accumulation point in $\mathbb{C}_+$. Let $\sigma = \sigma_{\{z_n\}_{n\geqslant 1}}$ denote the set of all accumulation points of $\{z_n\}_{n\geqslant 1}$ which are necessarily on the real line $\mathbb{R}$, and let $\Omega = \mathbb{C}\setminus(\sigma\cup \{\bar{z}_n : n \geqslant 1\})$. Then the partial products*

$$B_N(z) = \prod_{n=1}^{N} \frac{\left|\frac{i-z_n}{i-\bar{z}_n}\right|}{\frac{i-z_n}{i-\bar{z}_n}} \frac{z - z_n}{z - \bar{z}_n}$$

are uniformly convergent on compact subsets of Ω if and only if

$$\sum_{n=1}^{\infty} \frac{\Im z_n}{|i + z_n|^2} < \infty.$$

The limit of these partial products is called an infinite Blaschke product for the upper half plane and is denoted by

$$B(z) = \prod_{n=1}^{\infty} e^{i\alpha_n} \frac{z - z_n}{z - \bar{z}_n},$$

where the real number α_n is chosen such that

$$e^{i\alpha_n} = \frac{\left|\frac{i-z_n}{i-\bar{z}_n}\right|}{\frac{i-z_n}{i-\bar{z}_n}}.$$

Equivalently, we can say that α_n is a real number such that

$$e^{i\alpha_n} \frac{i - z_n}{i - \bar{z}_n} \geq 0.$$

Clearly $B \in H^\infty(\mathbb{C}_+)$ and $\|B\|_\infty \leq 1$. However, in the same manner, Theorem 7.4 is written as follows for the upper half plane. This result implies that $\|B\|_\infty = 1$.

Theorem 13.13 *Let B be a Blaschke product for the upper half plane. Let*

$$b(x) = \lim_{y\to 0} B(x + iy)$$

wherever the limit exists. Then

$$|b(x)| = 1$$

for almost all $x \in \mathbb{R}$.

13.7 The canonical factorization in $H^p(\mathbb{C}_+)$ $(0 < p \leq \infty)$

In this section we show that the zeros of a function in $H^p(\mathbb{C}_+)$ satisfy the Blaschke condition. Hence we form a Blaschke product to extract these zeros. This technique was discussed for Hardy spaces of the open unit disc in Section 7.4.

Lemma 13.14 *Let* $F \in H^p(\mathbb{C}_+)$, $0 < p \le \infty$, $F \not\equiv 0$. *Let* (z_n) *be the sequence of zeros of* F, *counting multiplicities, in* $\mathbb{C}_+$. *Then* (z_n) *is a Blaschke sequence for the upper half plane, i.e.*

$$\sum_n \frac{\Im z_n}{|i + z_n|^2} < \infty.$$

Proof. Let

$$G(w) = F\left(i\,\frac{1 - w}{1 + w}\right), \qquad (w \in \mathbb{D}).$$

By Lemma 13.9, $G \in H^p(\mathbb{D})$. Let

$$w_n = \frac{i - z_n}{i + z_n}.$$

The sequence (w_n) represents the set of zeros of G in $\mathbb{D}$. Hence, by Lemma 7.6,

$$\sum_n (1 - |w_n|^2) < \infty.$$

But

$$1 - |w_n|^2 = \frac{4\Im z_n}{|i + z_n|^2},$$

which implies that (z_n) satisfies the Blaschke condition for the upper half plane. $\qquad\square$

Knowing that the zeros of a function F in the Hardy class $H^p(\mathbb{C}_+)$ fulfil the Blaschke condition, we are able to construct a Blaschke product, say B. Then clearly $G = F/B$ is a well-defined analytic function in the upper half plane. To be more precise, we should say that G has removable singularities at the zeros of F. Moreover, as we discussed similarly in Theorem 7.7, G stays in the same class and has the same norm. However, we go further and provide the complete canonical factorization for elements of $H^p(\mathbb{C}_+)$.

A function $F \in H^\infty(\mathbb{C}_+)$ is called *inner* for the upper half plane if

$$|f(t)| = |\lim_{y \to 0} F(t + iy)| = 1$$

for almost all $t \in \mathbb{R}$. According to Theorem 13.13, any Blaschke product B is an inner function for the upper half plane. Let σ be a positive singular Borel measure on $\mathbb{R}$ satisfying

$$\int_{\mathbb{R}} \frac{d\sigma(t)}{1 + t^2} < \infty.$$

Then direct verification shows that

$$S_\sigma(z) = S(z) = \exp\left\{-\frac{i}{\pi} \int_{\mathbb{R}} \left(\frac{1}{z - t} + \frac{t}{1 + t^2}\right) d\sigma(t)\right\}, \qquad (z \in \mathbb{C}_+),$$

is an inner function for $\mathbb{C}_+$. For obvious reasons, S is called a singular inner function. Finally,

$$E_\alpha(z) = E(z) = e^{i\alpha z}, \qquad (z \in \mathbb{C}_+),$$

where $\alpha \geq 0$, is also an inner function for $\mathbb{C}_+$. The product of two inner functions is clearly inner. Hence, $I = EBS$ is inner. A part of the canonical factorization theorem says that each inner function has such a decomposition.

Let $h \geq 0$ and let

$$\int_{-\infty}^{\infty} \frac{|\log h(t)|}{1+t^2} dt < \infty.$$

Then

$$O_h(z) = O(z) = \exp\left\{ \frac{i}{\pi} \int_{-\infty}^{\infty} \left(\frac{1}{z-t} + \frac{t}{1+t^2} \right) \log h(t)\ dt \right\}, \qquad (z \in \mathbb{C}_+),$$

is called an *outer function* for the upper half plane. According to Theorem 13.10, if $F \in H^p(\mathbb{C}_+)$, $0 < p \leq \infty$, $F \not\equiv 0$ and $f(x) = \lim_{y \to 0} F(x+iy)$ wherever the limit exists, then

$$\int_{-\infty}^{\infty} \left| \log|f(t)| \right| \frac{dt}{1+t^2} < \infty.$$

Hence, we can define the outer part of F by

$$O_F(z) = \exp\left\{ \frac{i}{\pi} \int_{-\infty}^{\infty} \left(\frac{1}{z-t} + \frac{t}{1+t^2} \right) \log|f(t)|\ dt \right\}, \qquad (z \in \mathbb{C}_+).$$

Theorem 13.15 *Let $F \in H^p(\mathbb{C}_+)$, $0 < p \leq \infty$, $F \not\equiv 0$. Let B be the Blaschke product formed with zeros of F in $\mathbb{C}_+$. Then there is a positive singular Borel measure σ satisfying*

$$\int_{\mathbb{R}} \frac{d\sigma(t)}{1+t^2} < \infty$$

and a constant $\alpha \geq 0$ such that

$$F = I_F\, O_F,$$

where the inner part of F is given by

$$I_F = E_\alpha\, B\, S_\sigma.$$

Proof. Let

$$G(w) = F\left(i \frac{1-w}{1+w} \right), \qquad (w \in \mathbb{D}).$$

By Lemma 13.9, $G \in H^p(\mathbb{D})$. Hence, according to the canonical factorization theorem for Hardy spaces on the open unit disc, we have

$$G = B\, S_\nu\, O_G,$$

where B is the Blaschke product formed with the zeros of G, ν is a finite positive singular Borel measure on $\mathbb{T}$,

$$S_\nu(w) = \exp\left\{ -\frac{1}{2\pi} \int_\mathbb{T} \frac{e^{i\tau} + w}{e^{i\tau} - w} \, d\nu(e^{i\tau}) \right\}$$

and

$$O_G(w) = \exp\left\{ \frac{1}{2\pi} \int_\mathbb{T} \frac{e^{i\tau} + w}{e^{i\tau} - w} \, \log |G(e^{i\tau})| \, d\tau \right\}$$

is the outer part of G. Write

$$S_\nu(w) = \exp\left\{ -\frac{\nu(\{-1\})}{2\pi} \frac{1 - w}{1 + w} \right\} \exp\left\{ -\frac{1}{2\pi} \int_{\mathbb{T}\backslash\{-1\}} \frac{e^{i\tau} + w}{e^{i\tau} - w} \, d\nu(e^{i\tau}) \right\}.$$

Let $\alpha = \nu(\{-1\})/(2\pi)$, and let λ be the pullback of ν to the real line. Now we apply again the change of variables $z = i\frac{1-w}{1+w}$ and $t = i\frac{1-e^{i\tau}}{1+e^{i\tau}}$. Note that

$$\frac{e^{i\tau} + w}{e^{i\tau} - w} = i \left(\frac{1}{z - t} + \frac{t}{1 + t^2} \right) (1 + t^2).$$

Finally, let $d\sigma(t) = (1 + t^2) \, d\lambda(t)$. The canonical decomposition for the upper half plane follows immediately from the corresponding decomposition on the open unit disc. □

Exactly the same technique that was applied in the proof of Theorem 7.8 can be utilized again to obtain a similar result for the upper half plane. A different proof of this result was sketched in Exercise 13.5.2. Moreover, the result has already been proved when $1 \leq p < \infty$.

Corollary 13.16 *Let* $F \in H^p(\mathbb{C}_+)$, $0 < p < \infty$, *and let*

$$f(t) = \lim_{y \to 0} F(t + iy)$$

wherever the limit exists. Then $f \in L^p(\mathbb{R})$ *and*

$$\lim_{r \to 1} \|F_r - f\|_p = 0.$$

Exercises

Exercise 13.7.1 Let $h \geq 0$, and let

$$\int_{-\infty}^{\infty} \frac{|\log h(t)|}{1 + t^2} \, dt < \infty.$$

Show that $O_h \in H^p(\mathbb{C}_+)$, $0 < p \leq \infty$, if and only if $h \in L^p(\mathbb{R})$. Moreover, $\|O_h\|_p = \|h\|_p$.

Exercise 13.7.2   Let σ be a nonzero positive singular Borel measure on $\mathbb{R}$ satisfying

$$\int_{\mathbb{R}} \frac{d\sigma(t)}{1+t^2} < \infty.$$

Let

$$S(z) = \exp\left\{ -\frac{i}{\pi} \int_{\mathbb{R}} \left(\frac{1}{z-t} + \frac{t}{1+t^2} \right) d\sigma(t) \right\}, \qquad (z \in \mathbb{C}_+).$$

Show that there is an $x_0 \in \mathbb{R}$ such that

$$\lim_{y \to 0} S(x_0 + iy) = 0.$$

Remark: Remember that $\lim_{y \to 0} |S(x+iy)| = 1$ for almost all $x \in \mathbb{R}$.

13.8   A correspondence between $H^p(\mathbb{C}_+)$ and $H^p(\mathbb{D})$

In Section 11.5 we introduced a conformal mapping between the open unit disc and the upper half plane. However, if we apply this mapping, we do not obtain a bijection between the Hardy spaces of the upper half plane and those of the open unit disc. Nevertheless, these spaces are not completely irrelevant. Lemma 13.9 provides a contraction from $H^p(\mathbb{C}_+)$ into $H^p(\mathbb{D})$. However, the map is not surjective. The following result gives an isometric isomorphism between these two families of Hardy spaces.

Theorem 13.17 *Let $0 < p \leq \infty$. Let F and G be analytic functions respectively on $\mathbb{C}_+$ and $\mathbb{D}$ which are related by the following equivalent equations:*

$$F(z) = \left(\frac{1}{\sqrt{\pi}(z+i)} \right)^{2/p} G\left(\frac{i-z}{i+z} \right), \qquad (z \in \mathbb{C}_+),$$

$$G(w) = \left(\frac{2\sqrt{\pi}\,i}{1+w} \right)^{2/p} F\left(i\frac{1-w}{1+w} \right), \qquad (w \in \mathbb{D}).$$

Then $G \in H^p(\mathbb{D})$ if and only if $F \in H^p(\mathbb{C}_+)$. Moreover,

$$\|F\|_{H^p(\mathbb{C}_+)} = \|G\|_{H^p(\mathbb{D})}.$$

Proof. The result is clear if $p = \infty$. Hence let $0 < p < \infty$. First, suppose that $G \in H^p(\mathbb{D})$. Thus G has boundary values almost everywhere on $\mathbb{T}$, which we represent by $G(e^{i\tau})$, and

$$\|G\|_{H^p(\mathbb{D})} = \left(\frac{1}{2\pi} \int_{-\pi}^{\pi} |G(e^{i\tau})|^p \, d\tau \right)^{1/p}.$$

Hence, by the first relation, F also has boundary values almost everywhere on $\mathbb{R}$, which we denote by $F(t)$, and

$$F(t) = \left(\frac{1 + e^{i\tau}}{2\sqrt{\pi}\, i} \right)^{2/p} G(e^{i\tau}), \qquad \left(t = i\frac{1 - e^{i\tau}}{1 + e^{i\tau}} \in \mathbb{R} \right).$$

We write this identity as

$$\left(\sqrt{\pi}(i + t) \right)^{2/p} F(t) = G(e^{i\tau}).$$

Thus, by (11.7),

$$\int_{-\infty}^{\infty} |F(t)|^p\, dt = \frac{1}{2\pi} \int_{-\pi}^{\pi} |G(e^{i\tau})|^p\, d\tau. \qquad (13.14)$$

By Corollary 7.10, we have

$$|G(re^{i\theta})|^p \le \frac{1}{2\pi} \int_{-\pi}^{\pi} \frac{1 - r^2}{1 + r^2 - 2r\cos(\theta - \tau)}\, |G(e^{i\tau})|^p\, d\tau.$$

Hence, making the change of variable $w = \frac{i-z}{i+z}$, we obtain

$$|F(x + iy)|^p \le \frac{1}{\pi} \int_{-\infty}^{\infty} \frac{y}{(x - t)^2 + y^2}\, |F(t)|^p\, dt.$$

Therefore, by Fatou's lemma,

$$\int_{-\infty}^{\infty} |F(x + iy)|^p\, dx \le \int_{-\infty}^{\infty} |F(t)|^p\, dt, \qquad (y > 0).$$

This inequality shows that $F \in H^p(\mathbb{C}_+)$ and, by (13.14), $\|F\|_{H^p(\mathbb{C}_+)} = \|G\|_{H^p(\mathbb{D})}$. The reverse implication is similar, except that we need Corollary 13.11. $\square$

Chapter 14

The Hilbert transform on $\mathbb{R}$

14.1   Various definitions of the Hilbert transform

Let F be analytic and bounded on some half plane

$$\Omega = \{\, x + iy : y > -Y \,\},$$

with $Y > 0$. Suppose that

$$\int_{-\infty}^{\infty} \frac{|F(t)|}{1 + |t|}\, dt < \infty \qquad (14.1)$$

and

$$\lim_{\substack{z \to \infty \\ \Im z \geq 0}} F(z) = 0. \qquad (14.2)$$

Fix $x \in \mathbb{R}$. Let $0 < \varepsilon < Y$ and let $\Gamma_{\varepsilon,R}$ be the curve shown in Figure 14.1.

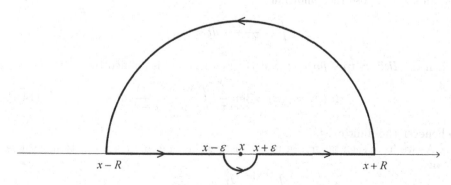

Fig. 14.1. The curve $\Gamma_{\varepsilon,R}$.

301

Then, by Cauchy's theorem,

$$
\begin{aligned}
F(x) &= \frac{1}{2\pi i} \int_{\Gamma_{\varepsilon,R}} \frac{F(\zeta)}{\zeta - x} \, d\zeta \\
&= \frac{1}{2\pi i} \int_{\varepsilon \leq |x-t| \leq R} \frac{F(t)}{t - x} \, dt + \frac{1}{2\pi} \int_{\pi}^{2\pi} F(x + \varepsilon e^{it}) \, dt \\
&\quad + \frac{1}{2\pi} \int_{0}^{\pi} F(x + Re^{it}) \, dt.
\end{aligned}
$$

Let $R \to \infty$. By (14.1) and (14.2) we obtain

$$
F(x) = \frac{1}{2\pi i} \int_{|x-t| \geq \varepsilon} \frac{F(t)}{t - x} \, dt + \frac{1}{2\pi} \int_{\pi}^{2\pi} F(x + \varepsilon e^{it}) \, dt.
$$

Thus

$$
F(x) = \lim_{\varepsilon \to 0} \frac{i}{\pi} \int_{|x-t| \geq \varepsilon} \frac{F(t)}{x - t} \, dt. \tag{14.3}
$$

Let us write $u = \Re F$ and $v = \Im F$ on the real line. Hence, (14.3) is equivalent to

$$
v(x) = \lim_{\varepsilon \to 0} \frac{1}{\pi} \int_{|x-t| \geq \varepsilon} \frac{u(t)}{x - t} \, dt
$$

and

$$
u(x) = -\lim_{\varepsilon \to 0} \frac{1}{\pi} \int_{|x-t| \geq \varepsilon} \frac{v(t)}{x - t} \, dt.
$$

Similar to the case of the unit circle, the preceding argument highlights the importance of the transform

$$
\lim_{\varepsilon \to 0} \frac{1}{\pi} \int_{|x-t| \geq \varepsilon} \frac{\phi(t)}{x - t} \, dt
$$

whenever ϕ is the real or the imaginary part of an analytic function satisfying certain properties. However, this transform is also well-defined on other classes of functions and has several interesting properties. To avoid any problem at infinity, we impose the condition

$$
\int_{-\infty}^{\infty} \frac{|\phi(t)|}{1 + |t|} \, dt < \infty.
$$

Then the *Hilbert transform* of ϕ at the point $x \in \mathbb{R}$ is defined by

$$
\mathcal{H}\phi(x) = \tilde{\phi}(x) = \lim_{\varepsilon \to 0} \frac{1}{\pi} \int_{|x-t| \geq \varepsilon} \frac{\phi(t)}{x - t} \, dt \tag{14.4}
$$

whenever the limit exists.

As we have seen before, in many cases we are faced with a function ϕ satisfying

$$
\int_{-\infty}^{\infty} \frac{|\phi(t)|}{1 + t^2} \, dt < \infty.
$$

In this case, the Hilbert transform is defined by

$$\mathcal{H}\phi(x) = \tilde{\phi}(x) = \lim_{\varepsilon \to 0} \frac{1}{\pi} \int_{|x-t|>\varepsilon} \left(\frac{1}{x-t} + \frac{t}{1+t^2} \right) \phi(t) \, dt. \tag{14.5}$$

The presence of $t/(1+t^2)$ ensures that for each fixed $x \in \mathbb{R}$ and $\varepsilon > 0$,

$$\int_{|x-t|>\varepsilon} \left(\frac{1}{x-t} + \frac{t}{1+t^2} \right) \phi(t) \, dt$$

is a well-defined Lebesgue integral. This fact is a consequence of

$$\frac{1}{x-t} + \frac{t}{1+t^2} = O(1/t^2)$$

as $|t| \to \infty$. If ϕ is a measurable function such that

$$\int_{-\infty}^{\infty} \frac{|\phi(t)|}{1+|t|^n} \, dt < \infty,$$

for a certain integer n, we are still able to add a specific term to $1/(x-t)$ such that the whole combination behaves like $1/|t|^n$, as $|t| \to \infty$, and thus obtain a meaningful Lebesgue integral which depends on ε. Then, by taking its limit as $\varepsilon \to 0$, we define the Hilbert transform of ϕ at x.

14.2 The Hilbert transform of $\mathcal{C}_c^1(\mathbb{R})$ functions

Let

$$\int_{-\infty}^{\infty} \frac{|\phi(t)|}{1+t^2} \, dt < \infty.$$

Fix $x \in \mathbb{R}$ and suppose that

$$\int_0^1 \frac{|\phi(x-t) - \phi(x+t)|}{t} \, dt < \infty. \tag{14.6}$$

We rewrite

$$\int_{|x-t|>\varepsilon} \left(\frac{1}{x-t} + \frac{t}{1+t^2} \right) \phi(t) \, dt$$

as

$$\int_\varepsilon^1 \frac{\phi(x-t) - \phi(x+t)}{t} \, dt + \int_{x-1}^{x+1} \frac{t\,\phi(t)}{1+t^2} \, dt + \int_{|x-t|>1} \left(\frac{1}{x-t} + \frac{t}{1+t^2} \right) \phi(t) \, dt.$$

Hence, $\tilde{\phi}(x)$, as defined in (14.5), exists and is given by

$$
\begin{aligned}
\pi\tilde{\phi}(x) &= \int_0^1 \frac{\phi(x-t) - \phi(x+t)}{t} \, dt + \int_{x-1}^{x+1} \frac{t\,\phi(t)}{1+t^2} \, dt \\
&\quad + \int_{|x-t|>1} \left(\frac{1}{x-t} + \frac{t}{1+t^2} \right) \phi(t) \, dt.
\end{aligned}
\tag{14.7}
$$

The condition (14.6) is fulfilled if ϕ is continuous on $[x-1, x+1]$ and differentiable at x. Clearly, the interval $[x-1, x+1]$ can be replaced by any other finite interval around x. Similarly, if $\phi \in \mathcal{C}_c^1(\mathbb{R})$, the space of continuously differentiable functions of compact support on $\mathbb{R}$, then $\tilde{\phi}$, as defined by (14.4), exists for all $x \in \mathbb{R}$ and is given by

$$\tilde{\phi}(x) = \frac{1}{\pi} \int_0^\infty \frac{\phi(x-t) - \phi(x+t)}{t}\, dt. \qquad (14.8)$$

Theorem 14.1 *Let* $\phi \in \mathcal{C}_c^1(\mathbb{R})$. *Then* $\tilde{\phi} \in \mathcal{C}_0(\mathbb{R})$.

Proof. Let

$$V(x+iy) = \frac{1}{\pi} \int_{-\infty}^\infty \frac{x-t}{(x-t)^2 + y^2}\, \phi(t)\, dt, \qquad (x+iy \in \mathbb{C}_+).$$

This formula can be rewritten as

$$V_y(x) = \frac{1}{\pi} \int_0^\infty \frac{t}{t^2 + y^2}\, (\phi(x-t) - \phi(x+t))\, dt.$$

Hence,

$$
\begin{aligned}
|V_y(x) - \tilde{\phi}(x)| &\leq \frac{1}{\pi} \int_0^\infty \left| \frac{t}{t^2 + y^2} - \frac{1}{t} \right| |\phi(x-t) - \phi(x+t)|\, dt \\
&\leq \frac{1}{\pi} \int_0^\infty \frac{y^2}{t^2 + y^2}\, \frac{|\phi(x-t) - \phi(x+t)|}{t}\, dt.
\end{aligned}
$$

However,

$$\frac{|\phi(x-t) - \phi(x+t)|}{t} \leq 2\, \|\phi'\|_\infty < \infty,$$

which implies

$$|V_y(x) - \tilde{\phi}(x)| \leq \frac{2\, \|\phi'\|_\infty}{\pi} \int_0^\infty \frac{y^2}{t^2 + y^2}\, dt = \|\phi'\|_\infty\, y$$

for all $x \in \mathbb{R}$. In other words, V_y converges uniformly to $\tilde{\phi}$ on $\mathbb{R}$. Since, for each $y > 0$, V_y is continuous, $\tilde{\phi}$ is also continuous on $\mathbb{R}$.

It remains to show that $\tilde{\phi}$ tends to zero at infinity. Let supp $\phi \subset [-M, M]$. Without loss of generality, suppose that $x > M$. Hence

$$\tilde{\phi}(x) = \frac{1}{\pi} \int_{x-M}^{x+M} \frac{\phi(x-t)}{t}\, dt,$$

and thus

$$|\tilde{\phi}(x)| \leq \frac{2M\, \|\phi\|_\infty}{x - M}.$$

Therefore, we indeed have

$$\tilde{\phi}(x) = O(1/|x|) \qquad (14.9)$$

as $|x| \to \infty$. $\square$

14.3 Almost everywhere existence of the Hilbert transform

In this section we follow a similar path as in Section 5.7 to show that the Hilbert transform is well-defined almost everywhere on $\mathbb{R}$.

Lemma 14.2 *Let*

$$\int_{-\infty}^{\infty} \frac{|u(t)|}{1+t^2} \, dt < \infty$$

and let

$$V(x+iy) = \frac{1}{\pi} \int_{-\infty}^{\infty} \left(\frac{x-t}{(x-t)^2 + y^2} + \frac{t}{1+t^2} \right) u(t) \, dt, \qquad (x+iy \in \mathbb{C}_+).$$

Then, for almost all $t \in \mathbb{R}$,

$$\lim_{y \to 0} \left\{ V(x+iy) - \frac{1}{\pi} \int_{|x-t|>y} \left(\frac{1}{x-t} + \frac{t}{1+t^2} \right) u(t) \, dt \right\} = 0.$$

Proof. By Lebegue's theorem

$$\lim_{\delta \to 0} \frac{1}{\delta} \int_0^\delta |u(x-t) - u(x+t)| \, dt = 0$$

for almost all $x \in \mathbb{R}$. We show that at such a point the lemma is valid.

Let

$$\Delta = V(x+iy) - \frac{1}{\pi} \int_{|x-t|>y} \left(\frac{1}{x-t} + \frac{t}{1+t^2} \right) u(t) \, dt.$$

Hence,

$$\Delta = I_1 + I_2 + I_3,$$

where

$$I_1 = \frac{1}{\pi} \int_{|x-t|\leq y} \frac{x-t}{(x-t)^2 + y^2} \, u(t) \, dt,$$

$$I_2 = \frac{1}{\pi} \int_{|x-t|\leq y} \frac{t}{1+t^2} \, u(t) \, dt$$

and

$$I_3 = \frac{1}{\pi} \int_{|x-t|>y} \left(\frac{x-t}{(x-t)^2 + y^2} - \frac{1}{x-t} \right) u(t) \, dt.$$

The first two integrals are easier to estimate:

$$|I_1| \leq \frac{1}{\pi} \int_0^y \frac{\tau}{\tau^2 + y^2} |u(x-\tau) - u(x+\tau)| \, d\tau$$

$$\leq \frac{1}{\pi y} \int_0^y |u(x-\tau) - u(x+\tau)| \, d\tau = o(1)$$

and

$$|I_2| \;\leq\; \frac{1}{\pi} \int_{|x-t| \leq y} \frac{|t|}{1+t^2} \, |u(t)| \, dt = o(1)$$

as $y \to 0$. To estimate the third integral, first note that

$$I_3 \;\leq\; \frac{1}{\pi} \int_y^\infty \left| \frac{\tau}{\tau^2 + y^2} - \frac{1}{\tau} \right| |u(x-\tau) - u(x+\tau)| \, d\tau$$

$$\leq\; \frac{y^2}{\pi} \int_y^\infty \frac{|u(x-\tau) - u(x+\tau)|}{\tau^3} \, d\tau.$$

Now, as was similarly shown in the proof of Lemma 5.16, the last expression also tends to zero when $y \to 0$. $\square$

In the preceding lemma, we saw that the difference of two functions of y tends to zero, as $y \to 0$, at almost all points of the real line. Now, we show that one of them has a finite limit almost everywhere. Hence, the other one necessarily has the same finite limit at almost all points of the real line.

Lemma 14.3 *Let*

$$\int_{-\infty}^\infty \frac{|u(t)|}{1+t^2} \, dt < \infty$$

and let

$$V(x + iy) = \frac{1}{\pi} \int_{-\infty}^\infty \left(\frac{x-t}{(x-t)^2 + y^2} + \frac{t}{1+t^2} \right) u(t) \, dt, \qquad (x + iy \in \mathbb{C}_+).$$

Then, for almost all $x \in \mathbb{R}$,

$$\lim_{y \to 0} V(x + iy)$$

exists.

Proof. Without loss of generality assume that $u \geq 0$, $u \not\equiv 0$. Since otherwise we can write $u = (u_1 - u_2) + i(u_3 - u_4)$, where each u_k is positive. Let

$$F(z) = \frac{i}{\pi} \int_{-\infty}^\infty \left(\frac{1}{z-t} + \frac{t}{1+t^2} \right) u(t) \, dt, \qquad (z \in \mathbb{C}_+).$$

Then

$$\Re F(z) = \frac{1}{\pi} \int_{-\infty}^\infty \frac{y}{(x-t)^2 + y^2} \, u(t) \, dt > 0, \qquad (z = x + iy \in \mathbb{C}_+).$$

Hence, using a conformal mapping φ between the unit disc and the upper half plane and considering $F \circ \varphi$, Lemma 5.14 ensures that

$$\lim_{y \to 0} F(x + iy)$$

exists and is finite for almost all $x \in \mathbb{R}$. Since $V = \Im F$, then

$$\lim_{y \to 0} V(x + iy)$$

also exists and is finite for almost all $x \in \mathbb{R}$. $\square$

A similar reasoning shows that if u satisfies the stronger condition

$$\int_{-\infty}^{\infty} \frac{|u(t)|}{1+|t|} \, dt < \infty$$

and we define

$$V(x+iy) = \frac{1}{\pi} \int_{-\infty}^{\infty} \frac{x-t}{(x-t)^2+y^2} \, u(t) \, dt, \qquad (x+iy \in \mathbb{C}_+),$$

then $\lim_{y \to 0} V(x+iy)$ exists and is finite for almost all $x \in \mathbb{R}$. Note that in this case

$$V(x+iy) = \Im \left(\frac{i}{\pi} \int_{-\infty}^{\infty} \frac{u(t)}{z-t} \, dt \right), \qquad (z = x+iy \in \mathbb{C}_+).$$

Theorem 14.4 *Let*

$$\int_{-\infty}^{\infty} \frac{|u(t)|}{1+t^2} \, dt < \infty.$$

Then, for almost all $x \in \mathbb{R}$,

$$\tilde{u}(x) = \lim_{\varepsilon \to 0} \frac{1}{\pi} \int_{|x-t|>\varepsilon} \left(\frac{1}{x-t} + \frac{t}{1+t^2} \right) u(t) \, dt$$

and

$$\lim_{y \to 0} \frac{1}{\pi} \int_{-\infty}^{\infty} \left(\frac{x-t}{(x-t)^2+y^2} + \frac{t}{1+t^2} \right) u(t) \, dt$$

exist and they are equal.

Proof. Apply Lemmas 14.2 and 14.3. $\qquad\square$

If

$$\int_{-\infty}^{\infty} \frac{|u(t)|}{1+|t|} \, dt < \infty$$

and we define

$$V(x+iy) = \frac{1}{\pi} \int_{-\infty}^{\infty} \frac{x-t}{(x-t)^2+y^2} \, u(t) \, dt, \qquad (x+iy \in \mathbb{C}_+),$$

then, for almost all $x \in \mathbb{R}$,

$$\lim_{y \to 0} \left\{ V(x+iy) - \frac{1}{\pi} \int_{|x-t|>y} \frac{u(t)}{x-t} \, dt \right\} = 0.$$

Therefore, we conclude that in this case,

$$\tilde{u}(x) = \lim_{\varepsilon \to 0} \frac{1}{\pi} \int_{|x-t|>\varepsilon} \frac{u(t)}{x-t} \, dt = \lim_{y \to 0} V(x+iy)$$

for almost all $x \in \mathbb{R}$.

14.4   Kolmogorov's theorem

Similar to the case of the unit circle, the assumption

$$\int_{-\infty}^{\infty} \frac{|u(t)|}{1+t^2}\, dt < \infty$$

is not enough to ensure that

$$\int_{-\infty}^{\infty} \frac{|\tilde{u}(t)|}{1+t^2}\, dt < \infty.$$

Nevertheless, a slight modification of the Kolmogorov theorem for the unit circle gives the following result for the Hilbert transform of functions defined on the real line.

Theorem 14.5 (Kolmogorov) *Let u be a real function satisfying*

$$\int_{-\infty}^{\infty} \frac{|u(t)|}{1+t^2}\, dt < \infty.$$

Then, for each $\lambda > 0$,

$$\int_{\{|\tilde{u}|>\lambda\}} \frac{dt}{1+t^2} \leq \frac{4}{\lambda} \int_{-\infty}^{\infty} \frac{|u(t)|}{1+t^2}\, dt.$$

Proof. (Carleson) Fix $\lambda > 0$. First suppose that $u \geq 0$ and $u \not\equiv 0$. Therefore, the analytic function

$$F(z) = \frac{i}{\pi} \int_{-\infty}^{\infty} \left(\frac{1}{z-t} + \frac{t}{1+t^2} \right) u(t)\, dt$$

maps the upper half plane into the right half plane, and moreover we have

$$F(i) = \frac{1}{\pi} \int_{-\infty}^{\infty} \frac{u(t)}{1+t^2}\, dt > 0. \tag{14.10}$$

Then, as $y \to 0$, by Corollary 11.11,

$$\Re F(x+iy) = \frac{1}{\pi} \int_{-\infty}^{\infty} \frac{y}{(x-t)^2+y^2}\, u(t)\, dt \longrightarrow u(x)$$

and by Lemma 14.3,

$$\Im F(x+iy) = \frac{1}{\pi} \int_{-\infty}^{\infty} \left(\frac{x-t}{(x-t)^2+y^2} + \frac{t}{1+t^2} \right) u(t)\, dt \longrightarrow \tilde{u}(x),$$

for almost everywhere $x \in \mathbb{R}$. Now, we apply the conformal mapping used in Theorem 6.1, and define

$$G(z) = \frac{2F(z)}{F(z)+\lambda}.$$

Since $G \in H^\infty(\mathbb{C}_+)$, $g(t) = \lim_{y \to 0} G(t + iy)$ exists and

$$g(t) = \frac{2(u(t) + i\tilde{u}(t))}{u(t) + i\tilde{u}(t) + \lambda}$$

for almost all $t \in \mathbb{R}$. Hence

$$\Re g = 1 + \frac{u^2 + \tilde{u}^2 - \lambda^2}{(u + \lambda)^2 + \tilde{u}^2} \geq 1$$

provided that $\tilde{u} > \lambda$. On the other hand, by Theorem 11.6,

$$\frac{1}{\pi} \int_{-\infty}^{\infty} \frac{\Re g(t)}{1 + t^2} \, dt = \Re G(i) = \frac{2F(i)}{F(i) + \lambda}$$

and, by (14.10),

$$\frac{2F(i)}{F(i) + \lambda} \leq \frac{2}{\lambda} \int_{-\infty}^{\infty} \frac{|u(t)|}{1 + t^2} \, dt.$$

Therefore,

$$\int_{\{|\tilde{u}| > \lambda\}} \frac{dt}{1 + t^2} \leq \int_{\{|\tilde{u}| > \lambda\}} \frac{\Re g(t)}{1 + t^2} \, dt \leq \frac{2}{\lambda} \int_{-\infty}^{\infty} \frac{|u(t)|}{1 + t^2} \, dt.$$

For an arbitrary real $u \in L^1(\mathbb{R})$, write $u = u_1 - u_2$, where $u_k \geq 0$ and $u_1 u_2 \equiv 0$. The rest of the proof is similar to the one given for Theorem 6.1. $\square$

Kolmogorov's theorem says

$$\int_{\{|\tilde{u}| > \lambda\}} \frac{dt}{1 + t^2} = O(1/\lambda)$$

as $\lambda \to \infty$. However, using the fact that in $L^1(dt/(1 + t^2))$, u can be approximated by functions in $C_c^\infty(\mathbb{R})$, the space of infinitely differentiable functions of compact support, we slightly generalize this result.

Corollary 14.6 *Let u be a real function satisfying*

$$\int_{-\infty}^{\infty} \frac{|u(t)|}{1 + t^2} \, dt < \infty.$$

Then

$$\int_{\{|\tilde{u}| > \lambda\}} \frac{dt}{1 + t^2} = o(1/\lambda)$$

as $\lambda \to \infty$.

Proof. Fix $\varepsilon > 0$. There is a $\phi \in C_c^\infty(\mathbb{R})$ such that

$$\int_{-\infty}^{\infty} \frac{|u(t) - \phi(t)|}{1 + t^2} \, dt < \varepsilon.$$

By Theorem 14.1, $\tilde{\phi}$ is bounded on $\mathbb{R}$. Note that we used (14.5) as the definition of $\tilde{\phi}$. If

$$\lambda > 2\|\tilde{\phi}\|_\infty,$$

then

$$\{\,|\tilde{u}| > \lambda\,\} \subset \{\,|\tilde{u} - \tilde{\phi}| > \lambda/2\,\}.$$

Hence, by Theorem 14.5,

$$\int_{\{|\tilde{u}|>\lambda\}} \frac{dt}{1+t^2} \le \int_{\{|\tilde{u}-\tilde{\phi}|>\lambda/2\}} \frac{dt}{1+t^2} \le \frac{8}{\lambda} \int_{-\infty}^{\infty} \frac{|u(t) - \phi(t)|}{1+t^2}\, dt \le \frac{8\varepsilon}{\lambda}.$$

$\square$

Exercises

Exercise 14.4.1 Find a measurable function u such that

$$\int_{-\infty}^{\infty} \frac{|u(t)|}{1+t^2}\, dt < \infty,$$

but

$$\int_{-\infty}^{\infty} \frac{|\tilde{u}(t)|}{1+t^2}\, dt = \infty.$$

Exercise 14.4.2 Let

$$\int_{-\infty}^{\infty} \frac{|u(t)|}{1+t^2}\, dt < \infty.$$

Suppose that there is an $\alpha > 0$ such that

$$\alpha x - \tilde{u}(x)$$

is increasing on $\mathbb{R}$. Show that

$$\lim_{x \to \pm\infty} \frac{\tilde{u}(x)}{x} = 0.$$

Hint: Fix $\varepsilon > 0$ and let $\gamma = 1 + \varepsilon$. By Corollary 14.6,

$$\lim_{n \to \infty} \gamma^n \int_{\{|\tilde{u}|>\varepsilon\gamma^n\}} \frac{dt}{1+t^2} = 0,$$

for all large integers n. Hence, there is an $x_n \in [\gamma^n, \gamma^{n+1}]$ such that

$$|\tilde{u}(x_n)| \le \varepsilon\gamma^n.$$

Note also that

$$\alpha x_n - \tilde{u}(x_n) \le \alpha x - \tilde{u}(x) \le \alpha x_{n+1} - \tilde{u}(x_{n+1})$$

for all $x \in [x_n, x_{n+1}]$.

14.5 M. Riesz's theorem

Let $u \in L^p(\mathbb{R})$, $1 < p < \infty$, and let

$$U(x + iy) = \frac{1}{\pi} \int_{-\infty}^{\infty} \frac{y}{(x - t)^2 + y^2} \, u(t) \, dt, \qquad (x + iy \in \mathbb{C}_+),$$

$$V(x + iy) = \frac{1}{\pi} \int_{-\infty}^{\infty} \frac{x - t}{(x - t)^2 + y^2} \, u(t) \, dt, \qquad (x + iy \in \mathbb{C}_+).$$

Then, by Corollary 10.12, $U \in h^p(\mathbb{C}_+)$,

$$\lim_{y \to 0} \|U_y - u\|_p = 0$$

and, as $y \to 0$, by Corollary 11.11,

$$U(x + iy) \longrightarrow u(x)$$

for almost everywhere $x \in \mathbb{R}$. For the conjugate function V, Theorem 14.4 says that

$$\lim_{y \to 0} V(x + iy) = \tilde{u}(x)$$

for almost all $x \in \mathbb{R}$. In this section our main goal is to show that $V \in h^p(\mathbb{C}_+)$, $\tilde{u} \in L^p(\mathbb{R})$ and $\lim_{y \to 0} \|V_y - \tilde{u}\|_p = 0$.

Theorem 14.7 (M. Riesz's theorem) *Let $u \in L^p(\mathbb{R})$, $1 < p < \infty$, and let*

$$V(x + iy) = \frac{1}{\pi} \int_{-\infty}^{\infty} \frac{x - t}{(x - t)^2 + y^2} \, u(t) \, dt, \qquad (x + iy \in \mathbb{C}_+).$$

Then $V \in h^p(\mathbb{C}_+)$, $\tilde{u} \in L^p(\mathbb{R})$ and there is a constant C_p, just depending on p, such that

$$\|V\|_p = \|\tilde{u}\|_p \leq C_p \|u\|_p.$$

Moreover,

$$V(x + iy) = \frac{1}{\pi} \int_{-\infty}^{\infty} \frac{y}{(x - t)^2 + y^2} \, \tilde{u}(t) \, dt, \qquad (x + iy \in \mathbb{C}_+),$$

and

$$\lim_{y \to 0} \|V_y - \tilde{u}\|_p = 0.$$

Proof. Without loss of generality, assume that u is real. Let $U = P * u$. The proof has several steps.

Case 1: $u \in L^p(\mathbb{R})$, $1 < p \leq 2$, $u \geq 0$, $u \not\equiv 0$ and u has compact support.

Let $F = U + iV$. Clearly F is analytic and $\Re F > 0$ in the upper half plane. By Corollary 10.12, for all $y > 0$,

$$\|U_y\|_p \leq \|u\|_p.$$

On the other hand, if supp $u \subset [-M, M]$, then

$$|V(x + iy)| \leq \frac{1}{\pi} \int_{-\infty}^{\infty} \frac{|x - t|}{(x - t)^2 + y^2} \, |u(t)| \, dt \leq \left(\frac{1}{2\pi} \int_{-M}^{M} |u(t)| \, dt \right) \frac{1}{y}$$

and

$$|V(x + iy)| \leq \frac{1}{\pi} \int_{-\infty}^{\infty} \frac{|x - t|}{(x - t)^2 + y^2} \, |u(t)| \, dt \leq \frac{|z| + M}{\pi(|z| - M)^2} \int_{-M}^{M} |u(t)| \, dt,$$

which implies

$$|V(z)| = \left(\frac{6}{\pi} \int_{-M}^{M} |u(t)| \, dt \right) \frac{1}{|z|}$$

for all $z \in \mathbb{C}_+$, $|z| > 2M$. Put

$$M' = \frac{1}{2\pi} \int_{-M}^{M} |u(t)| \, dt.$$

Fix $\beta > 0$. Therefore, for all $x + iy \in \mathbb{C}_+$,

$$\int_{-\infty}^{\infty} |V(x + iy + i\beta)|^p \, dx \leq 4M \, (M'/\beta)^p + 2(12M')^p \int_{2M}^{\infty} \frac{dx}{x^p} < \infty.$$

In other words, $V_\beta \in h^p(\mathbb{C}_+)$. Define

$$G(z) = F^p(z + i\beta).$$

Clearly, G is analytic in $\mathbb{C}_+$ and since

$$|G(x + iy)| \leq 2^p \left(|U(x + iy + i\beta)|^p + |V(x + iy + i\beta)|^p \right),$$

we have $G \in H^1(\mathbb{C}_+)$. Thus, by Corollary 13.7,

$$\int_{-\infty}^{\infty} G(x) \, dx = 0. \tag{14.11}$$

Choose $\alpha \in (\pi/2p, \pi/2)$. Since $1 < p \leq 2$, such a selection is possible. Let

$$S_\alpha = \{ x + iy : x > 0, \ |y/x| \leq \tan \alpha \}$$

and

$$\Omega_\alpha = \{ x + iy : x < 0, \ |y/x| \leq -\tan(p\alpha) \}.$$

(See Figures 14.2 and 14.3.)

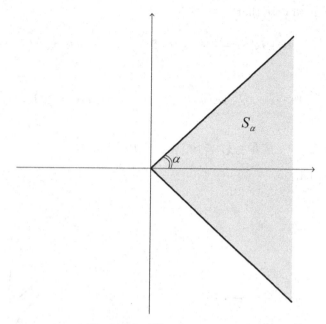

Fig. 14.2. The domain S_α.

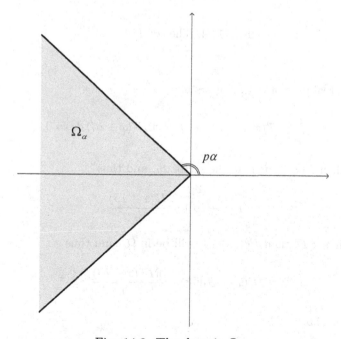

Fig. 14.3. The domain Ω_α.

It is important to note that

$$|x + iy| \leq \frac{x}{\cos \alpha} \qquad \text{if} \qquad x + iy \in S_\alpha$$

and

$$-x \geq |x + iy| \, |\cos(p\alpha)| \qquad \text{if} \qquad x + iy \in \Omega_\alpha.$$

Define

$$E = \{ \, x \in \mathbb{R} \, : \, F(x + i\beta) \in S_\alpha \, \}.$$

(See Figure 14.4.)

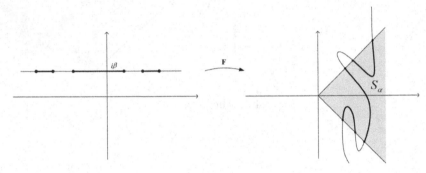

Fig. 14.4. The set $E + i\beta$.

Taking the real part of (14.11) gives us

$$\int_E \Re F^p(x + i\beta) \, dx + \int_{\mathbb{R} \backslash E} \Re F^p(x + i\beta) \, dx = 0. \qquad (14.12)$$

If $x \in E$, then $F(x + i\beta)$ is in the sector S_α and thus

$$|F(x + i\beta)| \leq \frac{U(x + i\beta)}{\cos \alpha}.$$

However, if $x \notin E$, then $F^p(x + i\beta)$ will be in Ω_α and thus

$$|F(x + i\beta)|^p \leq -\frac{\Re F^p(x + i\beta)}{|\cos(p\alpha)|}.$$

(See Figure 14.5.)

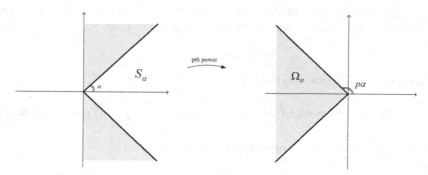

Fig. 14.5. The image under the pth power.

The first inequality implies

$$\int_E |F(x+i\beta)|^p \, dx \le \frac{1}{\cos^p \alpha} \int_E U^p(x+i\beta) \, dx$$

and the second one, along with (14.12), gives

$$
\begin{aligned}
\int_{\mathbb{R}\setminus E} |F(x+i\beta)|^p \, dx
&\le -\frac{1}{|\cos(p\alpha)|} \int_{\mathbb{R}\setminus E} \Re F^p(x+i\beta) \, dx \\
&= \frac{1}{|\cos(p\alpha)|} \int_E \Re F^p(x+i\beta) \, dx \\
&\le \frac{1}{|\cos(p\alpha)|} \int_E |F(x+i\beta)|^p \, dx \\
&\le \frac{1}{\cos^p \alpha \, |\cos(p\alpha)|} \int_E U^p(x+i\beta) \, dx.
\end{aligned}
$$

Hence,

$$\int_{-\infty}^{\infty} |F(x+i\beta)|^p \, dx \le \frac{1}{\cos^p \alpha}\left(1+\frac{1}{|\cos(p\alpha)|}\right) \int_{-\infty}^{\infty} U^p(x+i\beta) \, dx.$$

Since

$$|V(x+i\beta)| \le |F(x+i\beta)|,$$

the desired result follows for this case with the constant

$$K_p = \frac{1}{\cos \alpha}\left(1+\frac{1}{|\cos(p\alpha)|}\right)^{\frac{1}{p}}.$$

Case 2: $u \in L^p(\mathbb{R})$, $1 < p \le 2$, and u has compact support.

We prove a weaker form of the required inequality. Write $u = u_1 - u_2$ where $u_1, u_2 \geq 0$, $u_1 u_2 \equiv 0$ and they have compact support. Thus

$$|u|^p = |u_1|^p + |u_2|^p.$$

Moreover, with obvious notations, $V = V_1 - V_2$ and thus

$$\int_{-\infty}^{\infty} |V(x+i\beta)|^p \, dx \leq 2^p \left(\int_{-\infty}^{\infty} |V_1(x+i\beta)|^p \, dx + \int_{-\infty}^{\infty} |V_2(x+i\beta)|^p \, dx \right).$$

Hence, by the result of the preceding case, we obtain

$$
\begin{aligned}
\int_{-\infty}^{\infty} |V(x+i\beta)|^p \, dx \;&\leq\; 2^p \, K_p^p \left(\int_{-\infty}^{\infty} |U_1(x+i\beta)|^p \, dx + \int_{-\infty}^{\infty} |U_2(x+i\beta)|^p \, dx \right) \\
&\leq\; 2^p \, K_p^p \left(\int_{-\infty}^{\infty} |u_1(x)|^p \, dx + \int_{-\infty}^{\infty} |u_2(x)|^p \, dx \right) \\
&=\; 2^p \, K_p^p \int_{-\infty}^{\infty} |u(x)|^p \, dx.
\end{aligned}
$$

Case 3: $u \in L^p(\mathbb{R})$, $1 < p \leq 2$.

First, we prove a weaker form of the required inequality. For an arbitrary $u \in L^p(\mathbb{R})$, $1 < p \leq 2$, take a sequence $u_n \in L^p(\mathbb{R})$, u_n of compact support, such that

$$\lim_{y \to 0} \|u_n - u\|_p = 0.$$

Let $V_n = Q * u_n$. By Case 2, for each $\beta > 0$,

$$\int_{-\infty}^{\infty} |V_n(x+i\beta)|^p \, dx \leq 2^p \, K_p^p \int_{-\infty}^{\infty} |u_n(x)|^p \, dx.$$

On the other hand, by Hölder's inequality, for each $x + i\beta \in \mathbb{C}_+$,

$$V_n(x+i\beta) \longrightarrow V(x+i\beta)$$

as $n \to \infty$. Hence, by Fatou's lemma,

$$
\begin{aligned}
\int_{-\infty}^{\infty} |V(x+i\beta)|^p \, dx \;&\leq\; \liminf_{n \to \infty} \int_{-\infty}^{\infty} |V_n(x+i\beta)|^p \, dx \\
&\leq\; \lim_{n \to \infty} 2^p \, K_p^p \int_{-\infty}^{\infty} |u_n(x)|^p \, dx \\
&=\; 2^p \, K_p^p \int_{-\infty}^{\infty} |u(x)|^p \, dx.
\end{aligned}
$$

Second, we prove the general case. Fix $x_0 + iy_0 \in \mathbb{C}_+$. Let

$$\phi(x+iy) = \frac{x_0 - x}{(x_0 - x)^2 + (y + y_0)^2}, \qquad (x + iy \in \mathbb{C}_+).$$

Clearly, ϕ is a bounded harmonic function on the closed upper half plane, and thus, by Theorem 11.2,

$$\phi(x + iy) = \frac{1}{\pi} \int_{-\infty}^{\infty} \frac{y}{(x-t)^2 + y^2} \phi(t) \, dt.$$

More explicitly, we have

$$\frac{x_0 - x}{(x_0 - x)^2 + (y + y_0)^2} = \frac{1}{\pi} \int_{-\infty}^{\infty} \frac{y}{(x-t)^2 + y^2} \frac{x_0 - t}{(x_0 - t)^2 + y_0^2} \, dt$$

for all $x + iy \in \mathbb{C}_+$. Therefore, for all $y > 0$,

$$
\begin{aligned}
V(x_0 + iy_0 + iy) &= \frac{1}{\pi} \int_{-\infty}^{\infty} \frac{x_0 - x}{(x_0 - x)^2 + (y + y_0)^2} u(x) \, dx \\
&= \frac{1}{\pi} \int_{-\infty}^{\infty} \left(\frac{1}{\pi} \int_{-\infty}^{\infty} \frac{y}{(x-t)^2 + y^2} \frac{x_0 - t}{(x_0 - t)^2 + y_0^2} \, dt \right) u(x) \, dx \\
&= \frac{1}{\pi} \int_{-\infty}^{\infty} \left(\frac{1}{\pi} \int_{-\infty}^{\infty} \frac{y}{(x-t)^2 + y^2} u(x) \, dx \right) \frac{x_0 - t}{(x_0 - t)^2 + y_0^2} \, dt \\
&= \frac{1}{\pi} \int_{-\infty}^{\infty} \frac{x_0 - t}{(x_0 - t)^2 + y_0^2} U(t + iy) \, dt.
\end{aligned}
$$

Hence, by the result of Case 2,

$$\int_{-\infty}^{\infty} |V(x_0 + iy_0 + iy)|^p \, dx_0 \leq 2^p K_p^p \int_{-\infty}^{\infty} |U(t + iy)|^p \, dt.$$

Let $y_0 \to 0$ and apply Fatou's lemma to get the required inequality.

Case 4: $u \in L^p(\mathbb{R})$, $2 < p < \infty$.

The proof is by duality. Let q be the conjugate exponent of p. Hence $1 < q < 2$. Let $\phi \in L^q(\mathbb{R})$ be of compact support and let

$$\Phi(x + iy) = \frac{1}{\pi} \int_{-\infty}^{\infty} \frac{x - t}{(x-t)^2 + y^2} \phi(t) \, dt, \qquad (x + iy \in \mathbb{C}_+).$$

Hence, by Fubini's theorem,

$$
\begin{aligned}
\int_{-\infty}^{\infty} V(x + iy) \phi(x) \, dx &= \int_{-\infty}^{\infty} \left(\frac{1}{\pi} \int_{-\infty}^{\infty} \frac{x - t}{(x-t)^2 + y^2} u(t) \, dt \right) \phi(x) \, dx \\
&= -\int_{-\infty}^{\infty} \left(\frac{1}{\pi} \int_{-\infty}^{\infty} \frac{t - x}{(t-x)^2 + y^2} \phi(x) \, dx \right) u(t) \, dt \\
&= -\int_{-\infty}^{\infty} \Phi(t + iy) u(t) \, dt.
\end{aligned}
$$

Therefore, by Hölder's inequality and the result of Case 3,

$$\left| \int_{-\infty}^{\infty} V(x + iy) \phi(x) \, dx \right| \leq \|\Phi_y\|_q \|u\|_p \leq C_q \|\phi\|_q \|u\|_p.$$

Divide by $\|\phi\|_q$ and then take the supremum with respect to all ϕ to obtain

$$\|V_y\|_p \leq C_q \|u\|_p.$$

Now, the passage from this inequality to

$$\|V_y\|_p \leq C_q \|U_y\|_p$$

is exactly as explained in Case 3 for the case $1 < p \leq 2$.

Case 5: Poisson representation.

Knowing that $V \in h^p(\mathbb{C}_+)$, $1 < p < \infty$, Theorem 11.6 ensures the existence of $v \in L^p(\mathbb{R})$ such that

$$V(x + iy) = \frac{1}{\pi} \int_{-\infty}^{\infty} \frac{y}{(x-t)^2 + y^2} \, v(t) \, dt, \qquad (x + iy \in \mathbb{C}_+),$$

and

$$\lim_{y \to 0} \|V_y - v\|_p.$$

On the other hand, by Corollary 11.11,

$$\lim_{y \to 0} V(x + iy) = v(x)$$

for almost all $x \in \mathbb{R}$. Moreover, by Corollary 14.4, we also have

$$\lim_{y \to 0} V(x + iy) = \tilde{u}$$

for almost all $x \in \mathbb{R}$. Hence, $\tilde{u} = v$ and we are done. $\square$

Corollary 14.8 *Let u be real and let $u \in L^p(\mathbb{R})$, $1 < p < \infty$. Let*

$$F(z) = \frac{1}{\pi i} \int_{-\infty}^{\infty} \frac{u(t)}{t - z} \, dt, \qquad (z \in \mathbb{C}_+).$$

Then $F \in H^p(\mathbb{C}_+)$ and

$$\|F\|_p \leq K_p \|u\|_p,$$

where the constant K_p just depends on p. Moreover,

$$f(t) = \lim_{y \to 0} F(t + iy) = u(t) + i\tilde{u}(t)$$

for almost all $t \in \mathbb{R}$.

Proof. Write $F = U + iV$. Hence $U = P * u$ and $V = Q * u$. By Corollary 10.12, $U \in h^p(\mathbb{C}_+)$. Moreover, the M. Riesz theorem ensures that $V \in h^p(\mathbb{C}_+)$ and $\|V_y\|_p \leq C_p \|U_y\|_p$. Hence,

$$\|F_y\|_p \leq \|U_y\|_p + \|V_y\|_p \leq (1 + C_p) \|U_y\|_p \leq (1 + C_p) \|u\|_p.$$

Since $U \longrightarrow u$ and $V \longrightarrow \tilde{u}$, almost everywhere on $\mathbb{R}$, the last assertion immediately follows. $\square$

Let u be real, $u \in L^p(\mathbb{R})$, $1 < p < \infty$ and let

$$F(z) = \frac{1}{\pi i} \int_{-\infty}^{\infty} \frac{u(t)}{t - z} \, dt, \qquad (z \in \mathbb{C}_+).$$

By the preceding result, $F \in H^p(\mathbb{C}_+)$, whose boundary values on the real line are given by $f = u + i\tilde{u}$. Thus, by Theorem 13.2, we also have

$$F(z) = \frac{1}{2\pi i} \int_{-\infty}^{\infty} \frac{u(t) + i\tilde{u}(t)}{t - z} \, dt, \qquad (z \in \mathbb{C}_+).$$

Therefore,

$$F(z) = \frac{1}{\pi} \int_{-\infty}^{\infty} \frac{\tilde{u}(t)}{t - z} \, dt, \qquad (z \in \mathbb{C}_+),$$

which is equivalent to

$$F(z) = -\frac{1}{\pi} \int_{-\infty}^{\infty} \frac{x - t}{(x - t)^2 + y^2} \, \tilde{u}(t) \, dt + \frac{i}{\pi} \int_{-\infty}^{\infty} \frac{y}{(x - t)^2 + y^2} \, \tilde{u}(t) \, dt.$$

Hence, for almost all $t \in \mathbb{R}$,

$$\lim_{y \to 0} F(t + iy) = -\tilde{\tilde{u}}(t) + i\tilde{u}(t).$$

But, we also know that

$$\lim_{y \to 0} F(t + iy) = u(t) + i\tilde{u}(t)$$

for almost all $t \in \mathbb{R}$. Therefore, we obtain the following result.

Corollary 14.9 *Let $u \in L^p(\mathbb{R})$, $1 < p < \infty$. Then*

$$\tilde{\tilde{u}} = -u.$$

Combining this result with M. Riesz's theorem immediately implies the following.

Corollary 14.10 *Let $u \in L^p(\mathbb{R})$, $1 < p < \infty$. Then*

$$c_p \|u\|_p \le \|\tilde{u}\|_p \le C_p \|u\|_p,$$

where c_p and C_p are constants just depending on p.

Corollaries 14.9 and 14.10 show that the map

$$
\begin{array}{ccc}
L^p(\mathbb{R}) & \longrightarrow & L^p(\mathbb{R}) \\
u & \longmapsto & \tilde{u}
\end{array}
$$

is an automorphism of the Banach space $L^p(\mathbb{R})$, $1 < p < \infty$. In the following we show that if $p = 2$, then this map is an isometry.

Corollary 14.11 *Let $1 < p < \infty$, $u \in L^p(\mathbb{R})$ and $v \in L^q(\mathbb{R})$, where q is the conjugate exponent of p. Then*

$$\int_{-\infty}^{\infty} \tilde{u}(x)\, \tilde{v}(x)\, dx = \int_{-\infty}^{\infty} u(x)\, v(x)\, dx$$

and

$$\int_{-\infty}^{\infty} \tilde{u}(x)\, v(x)\, dx = -\int_{-\infty}^{\infty} u(x)\, \tilde{v}(x)\, dx.$$

In particular, if $u \in L^2(\mathbb{R})$, then

$$\int_{-\infty}^{\infty} |u(x)|^2\, dx = \int_{-\infty}^{\infty} |\tilde{u}(x)|^2\, dx,$$

$$\int_{-\infty}^{\infty} u(x)\, \tilde{u}(x)\, dx = 0.$$

Proof. Without loss of generality, assume that u and v are real-valued. Let

$$F(z) = \frac{1}{\pi i} \int_{-\infty}^{\infty} \frac{u(t)}{t - z}\, dt, \qquad (z \in \mathbb{C}_+),$$

and

$$G(z) = \frac{1}{\pi i} \int_{-\infty}^{\infty} \frac{v(t)}{t - z}\, dt, \qquad (z \in \mathbb{C}_+).$$

By Corollary 14.8, $F \in H^p(\mathbb{C}_+)$ and $G \in H^q(\mathbb{C}_+)$. Hence

$$FG \in H^1(\mathbb{C}_+)$$

with boundary values

$$F(t)\, G(t) = \big(u(t) + i\tilde{u}(t)\big)\, \big(v(t) + i\tilde{v}(t)\big).$$

Hence, by Corollary 13.7,

$$\int_{-\infty}^{\infty} F(t)\, G(t)\, dt = 0.$$

The real and imaginary parts of this equality are the required identities. □

Exercises

Exercise 14.5.1 Let $u \in L^2(\mathbb{R})$, and let

$$U(x + iy) = \frac{1}{\pi} \int_{-\infty}^{\infty} \frac{y}{(x - t)^2 + y^2}\, u(t)\, dt, \qquad (x + iy \in \mathbb{C}_+),$$

$$V(x + iy) = \frac{1}{\pi} \int_{-\infty}^{\infty} \frac{x - t}{(x - t)^2 + y^2}\, u(t)\, dt, \qquad (x + iy \in \mathbb{C}_+).$$

Show that, for all $y > 0$,

$$\int_{-\infty}^{\infty} U^2(x + iy)\, dx = \int_{-\infty}^{\infty} V^2(x + iy)\, dx.$$

Exercise 14.5.2   Let $u,\, \tilde{u} \in L^1(\mathbb{R})$. Show that $\tilde{\tilde{u}} = -u$.
Hint: $f = u + i\tilde{u} \in H^1(\mathbb{R})$.

Exercise 14.5.3   Let $f \in H^p(\mathbb{R})$, $1 \leq p < \infty$. Show that $\tilde{f} = -if$ and thus $\tilde{f} \in H^p(\mathbb{R})$.
Hint: Use Exercise 14.5.2 and Corollary 14.9.

Exercise 14.5.4   Let $F \in H^p(\mathbb{R})$, $1 \leq p < \infty$, and let f denote its boundary value function on $\mathbb{R}$. Show that

$$F(z) = \frac{1}{2\pi} \int_{-\infty}^{\infty} \frac{\tilde{f}(t)}{t - z}\, dt, \qquad (z \in \mathbb{C}_+).$$

Hint: Use Exercise 14.5.3.

14.6   The Hilbert transform of $\mathrm{Lip}_{\alpha(t)}$ functions

Let $f : \mathbb{R} \longrightarrow \mathbb{C}$ be a continuous function. The *modulus of continuity* of f is, as similarly defined in (6.6),

$$\omega_f(t) = \sup_{|x - x'| \leq t} |f(x) - f(x')|.$$

Of course, the assumption $\omega_f(t) \longrightarrow 0$, as $t \to 0$, implies that f is uniformly continuous on $\mathbb{R}$. However, even for a bounded continuous function f on $\mathbb{R}$, $\omega_f(t)$ does not necessarily tend to zero as $t \to 0$.

As in the unit circle, f is called Lip_α, for some α with $0 < \alpha \leq 1$, if

$$\omega_f(t) = O(t^\alpha)$$

as $t \to 0^+$. In this section, our first goal is to replace t^α by $t^{\alpha(t)}$, where $\alpha(t)$ is a function defined for small values of t, which *properly* tends to α as $t \to 0^+$. Then we will study the Hilbert transform of these families. The contents of this section are from [14].

A *test function* $\alpha(t)$ is a real continuous function defined in a right neighborhood of zero, say $(0, t_0)$, such that, as $t \to 0^+$, we have

$$\alpha(t) = \alpha + o(1), \qquad\qquad\qquad (\alpha \in \mathbb{R}),$$

$$\int_0^t \tau^{\alpha(\tau) - \beta}\, d\tau = \frac{t^{\alpha(t) - \beta + 1}}{\alpha + 1 - \beta} + o(t^{\alpha(t) - \beta + 1}), \qquad (\beta < \alpha + 1),$$

$$\int_t^{t_0} \tau^{\alpha(\tau) - \beta}\, d\tau = \frac{t^{\alpha(t) - \beta + 1}}{\beta - \alpha - 1} + o(t^{\alpha(t) - \beta + 1}), \qquad (\beta > \alpha + 1).$$

The space of all continuous functions $f : \mathbb{R} \longrightarrow \mathbb{C}$ satisfying

$$\omega_f(t) = O(t^{\alpha(t)}), \qquad (t \to 0^+),$$

will be denoted by $\mathrm{Lip}_{\alpha(t)}$. Clearly, the classical space Lip_α is a special case corresponding to the test function $\alpha(t) \equiv \alpha$.

Given a test function $\alpha(t)$, let

$$\hat{\alpha}(t) = \alpha + \frac{\log\left(\int_t^{t_0} \tau^{\alpha(\tau)-\alpha-1}\, d\tau \right)}{\log t}.$$

We show that $\hat{\alpha}(t)$ is also a test function. We call it the test function *associated* with $\alpha(t)$. In the definition of a test function, we put two conditions dealing with cases $\beta < \alpha + 1$ and $\beta > \alpha + 1$. The associated test function is introduced to deal with the troublesome case $\beta = \alpha + 1$. Note that the associated test function is defined so that

$$t^{\hat{\alpha}(t)} = t^\alpha \int_t^{t_0} \tau^{\alpha(\tau)-\alpha-1}\, d\tau.$$

Lemma 14.12 *Let* $\alpha(t)$ *be a test function. Then* $\hat{\alpha}(t)$ *is also a test function.*

Proof. Clearly, $\hat{\alpha}(t)$ is a continuous function on $(0, t_0)$. Our first task is to show that $\lim_{t\to 0^+} \hat{\alpha}(t) = \alpha$. For each $\varepsilon > 0$, we have

$$t^{\hat{\alpha}(t)-\alpha} = \int_t^{t_0} \tau^{\alpha(\tau)-\alpha-1}\, d\tau = \int_t^{t_0} \tau^{\alpha(\tau)-\alpha-1-\varepsilon}\, \tau^\varepsilon\, d\tau$$

$$\geq t^\varepsilon \int_t^{t_0} \tau^{\alpha(\tau)-\alpha-1-\varepsilon}\, d\tau = t^\varepsilon \left(\frac{t^{\alpha(t)-\alpha-\varepsilon}}{\varepsilon} + o(t^{\alpha(t)-\alpha-\varepsilon}) \right).$$

Hence, $\liminf_{t\to 0^+} t^{\hat{\alpha}(t)-\alpha(t)} \geq \frac{1}{\varepsilon}$. Let $\varepsilon \to 0$ to get

$$\lim_{t\to 0^+} t^{\hat{\alpha}(t)-\alpha(t)} = \infty. \tag{14.13}$$

Let $\lambda \in \mathbb{R}$ and let

$$\varphi_\lambda(t) = t^{\hat{\alpha}(t)-\alpha+\lambda}.$$

Then, we have

$$\varphi_\lambda'(t) = t^{\lambda-1-\alpha+\alpha(t)} \left(\lambda t^{\hat{\alpha}(t)-\alpha(t)} - 1 \right).$$

If $\lambda \leq 0$, then $\varphi_\lambda' < 0$ and thus φ_λ is strictly decreasing. On the other hand, if $\lambda > 0$, by (14.13), $\varphi_\lambda'(t) > 0$ for small values of t and thus φ_λ is strictly increasing.

Fix $\varepsilon > 0$. For small enough τ, say $0 < \tau < \tau_0 < \min\{1, t_0\}$, we have

$$\alpha - \varepsilon \leq \alpha(\tau) \leq \alpha + \varepsilon.$$

Hence,

$$\tau^{\varepsilon-1} \leq \tau^{\alpha(\tau)-\alpha-1} \leq \tau^{-\varepsilon-1}.$$

For $0 < t < \tau_0$, we thus get

$$\frac{\tau_0^\varepsilon}{\varepsilon} - \frac{t^\varepsilon}{\varepsilon} = \int_t^{\tau_0} \tau^{\varepsilon-1} \, d\tau \leq \int_t^{\tau_0} \tau^{\alpha(\tau)-\alpha-1} \, d\tau \leq \int_t^{\tau_0} \tau^{-\varepsilon-1} \, d\tau = \frac{t^{-\varepsilon}}{\varepsilon} - \frac{\tau_0^{-\varepsilon}}{\varepsilon}.$$

Therefore, for $0 < t < \tau_0$,

$$\frac{\log\left(\frac{t^{-\varepsilon}}{\varepsilon} - \frac{\tau_0^{-\varepsilon}}{\varepsilon}\right)}{\log t} \leq \frac{\log\left(\int_t^{\tau_0} \tau^{\alpha(\tau)-\alpha-1} \, d\tau\right)}{\log t} \leq \frac{\log\left(\frac{\tau_0^\varepsilon}{\varepsilon} - \frac{t^\varepsilon}{\varepsilon}\right)}{\log t},$$

which implies

$$-\varepsilon \leq \liminf_{t\to 0^+} \frac{\log\left(\int_t^{t_0} \tau^{\alpha(\tau)-\alpha-1} \, d\tau\right)}{\log t} \leq \limsup_{t\to 0^+} \frac{\log\left(\int_t^{t_0} \tau^{\alpha(\tau)-\alpha-1} \, d\tau\right)}{\log t} \leq 0.$$

Now, let $\varepsilon \to 0$ to get

$$\lim_{t\to 0^+} \frac{\log\left(\int_t^{t_0} \tau^{\alpha(\tau)-\alpha-1} \, d\tau\right)}{\log t} = 0.$$

Hence,

$$\lim_{t\to 0^+} \hat{a}(t) = \alpha.$$

Let $\beta < \alpha + 1$. Then

$$\int_0^t \tau^{\hat{a}(\tau)-\beta} \, d\tau = \int_0^t \tau^{\alpha-\beta} \, \tau^{\hat{a}(\tau)-\alpha} \, d\tau$$

$$\geq t^{\hat{a}(t)-\alpha} \int_0^t \tau^{\alpha-\beta} \, d\tau = \frac{t^{\hat{a}(t)-\beta+1}}{\alpha - \beta + 1}.$$

On the other hand, for $0 < \varepsilon < 1 + \alpha - \beta$, we have

$$\int_0^t \tau^{\hat{a}(\tau)-\beta} \, d\tau = \int_0^t \tau^{\alpha-\beta-\varepsilon} \, \tau^{\hat{a}(\tau)-\alpha+\varepsilon} \, d\tau$$

$$\leq t^{\hat{a}(t)-\alpha+\varepsilon} \int_0^t \tau^{\alpha-\beta-\varepsilon} \, d\tau = \frac{t^{\hat{a}(t)-\beta+1}}{\alpha - \beta + 1 - \varepsilon}.$$

Hence,

$$\int_0^t \tau^{\hat{a}(\tau)-\beta} \, d\tau = \frac{t^{\hat{a}(t)-\beta+1}}{\alpha - \beta + 1} + o\left(t^{\hat{a}(t)-\beta+1}\right).$$

Finally, let $\beta > \alpha + 1$. Then

$$\int_t^{t_0} \tau^{\hat{a}(\tau)-\beta} \, d\tau = \int_t^{t_0} \tau^{\alpha-\beta} \, \tau^{\hat{a}(\tau)-\alpha} \, d\tau$$

$$\leq t^{\hat{a}(t)-\alpha} \int_t^{t_0} \tau^{\alpha-\beta} \, d\tau \leq \frac{t^{\hat{a}(t)-\beta+1}}{\beta - \alpha - 1}.$$

On the other hand, for each $\varepsilon > 0$, we have

$$\int_t^{\tau_0} \tau^{\hat{\alpha}(\tau)-\beta}\,d\tau = \int_t^{\tau_0} \tau^{\alpha-\beta-\varepsilon}\,\tau^{\hat{\alpha}(\tau)-\alpha+\varepsilon}\,d\tau$$

$$\geq t^{\hat{\alpha}(t)-\alpha+\varepsilon}\int_t^{\tau_0}\tau^{\alpha-\beta-\varepsilon}\,d\tau = \frac{t^{\hat{\alpha}(t)-\beta+1}}{\beta-\alpha-1+\varepsilon}\left(1-(t/\tau_0)^{\beta+\varepsilon-\alpha-1}\right).$$

Hence,

$$\int_t^{t_0}\tau^{\hat{\alpha}(\tau)-\beta}\,d\tau = \frac{t^{\hat{\alpha}(t)-\beta+1}}{\beta-\alpha-1}+o(\,t^{\hat{\alpha}(t)-\beta+1}\,).$$

$\square$

In the following, the notion $f(t)\asymp g(t)$, as $t\to 0^+$, means that there are constants $c, C > 0$ such that

$$c\,g(t)\leq f(t)\leq C\,g(t)$$

in a right neighborhood of zero.

Corollary 14.13 *Let* $\alpha(t)$ *be a test function with* $\lim_{t\to 0^+}\alpha(t)=\alpha$. *Let* β,γ,C *be real constants such that* $C > 0$, $\gamma > 0$ *and* $\alpha > \beta-1$. *Then*

$$\int_0^{t_0}\frac{\tau^{\alpha(\tau)-\beta}}{\tau^\gamma+C\,t^\gamma}\,d\tau \asymp \begin{cases} t^{\alpha(t)+1-\beta-\gamma} & \text{if } \alpha < \gamma+\beta-1, \\[2mm] t^{\hat{\alpha}(t)-\alpha} & \text{if } \alpha = \gamma+\beta-1, \\[2mm] 1 & \text{if } \alpha > \gamma+\beta-1, \end{cases}$$

as $t\to 0^+$.

Proof. We decompose the integral over two intervals $(0,t)$ and (t,t_0) and then we use our assumptions on $\alpha(t)$. Hence,

$$\int_0^{t_0}\frac{\tau^{\alpha(\tau)-\beta}}{\tau^\gamma+C\,t^\gamma}\,d\tau = \left(\int_0^t+\int_t^{t_0}\right)\frac{\tau^{\alpha(\tau)-\beta}}{\tau^\gamma+C\,t^\gamma}\,d\tau$$

$$\asymp \int_0^t\frac{\tau^{\alpha(\tau)-\beta}}{t^\gamma}\,d\tau + \int_t^{t_0}\frac{\tau^{\alpha(\tau)-\beta}}{\tau^\gamma}\,d\tau$$

$$\asymp \frac{t^{\alpha(\tau)-\beta+1}}{t^\gamma}+t^{\alpha(\tau)-\beta-\gamma+1}.$$

The first estimate is true since $\alpha > \beta-1$, and the second one holds since $\alpha < \beta+\gamma-1$. Similarly, by (14.13), we have

$$\int_0^{t_0}\frac{\tau^{\alpha(\tau)-\alpha-1+\gamma}}{\tau^\gamma+C\,t^\gamma}\,d\tau = \left(\int_0^t+\int_t^{t_0}\right)\frac{\tau^{\alpha(\tau)-\alpha-1+\gamma}}{\tau^\gamma+C\,t^\gamma}\,d\tau$$

$$\asymp \int_0^t\frac{\tau^{\alpha(\tau)-\alpha-1+\gamma}}{t^\gamma}\,d\tau + \int_t^{t_0}\frac{\tau^{\alpha(\tau)-\alpha-1+\gamma}}{\tau^\gamma}\,d\tau$$

$$\asymp t^{\alpha(t)-\alpha}+t^{\hat{\alpha}(t)-\alpha}\asymp t^{\hat{\alpha}(t)-\alpha}.$$

The last case is proved similarly. $\square$

Let

$$\alpha(t) = \alpha - \alpha_1 \frac{\log_2 1/t}{\log 1/t} - \alpha_2 \frac{\log_3 1/t}{\log 1/t} - \cdots - \alpha_n \frac{\log_{n+1} 1/t}{\log 1/t}, \qquad (14.14)$$

where $\alpha, \alpha_1, \alpha_2, \ldots, \alpha_n \in \mathbb{R}$ and $\log_n = \log \log \cdots \log$ (n times). For this function, we have

$$t^{\alpha(t)} = t^\alpha \, (\log 1/t)^{\alpha_1} \, (\log_2 1/t)^{\alpha_2} \cdots (\log_n 1/t)^{\alpha_n}.$$

We give a sufficient and simple criterion showing at least that the function defined in (14.14) is a test function.

Theorem 14.14 *Let* $\alpha \in \mathbb{R}$ *and let* $\alpha(t)$ *be a real continuously differentiable function defined on* $(0, t_0)$ *with* $\lim_{t \to 0+} \alpha(t) = \alpha$. *Suppose that*

$$\lim_{t \to 0+} \alpha'(t) \, t \log t = 0.$$

Then $\alpha(t)$ *is a test function.*

Proof. Let

$$\frac{d(\, t^{\alpha(t) - \gamma}\,)}{dt} = (\, \alpha(t) - \gamma + \alpha'(t) \, t \log t \,) \, t^{\alpha(t) - \gamma - 1},$$

and note that

$$\alpha(t) - \gamma + \alpha'(t) \, t \log t \longrightarrow \alpha - \gamma$$

as $t \to 0^+$. Hence, for $\beta < \alpha + 1$, we have

$$
\begin{aligned}
\int_0^t \tau^{\alpha(\tau) - \beta} \, d\tau &= \int_0^t \tau^{\alpha(\tau) - \alpha} \, \tau^{\alpha - \beta} \, d\tau \\
&= \frac{t^{\alpha(t) - \beta + 1}}{\alpha - \beta + 1} - \frac{1}{\alpha - \beta + 1} \int_0^t (\, \alpha(\tau) - \alpha + \alpha'(\tau) \, \tau \log \tau \,) \, \tau^{\alpha(\tau) - \beta} \, d\tau \\
&= \frac{t^{\alpha(t) - \beta + 1}}{\alpha - \beta + 1} + o(1) \int_0^t \tau^{\alpha(\tau) - \beta} \, d\tau.
\end{aligned}
$$

Therefore,

$$(\, 1 + o(1) \,) \int_0^t \tau^{\alpha(\tau) - \beta} \, d\tau = \frac{t^{\alpha(t) - \beta + 1}}{\alpha - \beta + 1}.$$

Thus the second required condition is satisfied. For $\beta > \alpha + 1$, we have

$$
\begin{aligned}
\int_t^{t_0} \tau^{\alpha(\tau) - \beta} \, d\tau &= \int_t^{t_0} \tau^{\alpha(\tau) - \alpha} \, \tau^{\alpha - \beta} \, d\tau \\
&= \frac{\tau^{\alpha(\tau) - \beta + 1}}{\alpha - \beta + 1} \Bigg|_{\tau = t}^{\tau = t_0} - \int_t^{t_0} (\, \alpha(\tau) - \alpha + \alpha'(\tau) \, \tau \log \tau \,) \, \tau^{\alpha(\tau) - \beta} \, d\tau \\
&= O(1) - \frac{t^{\alpha(t) - \beta + 1}}{\alpha - \beta + 1} + o(1) \int_t^{t_0} \tau^{\alpha(\tau) - \beta} \, d\tau.
\end{aligned}
$$

Hence,

$$(1 + o(1)) \int_t^{t_0} \tau^{\alpha(\tau) - \beta} \, d\tau = \frac{t^{\alpha(t) - \beta + 1}}{\beta - \alpha - 1}.$$

Thus the third condition is also fulfilled.   □

Corollary 14.15 *Let* $\alpha, \alpha_1, \alpha_2, \ldots, \alpha_n$ *be real constants, and let*

$$\alpha(t) = \alpha - \alpha_1 \frac{\log_2 1/t}{\log 1/t} - \alpha_2 \frac{\log_3 1/t}{\log 1/t} - \cdots - \alpha_n \frac{\log_{n+1} 1/t}{\log 1/t}.$$

Then $\alpha(t)$ *is a test function. Moreover,*

$$t^{\hat{\alpha}(t)} \asymp \begin{cases} t^\alpha & if \ \ \alpha_1 < -1, \\ t^\alpha \, (\log 1/t)^{1+\alpha_1} \, (\log_2 1/t)^{\alpha_2} \cdots (\log_n 1/t)^{\alpha_n} & if \ \ \alpha_1 > -1. \end{cases}$$

If $\alpha_1 = \alpha_2 = \cdots = \alpha_{k-1} = -1$, *then*

$$t^{\hat{\alpha}(t)} \asymp \begin{cases} t^\alpha & if \ \ \alpha_k < -1, \\ t^\alpha \, (\log_k 1/t)^{1+\alpha_k} \, (\log_{k+1} 1/t)^{\alpha_{k+1}} \cdots (\log_n 1/t)^{\alpha_n} & if \ \ \alpha_k > -1. \end{cases}$$

Finally, if $\alpha_1 = \alpha_2 = \cdots = \alpha_n = -1$, *then*

$$t^{\hat{\alpha}(t)} \asymp t^\alpha \log_{n+1} 1/t.$$

The following theorems are two celebrated results about the Hilbert transform of bounded Lipschitz functions.

Theorem 14.16 (Privalov [16]) *If* u *is a bounded* Lip_α $(0 < \alpha < 1)$ *function on* $\mathbb{R}$, *then* $\tilde{u}(x)$ *exists for all* $x \in \mathbb{R}$ *and besides* $\tilde{u}$ *is also* Lip_α.

Theorem 14.17 (Titchmarsh [20]) *If* u *is a bounded* Lip_1 *function on* $\mathbb{R}$, *then* $\tilde{u}(x)$ *exists for all* $x \in \mathbb{R}$ *and besides*

$$|\, \tilde{u}(x + t) - \tilde{u}(x) \,| \leq C \, t \log 1/t$$

for all $x \in \mathbb{R}$ *and* $0 < t < 1/2$.

We are now able to generalize both theorems. The following theorem is the main result of this section.

Theorem 14.18 *Let* $\alpha(t)$ *be a test function with*

$$0 < \lim_{t \to 0^+} \alpha(t) \leq 1.$$

Let u *be a bounded* $\mathrm{Lip}_{\alpha(t)}$ *function on* $\mathbb{R}$. *Then* $\tilde{u}(x)$ *exists for all* $x \in \mathbb{R}$, *and besides*

$$\tilde{u} \ is \ \begin{cases} \mathrm{Lip}_{\alpha(t)} & if \ \ 0 < \lim_{t \to 0^+} \alpha(t) < 1, \\ \mathrm{Lip}_{\hat{\alpha}(t)} & if \ \ \ \ \lim_{t \to 0^+} \alpha(t) = 1. \end{cases}$$

Proof. Let

$$V(x+iy) = \frac{1}{\pi} \int_{-\infty}^{\infty} \left(\frac{x-t}{(x-t)^2+y^2} + \frac{t}{1+t^2} \right) u(t) \, dt, \qquad (x+iy \in \mathbb{C}_+).$$

To estimate $|\tilde{u}(x+t) - \tilde{u}(x)|$, instead of taking the real line as our straight path to go from x to $x+t$, we go from x up to $x+it$, then to $x+t+it$ and finally down to $x+t$:

$$\begin{aligned}
|\tilde{u}(x) - \tilde{u}(x+t)| &\leq |\tilde{u}(x) - V(x+it)| \qquad\qquad (14.15) \\
&+ |V(x+it) - V(x+t+it)| \\
&+ |V(x+t+it) - \tilde{u}(x+t)|.
\end{aligned}$$

Hence we proceed to study each term of the right side.

By (14.5), we have

$$\begin{aligned}
|\tilde{u}(x) - V(x+it)| &= \left| \lim_{\varepsilon \to 0} \frac{1}{\pi} \int_{|x-\tau|>\varepsilon} \left(\frac{1}{x-\tau} + \frac{\tau}{1+\tau^2} \right) u(\tau) \, d\tau \right. \\
&\quad \left. - \frac{1}{\pi} \int_{-\infty}^{\infty} \left(\frac{x-\tau}{(x-\tau)^2+t^2} + \frac{\tau}{1+\tau^2} \right) u(\tau) \, d\tau \right| \\
&= \left| \lim_{\varepsilon \to 0} \frac{1}{\pi} \int_{|x-\tau|>\varepsilon} \left(\frac{1}{x-\tau} - \frac{x-\tau}{(x-\tau)^2+t^2} \right) u(\tau) \, d\tau \right| \\
&= \left| \lim_{\varepsilon \to 0} \frac{t^2}{\pi} \int_{\varepsilon}^{\infty} \frac{u(x-\tau) - u(x+\tau)}{\tau\,(\tau^2+t^2)} \, d\tau \right| \\
&\leq \frac{t^2}{\pi} \int_{0}^{\infty} \frac{|u(x+\tau) - u(x-\tau)|}{\tau\,(\tau^2+t^2)} \, d\tau.
\end{aligned}$$

Since u is *bounded* and $Lip_{\alpha(t)}$, there is a constant C such that

$$|u(x+\tau) - u(x-\tau)| \leq C\,(2\tau)^{\alpha(2\tau)}$$

for *all* values of $x \in \mathbb{R}$, and small values of τ, say $0 < \tau < 1$. Hence, by Corollary 14.13, we have

$$\begin{aligned}
|\tilde{u}(x) - V(x+it)| &\leq \frac{t^2}{\pi} \int_{0}^{1} \frac{C\,(2\tau)^{\alpha(2\tau)}}{\tau\,(\tau^2+t^2)} \, d\tau + \frac{t^2}{\pi} \int_{1}^{\infty} \frac{2\|u\|_\infty}{\tau^3} \, d\tau \\
&\leq \frac{4C\,t^2}{\pi} \int_{0}^{2} \frac{\tau^{\alpha(\tau)-1}}{\tau^2+4t^2} \, d\tau + \frac{\|u\|_\infty}{\pi}\, t^2 \\
&\leq C'\, t^{\alpha(t)} + C''\, t^2.
\end{aligned}$$

Hence, for each $x \in \mathbb{R}$,

$$|\tilde{u}(x) - V(x+it)| = O(t^{\alpha(t)}). \qquad (14.16)$$

By the mean value theorem,

$$|V(x+it) - V(x+t+it)| \leq t \sup_{s \in \mathbb{R}} \left| \frac{\partial V}{\partial x}(s+it) \right|,$$

where

$$
\begin{aligned}
\frac{\partial V}{\partial x}(s+it) &= \frac{1}{\pi} \int_{-\infty}^{\infty} \frac{\partial}{\partial x}\left(\frac{x-\tau}{(x-\tau)^2+t^2} + \frac{\tau}{1+\tau^2}\right)\bigg|_{x=s} u(\tau)\, d\tau \\
&= \frac{1}{\pi} \int_{-\infty}^{\infty} \frac{t^2-(s-\tau)^2}{(t^2+(s-\tau)^2)^2} u(\tau)\, d\tau \\
&= \frac{1}{\pi} \int_{-\infty}^{\infty} \frac{t^2-\tau^2}{(t^2+\tau^2)^2} u(s+\tau)\, d\tau \\
&= \frac{1}{\pi} \int_{-\infty}^{\infty} \frac{t^2-\tau^2}{(t^2+\tau^2)^2} \left(u(s+\tau)-u(s)\right) d\tau.
\end{aligned}
$$

Hence,

$$
\begin{aligned}
|V(x+it)-V(x+t+it)| &\leq \sup_{s\in\mathbb{R}} \frac{t}{\pi} \int_{-\infty}^{\infty} \frac{|u(s+\tau)-u(s)|}{t^2+\tau^2}\, d\tau \\
&\leq Ct \int_{0}^{1} \frac{\tau^{\alpha(\tau)}}{t^2+\tau^2}\, d\tau + \frac{t}{\pi} \int_{1}^{\infty} \frac{2\,\|u\|_\infty}{\tau^2}\, d\tau.
\end{aligned}
$$

The asymptotic behavior of

$$
\int_{0}^{1} \frac{\tau^{\alpha(\tau)}}{t^2+\tau^2}\, d\tau
$$

depends on $\lim_{t\to 0+}\alpha(t)$. According to Corollary 14.13, we have

$$
|V(x+it)-V(x+t+it)| = \begin{cases} O(t^{\alpha(t)}) & \text{if } \lim_{t\to 0+}\alpha(t) < 1, \\ O(t^{\hat{\alpha}(t)}) & \text{if } \lim_{t\to 0+}\alpha(t) = 1. \end{cases} \tag{14.17}
$$

Finally, (14.15), (14.16) and (14.17) give the required result. $\square$

In the light of Corollary 14.15 and Theorem 14.18, we get the following special result.

Corollary 14.19 *Let $\alpha_1,\alpha_2,\ldots,\alpha_n \in \mathbb{R}$. Let u be a bounded function on $\mathbb{R}$ with*

$$
\omega_u(t) = O\big(t\,(\log 1/t)^{\alpha_1}\,(\log_2 1/t)^{\alpha_2}\cdots(\log_n 1/t)^{\alpha_n}\big).
$$

Then $\tilde{u}(x)$ exists for all values of $x \in \mathbb{R}$ and besides

$$
\omega_{\tilde{u}}(t) = \begin{cases} O(t) & \text{if } \alpha_1 < -1, \\ O\big(t\,(\log 1/t)^{1+\alpha_1}\,(\log_2 1/t)^{\alpha_2}\cdots(\log_n 1/t)^{\alpha_n}\big) & \text{if } \alpha_1 > -1. \end{cases}
$$

If $\alpha_1 = \alpha_2 = \cdots = \alpha_{k-1} = -1$, then

$$
\omega_{\tilde{u}}(t) = \begin{cases} O(t) & \text{if } \alpha_k < -1, \\ O\big(t\,(\log_k 1/t)^{1+\alpha_k}\,(\log_{k+1} 1/t)^{\alpha_{k+1}}\cdots(\log_n 1/t)^{\alpha_n}\big) & \text{if } \alpha_k > -1. \end{cases}
$$

Finally, if $\alpha_1 = \alpha_2 = \cdots = \alpha_n = -1$, then

$$
\omega_{\tilde{u}}(t) = O(t\log_{n+1} 1/t).
$$

Exercises

Exercise 14.6.1 Let $\alpha(t)$ be a test function and let

$$\beta(t) = \alpha(t) + \frac{\log\left(\int_t^{t_0} \tau^{\alpha(\tau)-\alpha-1}\, d\tau\right)}{\log t}.$$

Show that $\beta(t)$ is also a test function.

Exercise 14.6.2 Let

$$u(t) = \begin{cases} |t|\, \log 1/|t| & \text{if } |t| \le 1/e, \\ 1/e & \text{if } |t| \ge 1/e, \end{cases}$$

and let

$$\alpha(t) = 1 - \frac{\log_2 1/t}{\log 1/t}, \qquad (0 < t < 1/e),$$

or equivalently,

$$t^{\alpha(t)} = t \log 1/t.$$

Show that u is a bounded $\mathrm{Lip}_{\alpha(t)}$ function on $\mathbb{R}$ and that $\tilde{u}(t) - \tilde{u}(0)$ behaves asymptotically like $t^{\hat{\alpha}(t)} \asymp t\, (\log 1/t)^2$, as $t \to 0^+$.
Hint: Write

$$\tilde{u}(t) - \tilde{u}(0) = \left(\tilde{u}(t) - V(t+it)\right) + \left(V(t+it) - V(it)\right) + \left(V(it) - \tilde{u}(0)\right)$$

and show that

$$|\tilde{u}(t) - V(t+it)| = O(t^{\alpha(t)}), \qquad |V(it) - \tilde{u}(0)| = O(t^{\alpha(t)})$$

and

$$|V(t+it) - V(it)| = O(t^{\hat{\alpha}(t)}).$$

Remark: This exercise shows that Corollary 14.19 gives a *sharp* result.

14.7 Maximal functions

Let u be a measurable function on $\mathbb{R}$. Then the Hardy–Littlewood maximal function of u is defined by

$$Mu(t) = \sup_{k,k'>0} \frac{1}{k+k'} \int_{t-k'}^{t+k} |u(x)|\, dx, \qquad (t \in \mathbb{R}).$$

If U is defined on the upper half plane, we define a family of maximal functions related to U. Remember that a Stoltz domain anchored at $t \in \mathbb{R}$ is of the form

$$\Gamma_\alpha(t) = \{x + iy \in \mathbb{C}_+ : |x - t| < \alpha y\}, \qquad (\alpha > 0).$$

Then the nontangential maximal function of U is given by

$$M_\alpha U(t) = \sup_{z \in \Gamma_\alpha(t)} |U(z)|, \qquad (t \in \mathbb{R}),$$

and its radial maximal function by

$$M_0 U(t) = \sup_{y>0} |U(t+iy)|, \qquad (t \in \mathbb{R}).$$

If $u \in L^1(dt/(1+t^2))$, we first apply the Poisson integral formula to extend u to the upper half plane

$$U(x+iy) = \frac{1}{\pi} \int_{-\infty}^{\infty} \frac{y}{(x-t)^2 + y^2}\, u(t)\, dt, \qquad (x+iy \in \mathbb{C}_+).$$

Then we will use $M_\alpha u$ and $M_0 u$ respectively instead of $M_\alpha U$ and $M_0 U$. By definition, we clearly have

$$M_0 u(t) \le M_\alpha u(t), \qquad (t \in \mathbb{R}),$$

for all $\alpha > 0$. Moreover, since at almost all points of $\mathbb{R}$, $U(t+iy)$ tends to $u(t)$ as $y \to 0$ (Lemma 3.10), we also have

$$|u(t)| \le M_0 u(t)$$

for almost all $t \in \mathbb{R}$. We now obtain a more delicate relation between $M_\alpha u$ and Mu.

Lemma 14.20 *Let $u \in L^1(dt/(1+t^2))$ and let $\alpha \ge 0$. Then*

$$M_\alpha u(t) \le \left(1 + \frac{2\alpha}{\pi}\right) Mu(t)$$

for all $t \in \mathbb{R}$.

Proof. We give the proof for $\alpha > 0$. The case $\alpha = 0$ follows immediately.

Fix $t \in \mathbb{R}$. According to the definition of Mu

$$\int_{-\infty}^{\infty} \varphi(s)\, |u(s)|\, ds \le \left(\int_{-\infty}^{\infty} \varphi(s)\, ds \right) Mu(t), \qquad (14.18)$$

where φ is the step function

$$\varphi(s) = \begin{cases} 1 & \text{if } |t-s| < k, \\ 0 & \text{if } |t-s| \ge k. \end{cases}$$

By considering the linear combination of such step functions (with positive coefficients) we see that (14.18) still holds for any step function which is symmetric with respect to the vertical axis $s = t$ and is decreasing on $s > t$. (See Figure 14.6.)

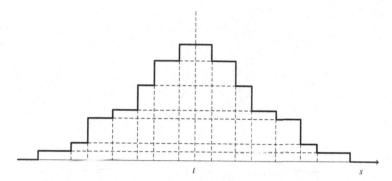

Fig. 14.6. A symmetric step function.

Finally, if φ is any positive and integrable function with the preceding properties, we can approximate it from below by a sequence of such step functions and thus a simple application of Fatou's lemma shows that (14.18) indeed holds for any positive and integrable function which is symmetric with respect to the vertical axis $s = t$ and is decreasing on $s > t$.

Now let $z = x + iy \in \Gamma_\alpha(t)$. Then

$$|U(x + iy)| \leq \frac{1}{\pi} \int_{-\infty}^{\infty} \frac{y}{(x - s)^2 + y^2} \, |u(s)| \, ds, \qquad (x + iy \in \mathbb{C}_+).$$

But, if $x \neq t$, the kernel $P_y(x - s)$ is not symmetric with respect to the axis $s = t$. (See Figure 14.7.) Nevertheless, we can overcome this difficulty. Without loss of generality assume that $x > t$. Let

$$\varphi(s) = \varphi_z(s) = \begin{cases} \dfrac{1}{\pi} \dfrac{y}{(x - s)^2 + y^2} & \text{if} \quad s \geq x, \\[2ex] \dfrac{1}{\pi y} & \text{if} \quad 2t - x < s < x, \\[2ex] \dfrac{1}{\pi} \dfrac{y}{(2t - x - s)^2 + y^2} & \text{if} \quad s \leq 2t - x. \end{cases}$$

(See Figure 14.8.) On the one hand, φ is positive and symmetric with respect to the axis $s = t$, and on the other hand it majorizes the Poisson kernel. Moreover,

$$\int_{-\infty}^{\infty} \varphi(s) \, ds = 1 + \frac{2|x - t|}{\pi y} \leq 1 + \frac{2\alpha}{\pi}.$$

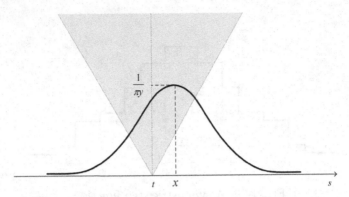

Fig. 14.7. The kernel $P_y(x - s)$.

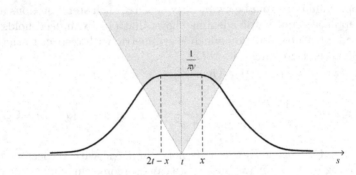

Fig. 14.8. The graph of $\varphi_z(s)$.

Hence, for each $z = x + iy \in \Gamma_\alpha(t)$,

$$|U(x + iy)| \le \int_{-\infty}^{\infty} \varphi(s)\, |u(s)|\, ds \le \left(1 + \frac{2\alpha}{\pi}\right) Mu(t).$$

$\square$

Considering the preceding chain of inequalities, we see that, for any function $u \in L^1(dt/(1 + t^2))$,

$$|u| \le M_0 u \le M_\alpha u \le \left(1 + \frac{2\alpha}{\pi}\right) Mu \qquad (14.19)$$

almost everywhere on $\mathbb{R}$. Hence, $Mu \in L^p(\mathbb{R})$ immediately implies that $M_\alpha u$, $M_0 u$ and u itself are also in $L^p(\mathbb{R})$. For the reverse implication (which holds for $1 < p < \infty$) we need the following celebrated result of Hardy–Littlewood. A measurable function u on $\mathbb{R}$ is locally integrable if it is integrable on each finite interval, i.e.

$$\int_a^b |f(t)|\, dt < \infty$$

for all $-\infty < a < b < \infty$. For the definition of distribution function m_φ, see Section A.5.

Theorem 14.21 (Hardy–Littlewood) *Let u be locally integrable on $\mathbb{R}$. Then, for each $\lambda > 0$,*

$$m_{Mu}(\lambda) \leq \frac{2}{\lambda} \int_{\{Mu > \lambda\}} |u(x)|\, dx.$$

Proof. (F. Riesz) Since $Mu = M|u|$, we may assume that $u \geq 0$. Let

$$M_r u(x) = \sup_{k>0} \frac{1}{k} \int_x^{x+k} u(t)\, dt$$

and

$$M_\ell u(x) = \sup_{k>0} \frac{1}{k} \int_{x-k}^{x} u(t)\, dt.$$

The notations M_r and M_ℓ are temporary for this proof and they do not mean M_α, as defined before, for a specific value of α.

Fix $\lambda > 0$ and consider

$$
\begin{aligned}
E &= \{\, x : Mu(x) > \lambda \,\}, \\
E_r &= \{\, x : M_r u(x) > \lambda \,\}, \\
E_\ell &= \{\, x : M_\ell u(x) > \lambda \,\}.
\end{aligned}
$$

Since $\int_a^b u$ is a continuous function of a and b we have $E_r \subset E$ and $E_\ell \subset E$. On the other hand, the identity

$$\frac{1}{h+k} \int_{x-k}^{x+h} u(t)\, dt = \frac{h}{h+k}\left(\frac{1}{h} \int_x^{x+h} u(t)\, dt \right) + \frac{k}{h+k}\left(\frac{1}{k} \int_{x-k}^{x} u(t)\, dt \right)$$

implies $E \subset E_r \cup E_\ell$. Hence,

$$E = E_r \cup E_\ell,$$

which yields

$$m_{Mu}(\lambda) \leq m_{M_r u}(\lambda) + m_{M_\ell u}(\lambda).$$

We show that

$$m_{M_r u}(\lambda) = \frac{1}{\lambda} \int_{E_r} |u(x)|\, dx \qquad (14.20)$$

and that

$$m_{M_\ell u}(\lambda) = \frac{1}{\lambda} \int_{E_\ell} |u(x)|\, dx \qquad (14.21)$$

and thus the theorem follows.

For each fixed $k > 0$, the function

$$\varphi_k(x) = \frac{1}{k} \int_x^{x+k} u(t)\, dt, \qquad (x \in \mathbb{R}),$$

is continuous on $\mathbb{R}$, and thus $M_r u = \sup_{k>0} \varphi_k$ is lower semicontinuous on $\mathbb{R}$. Hence, E_r is an open set. Let

$$\Psi(x) = \int_0^x u(t)\, dt, \qquad (x \in \mathbb{R}).$$

Thus Ψ is an increasing continuous function on $\mathbb{R}$ and the open set E_r is a countable union of disjoint intervals I_k. As F. Riesz observed, if one shines light downward on the graph of Ψ with slope λ, then the I_k are the parts of the x-axis where the graph of Ψ remains in shadow. From this geometric observation we immediately have

$$\lambda |I_k| = \Psi(b_k) - \Psi(a_k) = \int_{I_k} u(t)\, dt,$$

where $I_k = (a_k, b_k)$. Summing up over all I_k gives (14.20). The proof of (14.21) is similar. $\square$

Theorem 14.21 has several interesting consequences. The first corollary immediately follows from the theorem.

Corollary 14.22 *Let* $u \in L^1(\mathbb{R})$. *Then, for each* $\lambda > 0$,

$$m_{Mu}(\lambda) \le \frac{2}{\lambda} \|u\|_1.$$

Corollary 14.23 *Let* $u \in L^p(\mathbb{R})$, $1 < p < \infty$. *Then* $Mu \in L^p$ *and moreover*

$$\|Mu\|_p \le \frac{2p}{p-1} \|u\|_p.$$

Remark: The corollary clearly holds for $p = \infty$ with $\|Mu\|_\infty \le \|u\|_\infty$.

Proof. Without loss of generality, assume that $u \ge 0$. Let

$$u_n(t) = \begin{cases} u(t) & \text{if } |t| \le n \text{ and } |u(t)| \le n, \\ 0 & \text{otherwise.} \end{cases}$$

Hence,

$$0 \le u_1 \le u_2 \le \cdots \le u$$

and, for each $x \in \mathbb{R}$, $u_n(x) \to u(x)$ as $n \to \infty$. Therefore,

$$0 \le Mu_1 \le Mu_2 \le \cdots \le Mu$$

and, by the monotone convergence theorem, for each $x \in \mathbb{R}$, $Mu_n(x) \to Mu(x)$ as $n \to \infty$. Thus, if the theorem holds for each u_n, an application of Fatou's lemma immediately implies that it also holds for u.

Based on the observation made in the last paragraph, we assume that u is *bounded* and of *compact support*, say supp $u \subset [-A, A]$. Hence, the rough estimate

$$\frac{1}{k + k'} \int_{x-k'}^{x+k} u(t)\, dt \leq \frac{1}{|x| - A} \int_{-A}^{A} u(t)\, dt \leq \frac{4A\|u\|_\infty}{|x|}$$

for $|x| \geq 2A$ implies that

$$Mu(x) \leq \frac{C}{1 + |x|},$$

with an appropriate constant C. Hence, $Mu \in L^p$. Therefore, by Theorem 14.21 and by Fubini's theorem,

$$
\begin{aligned}
\|Mu\|_p^p &= p \int_0^\infty \lambda^{p-1}\, m_{Mu}(\lambda)\, d\lambda \\
&\leq 2p \int_0^\infty \lambda^{p-2} \left(\int_{\{Mu > \lambda\}} |u(x)|\, dx \right) d\lambda \\
&= 2p \int_{-\infty}^\infty \left(\int_0^{Mu(x)} \lambda^{p-2}\, d\lambda \right) |u(x)|\, dx \\
&= \frac{2p}{p-1} \int_{-\infty}^\infty (Mu(x))^{p-1}\, |u(x)|\, dx.
\end{aligned}
$$

Finally, by Hölder's inequality,

$$\|Mu\|_p^p \leq \frac{2p}{p-1} \|Mu\|_p^{p-1} \|u\|_p.$$

Since $\|Mu\|_p < \infty$ we can divide both sides by $\|Mu\|_p^{p-1}$. $\qquad\square$

Exercises

Exercise 14.7.1 Let $E \subset \mathbb{R}$ with $|E| < \infty$. Suppose that

$$\int_{-\infty}^\infty |u(x)|\, \log^+ |u(x)|\, dx < \infty.$$

Show that

$$\int_E Mu(x)\, dx \leq 2|E| + 4 \int_{-\infty}^\infty |u(x)|\, \log^+ |2u(x)|\, dx.$$

Exercise 14.7.2   Let $u \in L^1(\mathbb{R})$. Suppose that

$$\int_E Mu(x) \, dx < \infty$$

for any measurable E with $|E| < \infty$. Show that

$$\int_{-\infty}^{\infty} |u(x)| \, \log^+ |u(x)| \, dx < \infty.$$

14.8   The maximal Hilbert transform

The *maximal Hilbert transform* of a function $u \in L^1(dt/(1+|t|))$ is defined by

$$\breve{u}(x) = \sup_{\varepsilon>0} \left| \frac{1}{\pi} \int_{|x-t|>\varepsilon} \frac{u(t)}{x-t} \, dt \right|.$$

According to the M. Riesz theorem (Theorem 14.7), $\tilde{u}$ remains in $L^p(\mathbb{R})$ whenever $u \in L^p(\mathbb{R})$, $1 < p < \infty$. In this section, by showing that $\breve{u} \in L^p(\mathbb{R})$ we strengthen the M. Riesz result.

Lemma 14.24   *Let $u \in L^1(dt/(1+|t|))$, and let*

$$V(x+iy) = \frac{1}{\pi} \int_{-\infty}^{\infty} \frac{(x-t)}{(x-t)^2+y^2} \, u(t) \, dt, \qquad (x+iy \in \mathbb{C}_+).$$

Then

$$|\breve{u}(x)| \le (1+1/\pi) \, Mu(x) + M_0 V(x)$$

for all $x \in \mathbb{R}$.

Proof.   Let

$$\breve{u}_\varepsilon(x) = \frac{1}{\pi} \int_{|x-t|>\varepsilon} \frac{u(t)}{x-t} \, dt, \qquad (x \in \mathbb{R}).$$

Fix $y > 0$ and write

$$V(x+iy) = \frac{1}{\pi} \left(\int_{|x-t|<y} + \int_{|x-t|>y} \right) \frac{(x-t)}{(x-t)^2+y^2} \, u(t) \, dt.$$

Hence,

$$
\begin{aligned}
\pi |\breve{u}_y(x) - V(x+iy)| \;\le\; & \int_{|x-t|>y} \frac{y^2}{|x-t|((x-t)^2+y^2)} \, |u(t)| \, dt \\
& + \int_{|x-t|<y} \frac{|x-t|}{(x-t)^2+y^2} \, |u(t)| \, dt \\
\le\; & \int_{|x-t|>y} \frac{y}{(x-t)^2+y^2} \, |u(t)| \, dt + \frac{1}{2y} \int_{|x-t|<y} |u(t)| \, dt.
\end{aligned}
$$

Therefore, by Lemma 14.20,

$$\pi|\check{u}_y(x) - V(x+iy)| \le \pi M_0|u|(x) + Mu(x) \le (1+\pi)Mu(x).$$

Thus,

$$|\check{u}_y(x)| \le (1+1/\pi)\,Mu(x) + |V(x+iy)|.$$

Taking the supremum with respect to $y > 0$ gives the required result. $\qquad\square$

Corollary 14.25 *Let* $u \in L^p(\mathbb{R})$, $1 < p < \infty$. *Then*

$$|\check{u}(x)| \le (1+1/\pi)\,Mu(x) + M\tilde{u}(x)$$

for all $x \in \mathbb{R}$.

Proof. By Theorem 14.7 the conjugate harmonic function

$$V(x+iy) = \frac{1}{\pi}\int_{-\infty}^{\infty} \frac{(x-t)}{(x-t)^2 + y^2}\, u(t)\, dt, \qquad (x+iy \in \mathbb{C}_+),$$

is also the Poisson integral of $\tilde{u}$, i.e.

$$V(x+iy) = \frac{1}{\pi}\int_{-\infty}^{\infty} \frac{y}{(x-t)^2 + y^2}\, \tilde{u}(t)\, dt, \qquad (x+iy \in \mathbb{C}_+).$$

Hence $M_0 V = M_0\tilde{u}$ and, by Lemma 14.20, $M_0\tilde{u} \le M\tilde{u}$. Now, apply Lemma 14.24. $\qquad\square$

Corollary 14.26 *Let* $u \in L^p(\mathbb{R})$, $1 < p < \infty$. *Then* $\check{u} \in L^p(\mathbb{R})$, *and moreover*

$$\|\check{u}\|_p \le C_p\|u\|_p,$$

where C_p *is a constant just depending on* p.

Proof. By Corollary 14.23

$$\|Mu\|_p \le \frac{2p}{p-1}\|u\|_p$$

and

$$\|M\tilde{u}\|_p \le \frac{2p}{p-1}\|\tilde{u}\|_p.$$

However, by the M. Riesz theorem (Theorem 14.7),

$$\|\tilde{u}\|_p \le K_p\|u\|_p$$

where K_p is a constant just depending on p. Hence, by Corollary 14.25,

$$\begin{aligned}\|\check{u}\|_p &\le (1+1/\pi)\|Mu\|_p + \|M\tilde{u}\|_p \\ &\le \frac{2p}{p-1}((1+1/\pi)\|u\|_p + \|\tilde{u}\|_p) \\ &\le \frac{2p}{p-1}((1+1/\pi) + K_p)\|u\|_p.\end{aligned}$$

$\qquad\square$

Appendix A

Topics from real analysis

A.1 A very concise treatment of measure theory

A measure space has three components: the ambient space X, a family $\mathfrak{M}$ of subsets of X, and a measure μ. The family $\mathfrak{M}$ satisfies the following properties:

(a) $X \in \mathfrak{M}$;

(b) if $E \in \mathfrak{M}$, then E^c, the complement of E with respect to X, is also in $\mathfrak{M}$;

(c) if $E_n \in \mathfrak{M}$, $n \geq 1$, then $\cup_{n=1}^{\infty} E_n \in \mathfrak{M}$.

In technical terms, we say that $\mathfrak{M}$ is a σ-algebra and its elements are *measurable* subsets of X. The measure μ is a complex function defined on $\mathfrak{M}$ such that

(a) $\mu(\emptyset) = 0$,

(b) $\mu(\cup_{n=1}^{\infty} E_n) = \sum_{n=1}^{\infty} \mu(E_n)$, for any sequence $(E_n)_{n \geq 1}$ in $\mathfrak{M}$ whose elements are pairwise disjoint.

If μ is a positive measure, we usually allow $+\infty$ as a possible value in the range of μ. If μ is a complex measure, the smallest positive measure λ satisfying

$$|\mu(E)| \leq \lambda(E), \qquad (E \in \mathfrak{M}),$$

is called the *total variation* μ and is denoted by $|\mu|$.

A function $f : X \longrightarrow \mathbb{C}$ is measurable if $f^{-1}(V) \in \mathfrak{M}$ for any open set V in $\mathbb{C}$. In particular, each simple function

$$\varphi = \sum_{n=1}^{N} a_n \chi_{E_n},$$

where $a_n \in \mathbb{C}$ and $E_n \in \mathfrak{M}$, is measurable. We define

$$\int_X \varphi \, d\mu = \sum_{n=1}^{N} a_n \, \mu(E_n).$$

For a positive measurable function f, we define

$$\int_X f \, d\mu = \sup_\varphi \int_X \varphi \, d\mu,$$

where the supremum is taken over all simple functions φ such that $0 \le \varphi \le f$. A measurable function f is called *integrable* provided that

$$\int_X |f| \, d\mu < \infty.$$

For such a function, we define

$$\int_X f \, d\mu = \left(\int_X u^+ \, d\mu - \int_X u^- \, d\mu \right) + i \left(\int_X v^+ \, d\mu - \int_X v^- \, d\mu \right),$$

where $f = u + iv$, and for a real function ω, we put $\omega^+ = \max\{\omega, 0\}$ and $\omega^- = \max\{-\omega, 0\}$. The family of all integrable functions is denoted by $L^1(\mu)$. Sometimes we also use $L^1(X)$ if the measure μ is fixed throughout the discussion.

The great success of Lebesgue's theory of integration is mainly based on the following three convergence theorems.

Theorem A.1 (Monotone convergence theorem) *Let f_n, $n \ge 1$, be a sequence of measurable functions on X such that $0 \le f_1 \le f_2 \le \cdots$, and let*

$$f(x) = \lim_{n \to \infty} f_n(x), \qquad (x \in X).$$

Then f is measurable and

$$\lim_{n \to \infty} \int_X f_n \, d\mu = \int_X f \, d\mu.$$

Proof. If $f_n = \chi_{E_n}$, where $(E_n)_{n \ge 1}$ is an increasing sequence of measurable sets in X and $E = \cup_{n=1}^\infty E_n$, then the monotone convergence theorem reduces to

$$\mu(E) = \lim_{n \to \infty} \mu(E_n),$$

which is rather obvious. This special case in turn implies that

$$\lim_{n \to \infty} \int_X \chi_{E_n} \varphi \, d\mu = \int_X \chi_E \varphi \, d\mu,$$

for each simple function φ.

Now we are able to prove the general case. Since $f_n \le f_{n+1} \le f$,

$$\int_X f_n \, d\mu \le \int_X f_{n+1} \, d\mu \le \int_X f \, d\mu.$$

Thus $\alpha = \lim_{n \to \infty} \int_X f_n \, d\mu$ exists and

$$\alpha \le \int_X f \, d\mu.$$

If $\alpha = \infty$, we are done. Hence, suppose that $\alpha < \infty$. Let φ be any simple function such that $0 \leq \varphi \leq f$ and fix the constant $\gamma \in (0,1)$. Let

$$E_n = \{\, x \in X \,:\, f_n(x) \geq \gamma\varphi(x)\,\}.$$

Clearly, $(E_n)_{n\geq 1}$ is an increasing sequence of measurable sets with $\cup_{n=1}^{\infty} E_n = X$. Moreover,

$$\int_X f_n \, d\mu \geq \int_X \chi_{E_n} f_n \, d\mu \geq \gamma \int_X \chi_{E_n} \varphi \, d\mu.$$

Let $n \to \infty$ to obtain

$$\alpha \geq \gamma \int_X \varphi \, d\mu.$$

Taking the supremum with respect to φ gives

$$\alpha \geq \gamma \int_X f \, d\mu.$$

Finally, let $\gamma \to 1$. $\qquad\square$

Theorem A.2 (Fatou's lemma) *Let f_n, $n \geq 1$, be a sequence of positive measurable functions on X. Then*

$$\int_X (\liminf_{n\to\infty} f_n) \, d\mu \leq \liminf_{n\to\infty} \int_X f_n \, d\mu.$$

Proof. Let $g = \liminf_{n\to\infty} f_n$, and let

$$g_k = \inf_{n\geq k} f_n, \qquad (k \geq 1).$$

Hence, $0 \leq g_1 \leq g_2 \leq \cdots$ and

$$\lim_{k\to\infty} g_k(x) = g(x), \qquad (x \in X).$$

Therefore, by the monotone convergence theorem,

$$\int_X g \, d\mu = \lim_{k\to\infty} \int_X g_k \, d\mu.$$

However, $g_k \leq f_k$, which implies

$$\lim_{k\to\infty} \int_X g_k \, d\mu \leq \liminf_{k\to\infty} \int_X f_k \, d\mu.$$

$\qquad\square$

Theorem A.3 (Dominated convergence theorem) *Let f_n, $n \geq 1$, be a sequence of measurable functions on X such that*

$$f(x) = \lim_{n\to\infty} f_n(x)$$

exists for every $x \in X$. Suppose that there is a $g \in L^1(\mu)$ such that

$$|f_n(x)| \leq g(x), \qquad (x \in X, n \geq 1).$$

Then $f \in L^1(\mu)$ and

$$\lim_{n \to \infty} \int_X |f_n - f| \, d\mu = 0. \tag{A.1}$$

Proof. Since $|f| \leq g$, we clearly have $f \in L^1(\mu)$. Let $g_n = 2g - |f_n - f|$, $n \geq 1$. Hence, by Fatou's lemma,

$$
\begin{aligned}
\int_X 2g \, d\mu &= \int_X \liminf_{n \to \infty} g_n \, d\mu \\
&\leq \liminf_{n \to \infty} \int_X g_n \, d\mu \\
&= \int_X 2g \, d\mu - \limsup_{n \to \infty} \int_X |f_n - f| \, d\mu.
\end{aligned}
$$

Since $g \in L^1(\mu)$, we conclude that

$$\limsup_{n \to \infty} \int_X |f_n - f| \, d\mu \leq 0,$$

and thus we obtain (A.1). □

Two more advanced results in measure theory are used in the text, which we explain heuristically here. The first one that we have frequently applied is Fubini's theorem. Let $(X, \mathfrak{M}, \mu)$ and $(Y, \mathfrak{N}, \nu)$ be two measure spaces and let $f : X \times Y \longrightarrow [0, \infty]$ be a measurable function. Then the Fubini theorem says

$$\int_X \left(\int_Y f(x, y) \, d\nu(y) \right) d\mu(x) = \int_Y \left(\int_X f(x, y) \, d\mu(x) \right) d\nu(y).$$

However, if we only assume that f is a complex measurable function, then some extra assumption is needed. For example, the condition $f \in L^1(\mu \times \nu)$ is enough to ensure that the preceding identity holds.

Suppose that λ and μ are two finite positive measures defined on the same σ-algebra $\mathfrak{M}$ of an ambient space X. We say that λ is *absolutely continuous* with respect to μ and write $\lambda \ll \mu$ if $\lambda(E) = 0$ for every $E \in \mathfrak{M}$ for which $\mu(E) = 0$. At the other extreme, we say that λ and μ are *mutually singular* and write $\lambda \perp \mu$ if there is a partition of X, say $X = A \cup B$, with $A, B \in \mathfrak{M}$ and $A \cap B = \emptyset$, such that $\lambda(A) = \mu(B) = 0$. Naively speaking, this means that λ and μ are living on different territories; λ is concentrated on B and μ is concentrated on A. If λ and μ are complex measures, then, by definition, λ is *absolutely continuous* with respect to μ if $|\lambda|$ is *absolutely continuous* with respect to $|\mu|$. Mutually singular complex measures are defined similarly.

Given two complex measures λ and μ, the celebrated theorem of Lebesgue–Radon–Nikodym says that there is a unique $f \in L^1(\mu)$ and a unique measure $\sigma \perp \mu$ such that

$$d\lambda = f \, d\mu + d\sigma.$$

This notation is a shorthand to say

$$\lambda(E) = \int_X \chi_E f \, d\mu + \sigma(E)$$

for all $E \in \mathfrak{M}$. In particular, λ is *absolutely continuous* with respect to μ if and only if there is an $f \in L^1(\mu)$ such that $d\lambda = f \, d\mu$.

Exercises

Exercise A.1.1   Let $(\mathfrak{M}_\iota)_{\iota \in I}$ be any family of σ-algebras on X. Put

$$\mathfrak{M} = \bigcap_{\iota \in I} \mathfrak{M}_\iota.$$

Show that $\mathfrak{M}$ is a σ-algebra on X.

Exercise A.1.2   Let $\mathfrak{T}$ be any collection of subsets of X. Show that there is a smallest σ-algebra on X which contains $\mathfrak{T}$.
Hint: Use Exercise A.1.1.
Remark: If $\mathfrak{T}$ is a topology on X, the elements of the smallest σ-algebra on X which contains $\mathfrak{T}$ are called the *Borel* subsets of X.

Exercise A.1.3   Suppose that the simple function φ has two representations, i.e.

$$\varphi = \sum_{n=1}^{N} a_n \chi_{E_n} = \sum_{m=1}^{M} b_m \chi_{F_m}.$$

Show that

$$\sum_{n=1}^{N} a_n \mu(E_n) = \sum_{m=1}^{M} b_m \mu(F_m).$$

Hint: Without loss of generality, assume that $(E_n)_{1 \le n \le N}$ and $(F_m)_{1 \le m \le M}$ are pairwise disjoint. Then

$$\sum_{n=1}^{N} a_n \chi_{E_n} = \sum_{n=1}^{N} \sum_{m=1}^{M} a_n \chi_{E_n \cap F_m}$$

and similarly

$$\sum_{m=1}^{M} b_m \chi_{F_m} = \sum_{m=1}^{M} \sum_{n=1}^{N} b_m \chi_{F_m \cap E_n}.$$

Remark: This fact shows that $\int_X \varphi \, d\mu$ is well-defined.

A.2   Riesz representation theorems

Let $\mathfrak{X}$ be a Banach space. The *dual* space $\mathfrak{X}^*$ consists of all continuous linear functionals

$$\Lambda : \mathfrak{X} \longrightarrow \mathbb{C}.$$

The norm of Λ is defined by

$$\|\Lambda\| = \sup_{\substack{x \in \mathfrak{X} \\ x \neq 0}} \frac{|\Lambda(x)|}{\|x\|}.$$

For each Banach space $\mathfrak{X}$, an important question is to characterize its dual $\mathfrak{X}^*$. F. Riesz answered this question for certain spaces of functions. We discuss two such theorems.

Let $(X, \mathfrak{M}, \mu)$ be a measure space and consider $\mathfrak{X} = L^p(\mu)$, $1 \leq p < \infty$. Let q be the conjugate exponent of p, i.e. $1/p + 1/q = 1$. Then the first theorem says that

$$L^p(\mu)^* = L^q(\mu).$$

This identity means that given any $\Lambda \in L^p(\mu)^*$, there is a unique $g \in L^q(\mu)$ such that

$$\Lambda(f) = \int_X f\,g\,d\mu, \qquad (f \in L^p(\mu)),$$

and

$$\|\Lambda\| = \|g\|_q.$$

On the other hand, if $g \in L^q(\mu)$ is given and we define Λ by the given formula, then clearly Λ is a bounded linear functional on $L^p(\mu)$ and its norm is equal to $\|g\|_q$. Therefore, Riesz's theorem actually provides an isometric correspondence between $L^p(\mu)^*$ and $L^q(\mu)$. This assertion is not valid if $p = \infty$. As a matter of fact, using the above formula, $L^1(\mu)$ can be embedded in $L^\infty(\mu)^*$, but in the general case, this is a proper inclusion.

For the second theorem, let X be a locally compact Hausdorff space. In our applications, X was either the real line $\mathbb{R}$ or the unit circle $\mathbb{T}$. Put $\mathfrak{X} = \mathcal{C}_0(X)$, the space of all continuous functions vanishing at infinity equipped with the supremum norm. Note that if X is compact, then we have $\mathcal{C}_0(X) = \mathcal{C}(X)$. Then we have

$$\mathfrak{X}^* = \mathcal{M}(X),$$

where $\mathcal{M}(X)$ is the space of all complex regular Borel measures on X. Hence, if $\Lambda \in \mathcal{C}_0(X)^*$ is given, there is a unique $\mu \in \mathcal{M}(X)$ such that

$$\Lambda(f) = \int_X f\,d\mu, \qquad (f \in \mathcal{C}_0(X)),$$

and

$$\|\Lambda\| = \|\mu\|.$$

On the other hand, if $\mu \in \mathcal{M}(X)$ is given and we define Λ as above, then Λ is a bounded linear functional on $\mathcal{C}_0(X)$ whose norm is equal to $\|\mu\|$. Therefore, Riesz's second theorem provides an isometric correspondence between $\mathcal{C}_0(X)^*$ and $\mathcal{M}(X)$.

A.3 Weak* convergence of measures

Let X be a locally compact Hausdorff space. In the preceding section we saw that $\mathcal{M}(X)$ is the dual of $\mathcal{C}_0(X)$. Hence if μ and μ_n, $n \geq 1$, are measures in $\mathcal{M}(X)$, we say that μ_n converges to μ in the weak* topology if

$$\lim_{n \to \infty} \int_X \varphi \, d\mu_n = \int_X \varphi \, d\mu$$

for all $\varphi \in \mathcal{C}_0(X)$. For positive measures on separable spaces, there are several equivalent ways to define the weak* convergence.

The *cumulative distribution function* of the positive measure $\mu \in \mathcal{M}(\mathbb{R})$ is defined by

$$F(x) = \mu((-\infty, x]), \qquad (x \in \mathbb{R}).$$

Clearly, F is an increasing right continuous function on $\mathbb{R}$. Moreover,

$$\int_{\mathbb{R}} \varphi \, d\mu = \int_{-\infty}^{\infty} \varphi(x) \, dF(x)$$

for all $\varphi \in \mathcal{C}_0(\mathbb{R})$. In the last identity, the left side is a Lebesgue integral and the right side is a Riemann–Stieltjes integral. We can also define the cumulative distribution function for positive Borel measures on $\mathbb{T}$ and then obtain

$$\int_{\mathbb{T}} \varphi \, d\mu = \int_{-\pi}^{\pi} \varphi(x) \, dF(x)$$

for all $\varphi \in \mathcal{C}(\mathbb{T})$.

Theorem A.4 *Let X be a separable locally compact Hausdorff space. Let $\mu, \mu_1, \mu_2, \ldots$ be positive measures in $\mathcal{M}(X)$. Then the following are equivalent:*

(i) μ_n converges to μ in the weak* topology;

(ii) $\mu_n(X) \to \mu(X)$ and, for all open subsets U of X,

$$\liminf_{n \to \infty} \mu_n(U) \geq \mu(U);$$

(iii) $\mu_n(X) \to \mu(X)$ and, for all closed subsets F of X,

$$\limsup_{n \to \infty} \mu_n(F) \leq \mu(F);$$

(iv) for all Borel subsets E of X with $\mu(\partial E) = 0$, where ∂E is the boundary of E in X,

$$\lim_{n \to \infty} \mu_n(E) = \mu(E);$$

(v) in the case $X = \mathbb{R}$ (or $X = \mathbb{T}$), if F_n and F respectively denote the cumulative distribution functions of the measures μ_n and μ, then

$$\lim_{n \to \infty} F_n(x) = F(x)$$

at all points $x \in \mathbb{R}$ where F is continuous.

A.4 $\mathcal{C}(\mathbb{T})$ is dense in $L^p(\mathbb{T})$ $(0 < p < \infty)$

The space of continuous functions $\mathcal{C}(\mathbb{T})$ is *dense* in $L^p(\mathbb{T})$, $0 < p < \infty$. This assertion is not true if $p = \infty$, since the uniform limit of a sequence of continuous functions has to be continuous and a typical element of $L^\infty(\mathbb{T})$ is not necessarily continuous. As a matter of fact, we implicitly proved that $\mathcal{C}(\mathbb{T})$ is closed in $L^\infty(\mathbb{T})$. If we know that each $f \in L^p(\mathbb{T})$ can be approximated by *step* functions

$$\sum_{n=1}^{N} a_n \chi_{I_n},$$

where I_n is an open interval, then we can approximate each χ_{I_n} by a trapezoid-shaped continuous function, and thus the result is immediate. Even though approximation by step function is possible, it does not come from the first principles of measure theory. Based on the definition of a Lebesgue integral, each $f \in L^p(\mathbb{T})$ can be approximated by *simple* functions

$$s = \sum_{n=1}^{N} a_n \chi_{E_n},$$

where E_n is a Lebesgue measurable set. At this point, we need two deep results from real analysis. First, given E, a Lebesgue measurable set on $\mathbb{T}$, there is a compact set K and an open set V such that $K \subset E \subset V$, and the Lebesgue measure of $V \setminus K$ is as small as we want. Second, by Urysohn's lemma, there is a continuous function φ such that $0 \leq \varphi \leq 1$, $\varphi|_K \equiv 1$ and $\operatorname{supp} \varphi \subset V$. Hence,

$$\int_{\mathbb{T}} |\varphi - \chi_E|^p \, d\theta \leq \int_{V \setminus K} d\theta = |V \setminus K|.$$

Given $f \in L^p(\mathbb{T})$ and $\varepsilon > 0$, choose the simple function $s = \sum_{n=1}^{N} a_n \chi_{E_n}$ such that $\|f - s\|_p < \varepsilon$. Then, for each n, pick φ_n such that $\|\varphi_n - \chi_{E_n}\|_p < \varepsilon/(N(1 + |a_n|))$. Put

$$\varphi = \sum_{n=1}^{N} a_n \varphi_n.$$

Therefore,

$$\|f - \varphi\|_p \leq \|f - s\|_p + \|s - \varphi\|_p$$

$$\leq \varepsilon + \sum_{n=1}^{N} |a_n| \|\varphi_n - \chi_{E_n}\|_p \leq 2\varepsilon.$$

Exercises

Exercise A.4.1 Show that $C_c(\mathbb{R})$ is dense in $L^p(\mathbb{R})$, $1 \leq p < \infty$.

Exercise A.4.2 Show that $C_c^\infty(\mathbb{R})$ is dense in $L^p(\mathbb{R})$, $1 \leq p < \infty$.
Hint: Use Exercise A.4.1 and the convolution.

A.5   The distribution function

For a measurable set $E \subset \mathbb{T}$, let $|E|$ denote its Lebesgue measure. Let φ be a measurable function on $\mathbb{T}$. Then the *distribution function* of φ is defined by

$$m_\varphi(\lambda) = |\{e^{i\theta} : |\varphi(e^{i\theta})| > \lambda\}|, \qquad (\lambda > 0).$$

Clearly, m_φ is a decreasing function on $(0, \infty)$. The space weak-$L^1(\mathbb{T})$ consists of all measurable functions φ on $\mathbb{T}$ such that

$$m_\varphi(\lambda) \leq \frac{C}{\lambda},$$

where C is a constant (not depending on λ, but it may depend on φ). Since, for each $\varphi \in L^1(\mathbb{T})$,

$$m_\varphi(\lambda) \leq \frac{2\pi \|\varphi\|_1}{\lambda},$$

weak-$L^1(\mathbb{T})$ contains $L^1(\mathbb{T})$ as a subspace.

Let $0 < p < \infty$. On the measure space $\mathbb{T} \times [0, \infty)$ equipped with the product measure $dt \times d\lambda$, consider the measurable function

$$\Phi(e^{it}, \lambda) = \begin{cases} 1 & \text{if} \quad 0 \leq \lambda < |\varphi(e^{it})|, \\ 0 & \text{if} \quad \lambda \geq |\varphi(e^{it})|. \end{cases}$$

Now, using Fubini's theorem, we evaluate the integral

$$I = \int_{\mathbb{T} \times [0, \infty)} p \lambda^{p-1} \Phi(e^{it}, \lambda) \, dt \times d\lambda$$

in two ways. On the one hand, we have

$$I = \int_{-\pi}^{\pi} \left(\int_0^{|\varphi(e^{it})|} p\,\lambda^{p-1}\,d\lambda \right) dt = \int_{-\pi}^{\pi} |\varphi(e^{it})|^p\,dt$$

and on the other hand,

$$I = \int_0^{\infty} \left(\int_{\{|\varphi|>\lambda\}} dt \right) p\,\lambda^{p-1}\,d\lambda = \int_0^{\infty} p\,\lambda^{p-1}\,m_\varphi(\lambda)\,d\lambda.$$

Therefore, for all $0 < p < \infty$ and for all measurable functions φ, we have

$$\int_{-\pi}^{\pi} |\varphi(e^{it})|^p\,dt = \int_0^{\infty} p\,\lambda^{p-1}\,m_\varphi(\lambda)\,d\lambda. \tag{A.2}$$

Exercises

Exercise A.5.1   Let $(X, \mathfrak{M}, \mu)$ be a measure space with $\mu \geq 0$. Let $\varphi : X \longrightarrow \mathbb{C}$ be a measurable function on X. Define the distribution function of φ by

$$m_\varphi(\lambda) = \mu(\{x \in X : |\varphi(x)| > \lambda\}).$$

Show that

$$\int_X |\varphi(x)|^p\,d\mu(x) = \int_0^{\infty} p\,\lambda^{p-1}\,m_\varphi(\lambda)\,d\lambda$$

for all $0 < p < \infty$.

Exercise A.5.2   Let $0 \leq a < b \leq \infty$. Using the notation of Exercise A.5.1, show that

$$\int_{\{x \in X : \varphi(x) > a\}} \min\{ |\varphi(x)|^p - a^p, \; b^p - a^p \}\,d\mu(x) = \int_a^b p\,\lambda^{p-1}\,m_\varphi(\lambda)\,d\lambda.$$

A.6   Minkowski's inequality

Let (Ω, μ) and (Ω', ν) be two measure spaces with positive measures μ and ν, and let $f(x, t)$ be a positive measurable function on $\Omega' \times \Omega$. Define

$$F(x) = \int_\Omega f(x, t)\,d\mu(t).$$

Fix p, $1 \leq p < \infty$, and let $1/p + 1/q = 1$. Then, by Fubini's theorem, for all positive measurable functions G, we have

$$
\begin{aligned}
\int_{\Omega'} F(x)\, G(x)\, d\nu(x) &= \int_{\Omega'} \left(\int_{\Omega} f(x,t)\, d\mu(t) \right) G(x)\, d\nu(x) \\
&= \int_{\Omega} \left(\int_{\Omega'} f(x,t)\, G(x)\, d\nu(x) \right) d\mu(t) \\
&\leq \int_{\Omega} \left(\int_{\Omega'} |f(x,t)|^p\, d\nu(x) \right)^{\frac{1}{p}} \left(\int_{\Omega'} |G(x)|^q\, d\nu(x) \right)^{\frac{1}{q}} d\mu(t) \\
&= \|G\|_q \int_{\Omega} \left(\int_{\Omega'} |f(x,t)|^p\, d\nu(x) \right)^{\frac{1}{p}} d\mu(t).
\end{aligned}
$$

In particular, for $G = F^{p-1} \chi_E$, where E is any measurable set on which $|F|^p$ is summable, we have

$$
\left(\int_E |F(x)|^p\, d\nu(x) \right)^{\frac{1}{p}} \leq \int_{\Omega} \left(\int_{\Omega'} |f(x,t)|^p\, d\nu(x) \right)^{\frac{1}{p}} d\mu(t).
$$

Taking the supremum over E, we get

$$
\|F\|_p \leq \int_{\Omega} \left(\int_{\Omega'} |f(x,t)|^p\, d\nu(x) \right)^{\frac{1}{p}} d\mu(t).
$$

Equivalently, for every measurable function f,

$$
\left\{ \int_{\Omega'} \left(\int_{\Omega} |f(x,t)|\, d\mu(t) \right)^p d\nu(x) \right\}^{\frac{1}{p}} \leq \int_{\Omega} \left(\int_{\Omega'} |f(x,t)|^p\, d\nu(x) \right)^{\frac{1}{p}} d\mu(t).
$$

This is the most generalized form of Minkowski's inequality. If $\Omega = \{1, 2\}$ and μ is the counting measure, then the last inequality reduces to

$$
\begin{aligned}
\left(\int_{\Omega'} \Big(|f(x,1)| + |f(x,2)| \Big)^p d\nu(x) \right)^{\frac{1}{p}} &\leq \left(\int_{\Omega'} |f(x,1)|^p\, d\nu(x) \right)^{\frac{1}{p}} \\
&+ \left(\int_{\Omega'} |f(x,2)|^p\, d\nu(x) \right)^{\frac{1}{p}},
\end{aligned}
$$

which is equivalent to $\|f_1 + f_2\|_p \leq \|f_1\|_p + \|f_2\|_p$, where $f_n(x) = |f(x,n)|$.

A.7 Jensen's inequality

Let (a, b) be an interval on the real line $\mathbb{R}$. A real function φ on (a, b) is *convex* if the inequality

$$
\varphi\big((1-t)x + ty \big) \leq (1-t)\varphi(x) + t\varphi(y)
$$

holds for each $x, y \in (a,b)$ and $0 \le t \le 1$. It is easy to see that this condition is equivalent to

$$\frac{\varphi(z) - \varphi(x)}{z - x} \le \frac{\varphi(y) - \varphi(z)}{y - z} \tag{A.3}$$

whenever $a < x < z < y < b$.

Theorem A.5 (Jensen's inequality) *Let* $(X, \mathfrak{M}, \mu)$ *be a measure space with* $\mu \ge 0$ *and* $\mu(X) = 1$. *Let* $f \in L^1(\mu)$ *and suppose that*

$$-\infty \le a < f(x) < b \le \infty \tag{A.4}$$

for all $x \in X$. *Suppose that* φ *is convex on* (a,b). *Then*

$$\varphi\left(\int_X f d\mu\right) \le \int_X (\varphi \circ f)\, d\mu.$$

Proof. Let $z = \int_X f d\mu$. The condition (A.4) ensures that $z \in (a,b)$. According to (A.3), there is a constant c such that

$$\frac{\varphi(z) - \varphi(x)}{z - x} \le c \le \frac{\varphi(y) - \varphi(z)}{y - z}$$

for all $x \in (a, z)$ and all $y \in (z, b)$. This implies

$$\varphi(w) \ge \varphi(z) + c(w - z)$$

for each $w \in (a, b)$. Hence,

$$(\varphi \circ f)(x) \ge \varphi(z) + c(\,f(x) - z\,)$$

for every $x \in X$. Therefore,

$$\int_X (\varphi \circ f)(x)\, d\mu(x) \ge \int_X \left(\varphi(z) + c(\,f(x) - z\,)\right) d\mu(x)$$
$$= \varphi(z) + c\left(\int_X f(x)\, d\mu(x) - z\right)$$
$$= \varphi(z) = \varphi\left(\int_X f d\mu\right).$$

$\square$

Exercises

Exercise A.7.1 Let φ be convex on (a,b). Show that φ is continuous on (a,b).
Hint: Use (A.3).

Exercise A.7.2 Let φ be convex on (a, b). Show that at each $t \in (a,b)$, φ has a right and a left derivative.
Hint: Use (A.3).

Appendix B

A panoramic view of the representation theorems

We gather here a rather complete list of *representation theorems* for harmonic or analytic functions on the open unit disc or on the upper half plane. These theorems have been discussed in detail throughout the text. Hence, we do not provide any definition of proof in this appendix. However, we remember that our main objects are elements of the following spaces:

$$
\begin{aligned}
h^p(\mathbb{D}) &= \{\, U \in h(\mathbb{D}) : \|U\|_p = \sup_{0 \le r < 1} \|U_r\|_p < \infty \,\}, \\
H^p(\mathbb{D}) &= \{\, F \in H(\mathbb{D}) : \|F\|_p = \sup_{0 \le r < 1} \|F_r\|_p < \infty \,\}, \\
H^p(\mathbb{T}) &= \{\, f \in L^p(\mathbb{T}) : \exists F \in H^p(\mathbb{D}) \text{ such that } F_r \longrightarrow f \,\}
\end{aligned}
$$

and

$$
\begin{aligned}
h^p(\mathbb{C}_+) &= \{\, U \in h(\mathbb{C}_+) : \|U\|_p = \sup_{y>0} \|U_y\|_p < \infty \,\}, \\
H^p(\mathbb{C}_+) &= \{\, F \in H(\mathbb{C}_+) : \|F\|_p = \sup_{y>0} \|F_y\|_p < \infty \,\}, \\
H^p(\mathbb{R}) &= \{\, f \in L^p(\mathbb{R}) : \exists F \in H^p(\mathbb{C}_+) \text{ such that } F_y \longrightarrow f \,\}.
\end{aligned}
$$

B.1 $h^p(\mathbb{D})$

Let U be harmonic on the open unit disc $\mathbb{D}$.

B.1.1 $h^1(\mathbb{D})$

(a) $U \in h^1(\mathbb{D})$ if and only if there exists $\mu \in \mathcal{M}(\mathbb{T})$ such that

$$U(re^{i\theta}) = \int_{\mathbb{T}} \frac{1 - r^2}{1 + r^2 - 2r\cos(\theta - t)} \, d\mu(e^{it}), \qquad (re^{i\theta} \in \mathbb{D}).$$

(b) μ is unique and

$$U(re^{i\theta}) = \sum_{n=-\infty}^{\infty} \hat{\mu}(n) \, r^{|n|} \, e^{in\theta}, \qquad (re^{i\theta} \in \mathbb{D}).$$

The series is absolutely and uniformly convergent on compact subsets of $\mathbb{D}$.

(c) $\hat{\mu} \in \ell^\infty(\mathbb{Z})$ and
$$\|\hat{\mu}\|_\infty \le \|\mu\|.$$

(d) U is a positive harmonic function on $\mathbb{D}$ if and only if μ is a positive Borel measure on $\mathbb{T}$.

(e) As $r \to 1$, $d\mu_r(e^{it}) = U(re^{it}) \, dt/2\pi$ converges to $d\mu(e^{it})$ in the weak* topology, i.e.
$$\lim_{r \to 1} \int_{\mathbb{T}} \varphi(e^{it}) \, d\mu_r(e^{it}) = \int_{\mathbb{T}} \varphi(e^{it}) \, d\mu(e^{it})$$
for all $\varphi \in \mathcal{C}(\mathbb{T})$.

(f)
$$\|U\|_1 = \|\mu\| = |\mu|(\mathbb{T}).$$

(g) μ decomposes as
$$d\mu(e^{it}) = u(e^{it}) \, dt/2\pi + d\sigma(e^{it}),$$
where $u \in L^1(\mathbb{T})$ and σ is singular with respect to the Lebesgue measure. Then we have
$$\lim_{r \to 1} U(re^{i\theta}) = u(e^{i\theta})$$
for almost all $e^{i\theta} \in \mathbb{T}$.

(h) The harmonic conjugate of U is given by

$$
\begin{aligned}
V(re^{i\theta}) &= \int_{\mathbb{T}} \frac{2r\sin(\theta - t)}{1 + r^2 - 2r\cos(\theta - t)} \, d\mu(e^{it}) \\
&= \sum_{n=-\infty}^{\infty} -i\,\mathrm{sgn}(n)\,\hat{\mu}(n)\,r^{|n|}\,e^{in\theta}, \qquad (re^{i\theta} \in \mathbb{D}).
\end{aligned}
$$

The series is absolutely and uniformly convergent on compact subsets of $\mathbb{D}$.

(i) For almost all $e^{i\theta} \in \mathbb{T}$,

$$\lim_{r \to 1} V(re^{i\theta})$$

exists.

(j) $U \in h^1(\mathbb{D})$ does not imply that $V \in h^1(\mathbb{D})$.

(k) If

$$U(re^{i\theta}) = \frac{1}{2\pi} \int_{-\pi}^{\pi} \frac{1 - r^2}{1 + r^2 - 2r\cos(\theta - t)} u(e^{it})\, dt, \qquad (re^{i\theta} \in \mathbb{D}),$$

with $u \in L^1(\mathbb{T})$, then the following holds:

(i)

$$\lim_{r \to 1} \|U_r - u\|_1 = 0;$$

(ii)

$$\|U\|_1 = \|u\|_1;$$

(iii)

$$\hat{u} \in c_0(\mathbb{Z});$$

(iv)

$$\|\hat{u}\|_\infty \leq \|u\|_1;$$

(v) the harmonic conjugate of U is given by

$$V(re^{i\theta}) = \frac{1}{2\pi} \int_{-\pi}^{\pi} \frac{2r\sin(\theta - t)}{1 + r^2 - 2r\cos(\theta - t)} u(e^{it})\, dt$$

$$= \sum_{n=-\infty}^{\infty} -i\operatorname{sgn}(n)\, \hat{u}(n)\, r^{|n|}\, e^{in\theta}, \qquad (re^{i\theta} \in \mathbb{D});$$

(vi) for almost all $e^{i\theta} \in \mathbb{T}$,

$$\lim_{r \to 1} V(re^{i\theta}) = \tilde{u}(e^{i\theta});$$

(vii) for each $0 < p < 1$, $V \in h^p(\mathbb{D})$, $\tilde{u} \in L^p(\mathbb{T})$,

$$\lim_{r \to 1} \|V_r - \tilde{u}\|_p = 0$$

and

$$\|\tilde{u}\|_p = \|V\|_p \leq c_p \|u\|_1;$$

(viii) $u \in L^1(\mathbb{T})$ does not imply that $\tilde{u} \in L^1(\mathbb{T})$;

(ix) if $u, \tilde{u} \in L^1(\mathbb{T})$, then $V \in h^1(\mathbb{D})$,

$$\lim_{r \to 1} \|V_r - \tilde{u}\|_1 = 0$$

and the harmonic conjugate is also given by

$$V(re^{i\theta}) = \frac{1}{2\pi} \int_{-\pi}^{\pi} \frac{1 - r^2}{1 + r^2 - 2r\cos(\theta - t)} \tilde{u}(e^{it}) \, dt, \qquad (re^{i\theta} \in \mathbb{D});$$

(x) if $u, \tilde{u} \in L^1(\mathbb{T})$, then

$$\hat{\tilde{u}}(n) = -i \operatorname{sgn}(n) \hat{u}(n), \qquad (n \in \mathbb{Z}).$$

B.1.2 $h^p(\mathbb{D})$ $(1 < p < \infty)$

(a) $U \in h^p(\mathbb{D})$, $1 < p < \infty$, if and only if there exists $u \in L^p(\mathbb{T})$ such that

$$U(re^{i\theta}) = \frac{1}{2\pi} \int_{-\pi}^{\pi} \frac{1 - r^2}{1 + r^2 - 2r\cos(\theta - t)} u(e^{it}) \, dt, \qquad (re^{i\theta} \in \mathbb{D}).$$

(b) u is unique and

$$U(re^{i\theta}) = \sum_{n=-\infty}^{\infty} \hat{u}(n) \, r^{|n|} \, e^{in\theta}, \qquad (re^{i\theta} \in \mathbb{D}).$$

The series is absolutely and uniformly convergent on compact subsets of $\mathbb{D}$.

(c) If $1 < p \leq 2$, then $\hat{u} \in \ell^q(\mathbb{Z})$, where $1/p + 1/q = 1$ and

$$\|\hat{u}\|_q \leq \|u\|_p.$$

(d)

$$\lim_{r \to 1} \|U_r - u\|_p = 0.$$

(e)

$$\|U\|_p = \|u\|_p.$$

(f) For almost all $e^{i\theta} \in \mathbb{T}$,

$$\lim_{r \to 1} U(re^{i\theta}) = u(e^{i\theta}).$$

(g) $U \in h^2(\mathbb{D})$ if and only if $u \in L^2(\mathbb{T})$. If so,

$$\|U\|_2 = \|u\|_2 = \|\hat{u}\|_2$$

and if $U(0) = 0$, then

$$\|U\|_2 = \|\tilde{u}\|_2.$$

(h)
$$\hat{\tilde{u}}(n) = -i\,\mathrm{sgn}(n)\,\hat{u}(n), \qquad (n \in \mathbb{Z}).$$

(i) The harmonic conjugate of U is given by

$$
\begin{aligned}
V(re^{i\theta}) &= \frac{1}{2\pi} \int_{-\pi}^{\pi} \frac{2r\sin(\theta - t)}{1 + r^2 - 2r\cos(\theta - t)}\, u(e^{it})\, dt \\
&= \frac{1}{2\pi} \int_{-\pi}^{\pi} \frac{1 - r^2}{1 + r^2 - 2r\cos(\theta - t)}\, \tilde{u}(e^{it})\, dt \\
&= \sum_{n=-\infty}^{\infty} -i\,\mathrm{sgn}(n)\,\hat{u}(n)\, r^{|n|}\, e^{in\theta}, \qquad (re^{i\theta} \in \mathbb{D}).
\end{aligned}
$$

The series is absolutely and uniformly convergent on compact subsets of $\mathbb{D}$.

(j) For almost all $e^{i\theta} \in \mathbb{T}$,

$$\lim_{r \to 1} V(re^{i\theta}) = \tilde{u}(e^{i\theta}).$$

(k) $V \in h^p(\mathbb{D})$, $\tilde{u} \in L^p(\mathbb{T})$,

$$\lim_{r \to 1} \|V_r - \tilde{u}\|_p = 0$$

and

$$\|V\|_p = \|\tilde{u}\|_p \le c_p \|U\|_p.$$

B.1.3 $h^\infty(\mathbb{D})$

(a) $U \in h^\infty(\mathbb{D})$ if and only if there exists $u \in L^\infty(\mathbb{T})$ such that

$$U(re^{i\theta}) = \frac{1}{2\pi} \int_{-\pi}^{\pi} \frac{1 - r^2}{1 + r^2 - 2r\cos(\theta - t)}\, u(e^{it})\, dt, \qquad (re^{i\theta} \in \mathbb{D}).$$

(b) u is unique and

$$U(re^{i\theta}) = \sum_{n=-\infty}^{\infty} \hat{u}(n)\, r^{|n|}\, e^{in\theta}, \qquad (re^{i\theta} \in \mathbb{D}).$$

The series is absolutely and uniformly convergent on compact subsets of $\mathbb{D}$.

(c) As $r \to 1$, U_r converges to u in the weak* topology, i.e.

$$\lim_{r \to 1} \int_{-\pi}^{\pi} \varphi(e^{it})\, U(re^{it})\, dt = \int_{-\pi}^{\pi} \varphi(e^{it})\, u(e^{it})\, dt$$

for all $\varphi \in L^1(\mathbb{T})$.

(d)
$$\|U\|_\infty = \|u\|_\infty.$$

(e) But, we have

$$\lim_{r \to 1} \|U_r - u\|_\infty = 0$$

if and only if $u \in \mathcal{C}(\mathbb{T})$.

(f) For almost all $e^{i\theta} \in \mathbb{T}$,

$$\lim_{r \to 1} U(re^{i\theta}) = u(e^{i\theta}).$$

(g) The harmonic conjugate of U is given by

$$
\begin{aligned}
V(re^{i\theta}) &= \frac{1}{2\pi} \int_{-\pi}^{\pi} \frac{2r\sin(\theta - t)}{1 + r^2 - 2r\cos(\theta - t)} u(e^{it})\, dt \\
&= \frac{1}{2\pi} \int_{-\pi}^{\pi} \frac{1 - r^2}{1 + r^2 - 2r\cos(\theta - t)} \tilde{u}(e^{it})\, dt \\
&= \sum_{n=-\infty}^{\infty} -i\,\mathrm{sgn}(n)\,\hat{u}(n)\, r^{|n|}\, e^{in\theta}, \qquad (re^{i\theta} \in \mathbb{D}).
\end{aligned}
$$

The series is absolutely and uniformly convergent on compact subsets of $\mathbb{D}$.

(h) For almost all $e^{i\theta} \in \mathbb{T}$,

$$\lim_{r \to 1} V(re^{i\theta}) = \tilde{u}(e^{i\theta}).$$

(i)
$$\hat{\tilde{u}}(n) = -i\,\mathrm{sgn}(n)\,\hat{u}(n), \qquad (n \in \mathbb{Z}).$$

(j) $U \in h^\infty(\mathbb{D})$ does not imply that $V \in h^\infty(\mathbb{D})$.

(k) $u \in L^\infty(\mathbb{T})$ does not imply that $\tilde{u} \in L^\infty(\mathbb{T})$. However, for real u,

$$\frac{1}{2\pi} \int_{-\pi}^{\pi} e^{\lambda |\tilde{u}(e^{it})|}\, dt \leq 2\sec(\lambda \|u\|_\infty), \qquad (0 \leq \lambda < \pi/2\, \|u\|_\infty).$$

B.2 $\quad H^p(\mathbb{D})$

Let F be analytic on the open unit disc $\mathbb{D}$.

B.2.1 $\quad H^p(\mathbb{D})\ (1 \leq p < \infty)$

(a) $F \in H^p(\mathbb{D})$, $1 \leq p < \infty$, if and only if there exists $f \in H^p(\mathbb{T})$ such that

$$F(re^{i\theta}) = \frac{1}{2\pi} \int_{-\pi}^{\pi} \frac{1 - r^2}{1 + r^2 - 2r\cos(\theta - t)} f(e^{it})\, dt, \qquad (re^{i\theta} \in \mathbb{D}).$$

(b) f is unique and

$$
\begin{aligned}
F(z) &= \frac{1}{2\pi} \int_{-\pi}^{\pi} \frac{f(e^{it})}{1 - e^{-it}z}\, dt \\
&= \frac{i}{2\pi} \int_{-\pi}^{\pi} \frac{2r\sin(\theta - t)}{1 + r^2 - 2r\cos(\theta - t)} f(e^{it})\, dt + F(0) \\
&= \frac{1}{2\pi} \int_{-\pi}^{\pi} \frac{e^{it} + z}{e^{it} - z} \Re f(e^{it})\, dt + i\Im F(0) \\
&= \frac{i}{2\pi} \int_{-\pi}^{\pi} \frac{e^{it} + z}{e^{it} - z} \widetilde{\Re f}(e^{it})\, dt + F(0) \\
&= \sum_{n=0}^{\infty} \hat{f}(n) z^n, \qquad (z \in \mathbb{D}).
\end{aligned}
$$

The series is absolutely and uniformly convergent on compact subsets of $\mathbb{D}$.

(c) If $\hat{f}(0) = 0$, then $\tilde{f} = -if$ or equivalently,

$$\widetilde{\Re f} = \Im f \qquad \text{and} \qquad \widetilde{\Im f} = -\Re f.$$

(d) If $1 \le p \le 2$, then $\hat{f} \in \ell^q(\mathbb{Z}^+)$, where $1/p + 1/q = 1$ and

$$\|\hat{f}\|_q \le \|f\|_p.$$

(e)

$$\lim_{r \to 1} \|F_r - f\|_p = 0.$$

(f)

$$\|F\|_p = \|f\|_p.$$

(g) For almost all $e^{i\theta} \in \mathbb{T}$,

$$\lim_{r \to 1} F(re^{i\theta}) = f(e^{i\theta}).$$

(h) $F \in H^2(\mathbb{D})$ if and only if $f \in H^2(\mathbb{T})$. If so,

$$\|F\|_2 = \|f\|_2 = \|\hat{f}\|_2$$

and moreover,

$$\|F\|_2 = \|\tilde{f}\|_2$$

provided that $F(0) = 0$.

B.2.2 $H^\infty(\mathbb{D})$

(a) $F \in H^\infty(\mathbb{D})$ if and only if there exists $f \in H^\infty(\mathbb{T})$ such that

$$F(re^{i\theta}) = \frac{1}{2\pi} \int_{-\pi}^{\pi} \frac{1-r^2}{1+r^2-2r\cos(\theta-t)} f(e^{it})\, dt, \qquad (re^{i\theta} \in \mathbb{D}).$$

(b) f is unique and

$$
\begin{aligned}
F(z) &= \frac{1}{2\pi} \int_{-\pi}^{\pi} \frac{f(e^{it})}{1-e^{-it}z}\, dt \\
&= \frac{i}{2\pi} \int_{-\pi}^{\pi} \frac{2r\sin(\theta-t)}{1+r^2-2r\cos(\theta-t)} f(e^{it})\, dt + F(0) \\
&= \frac{1}{2\pi} \int_{-\pi}^{\pi} \frac{e^{it}+z}{e^{it}-z} \Re f(e^{it})\, dt + i\Im F(0) \\
&= \frac{i}{2\pi} \int_{-\pi}^{\pi} \frac{e^{it}+z}{e^{it}-z} \widetilde{\Re} f(e^{it})\, dt + F(0) \\
&= \sum_{n=0}^{\infty} \hat{f}(n)\, z^n, \qquad (z \in \mathbb{D}).
\end{aligned}
$$

The series is absolutely and uniformly convergent on compact subsets of $\mathbb{D}$.

(c) If $\hat{f}(0) = 0$, then $\tilde{f} = -if$ or equivalently,

$$\widetilde{\Re} f = \Im f \qquad \text{and} \qquad \widetilde{\Im} f = -\Re f.$$

(d) As $r \to 1$, F_r converges to f in the weak* topology, i.e.

$$\lim_{r \to 1} \int_{-\pi}^{\pi} \varphi(e^{it})\, F(re^{it})\, dt = \int_{-\pi}^{\pi} \varphi(e^{it})\, f(e^{it})\, dt$$

for all $\varphi \in L^1(\mathbb{T})$.

(e)
$$\|F\|_\infty = \|f\|_\infty.$$

(f) But, we have
$$\lim_{r \to 1} \|F_r - f\|_\infty = 0$$

if and only if $f \in \mathcal{A}(\mathbb{T}) = \mathcal{C}(\mathbb{T}) \cap H^1(\mathbb{T})$.

(g) For almost all $e^{i\theta} \in \mathbb{T}$,

$$\lim_{r \to 1} F(re^{i\theta}) = f(e^{i\theta}).$$

B.3 $h^p(\mathbb{C}_+)$

Let U be harmonic in the upper half plane $\mathbb{C}_+$.

B.3.1 $h^1(\mathbb{C}_+)$

(a) $U \in h^1(\mathbb{C}_+)$ if and only if there exists $\mu \in \mathcal{M}(\mathbb{R})$ such that

$$U(x+iy) = \frac{1}{\pi} \int_{\mathbb{R}} \frac{y}{(x-t)^2 + y^2} \, d\mu(t), \qquad (x+iy \in \mathbb{C}_+).$$

(b) μ is unique and

$$U(x+iy) = \int_{-\infty}^{\infty} \hat{\mu}(t) \, e^{-2\pi y|t|+i2\pi xt} \, dt, \qquad (x+iy \in \mathbb{C}_+).$$

The integral is absolutely and uniformly convergent on compact subsets of $\mathbb{C}_+$.

(c) $\hat{\mu} \in \mathcal{C}(\mathbb{R}) \cap L^\infty(\mathbb{R})$ and

$$\|\hat{\mu}\|_\infty \leq \|\mu\|.$$

(d) As $y \to 0$, $d\mu_y(t) = U(t+iy) \, dt$ converges to $d\mu(t)$ in the weak* topology, i.e.,

$$\lim_{y \to 0} \int_{\mathbb{R}} \varphi(t) \, d\mu_y(t) = \int_{\mathbb{R}} \varphi(t) \, d\mu(t)$$

for all $\varphi \in \mathcal{C}_0(\mathbb{R})$.

(e)
$$\|U\|_1 = \|\mu\| = |\mu|(\mathbb{R}).$$

(f) μ decomposes as

$$d\mu(t) = u(t) \, dt + d\sigma(t),$$

where $u \in L^1(\mathbb{R})$ and σ is singular with respect to the Lebesgue measure. Then we have

$$\lim_{y \to 0} U(x+iy) = u(x)$$

for almost all $x \in \mathbb{R}$.

(g) The harmonic conjugate of U is given by

$$\begin{aligned}
V(x+iy) &= \frac{1}{\pi} \int_{\mathbb{R}} \frac{x-t}{(x-t)^2 + y^2} \, d\mu(t) \\
&= \int_{-\infty}^{\infty} -i \, \mathrm{sgn}(t) \, \hat{\mu}(t) \, e^{-2\pi y|t|+i\,2\pi xt} \, dt, \qquad (x+iy \in \mathbb{C}_+).
\end{aligned}$$

Each integral is absolutely and uniformly convergent on compact subsets of $\mathbb{C}_+$.

(h) For almost all $x \in \mathbb{R}$,

$$\lim_{y \to 0} V(x + iy)$$

exists.

(i) $U \in h^1(\mathbb{C}_+)$ does not imply that $V \in h^1(\mathbb{C}_+)$.

(j) If

$$U(x + iy) = \frac{1}{\pi} \int_{-\infty}^{\infty} \frac{y}{(x-t)^2 + y^2} \, u(t) \, dt, \qquad (x + iy \in \mathbb{C}_+),$$

with $u \in L^1(\mathbb{R})$, then the following holds:

(i)
$$\lim_{y \to 0} \|U_y - u\|_1 = 0;$$

(ii)
$$\|U\|_1 = \|u\|_1;$$

(iii)
$$\hat{u} \in C_0(\mathbb{R});$$

(iv)
$$\|\hat{u}\|_\infty \le \|u\|_1;$$

(v) the harmonic conjugate of U is given by

$$
\begin{aligned}
V(x + iy) &= \frac{1}{\pi} \int_{\mathbb{R}} \frac{x-t}{(x-t)^2 + y^2} \, u(t) \, dt \\
&= \int_{-\infty}^{\infty} -i \operatorname{sgn}(t) \, \hat{u}(t) \, e^{-2\pi y |t| + i \, 2\pi x t} \, dt, \qquad (x + iy \in \mathbb{C}_+).
\end{aligned}
$$

Each integral is uniformly convergent on compact subsets of $\mathbb{C}_+$;

(vi) for almost all $x \in \mathbb{R}$,

$$\lim_{y \to 0} V(x + iy) = \tilde{u}(x).$$

(k) $u \in L^1(\mathbb{R})$ does not imply that $\tilde{u} \in L^1(\mathbb{R})$.

(l) If $u, \tilde{u} \in L^1(\mathbb{R})$, then $V \in h^1(\mathbb{C}_+)$,

$$\lim_{y \to 0} \|V_y - \tilde{u}\|_1 = 0$$

and

$$V(x + iy) = \frac{1}{\pi} \int_{\mathbb{R}} \frac{y}{(x-t)^2 + y^2} \, \tilde{u}(t) \, dt, \qquad (x + iy \in \mathbb{C}_+).$$

(m) If $u, \tilde{u} \in L^1(\mathbb{R})$, then

$$\hat{\tilde{u}}(t) = -i \operatorname{sgn}(t) \, \hat{u}(t), \qquad (t \in \mathbb{R}).$$

B.3.2 $h^p(\mathbb{C}_+)$ $(1 < p \le 2)$

(a) $U \in h^p(\mathbb{C}_+)$, $1 < p \le 2$, if and only if there exists $u \in L^p(\mathbb{R})$ such that

$$U(x+iy) = \frac{1}{\pi} \int_{-\infty}^{\infty} \frac{y}{(x-t)^2+y^2}\, u(t)\, dt, \qquad (x+iy \in \mathbb{C}_+).$$

(b) u is unique and

$$U(x+iy) = \int_{-\infty}^{\infty} \hat{u}(t)\, e^{-2\pi y|t|+i2\pi x t}\, dt, \qquad (x+iy \in \mathbb{C}_+).$$

The integral is absolutely and uniformly convergent on compact subsets of $\mathbb{C}_+$.

(c) $\hat{u} \in L^q(\mathbb{R})$, where $1/p + 1/q = 1$ and

$$\|\hat{u}\|_q \le \|u\|_p.$$

(d)
$$\lim_{y\to 0} \|U_y - u\|_p = 0.$$

(e)
$$\|U\|_p = \|u\|_p.$$

(f) For almost all $x \in \mathbb{R}$,
$$\lim_{y\to 0} U(x+iy) = u(x).$$

(g) $U \in h^2(\mathbb{C}_+)$ if and only if $u \in L^2(\mathbb{R})$. If so,
$$\|U\|_2 = \|u\|_2 = \|\hat{u}\|_2 = \|\tilde{u}\|_2.$$

(h) The harmonic conjugate of U is given by

$$\begin{aligned}
V(x+iy) &= \frac{1}{\pi} \int_{-\infty}^{\infty} \frac{x-t}{(x-t)^2+y^2}\, u(t)\, dt \\
&= \frac{1}{\pi} \int_{-\infty}^{\infty} \frac{y}{(x-t)^2+y^2}\, \tilde{u}(t)\, dt \\
&= \int_{-\infty}^{\infty} -i\,\mathrm{sgn}(t)\, \hat{u}(t)\, e^{-2\pi y|t|+i\,2\pi x t}\, dt, \qquad (x+iy \in \mathbb{C}_+).
\end{aligned}$$

Each integral is absolutely and uniformly convergent on compact subsets of $\mathbb{C}_+$.

(i) For almost all $x \in \mathbb{R}$,
$$\lim_{y\to 0} V(x+iy) = \tilde{u}(x).$$

(j) $V \in h^p(\mathbb{C}_+)$, $\tilde{u} \in L^p(\mathbb{R})$,

$$\lim_{y \to 0} \|V_y - \tilde{u}\|_p = 0$$

and

$$\|V\|_p = \|\tilde{u}\|_p \le c_p \|U\|_p.$$

(k)

$$\hat{\tilde{u}}(t) = -i \operatorname{sgn}(t)\, \hat{u}(t), \qquad (t \in \mathbb{R}).$$

B.3.3 $h^p(\mathbb{C}_+)$ $(2 < p < \infty)$

(a) $U \in h^p(\mathbb{C}_+)$, $2 < p < \infty$, if and only if there exists $u \in L^p(\mathbb{R})$ such that

$$U(x + iy) = \frac{1}{\pi} \int_{-\infty}^{\infty} \frac{y}{(x - t)^2 + y^2}\, u(t)\, dt, \qquad (x + iy \in \mathbb{C}_+).$$

(b) u is unique.

(c)

$$\lim_{y \to 0} \|U_y - u\|_p = 0.$$

(d)

$$\|U\|_p = \|u\|_p.$$

(e) The harmonic conjugate of U is given by

$$\begin{aligned}
V(x + iy) &= \frac{1}{\pi} \int_{-\infty}^{\infty} \frac{x - t}{(x - t)^2 + y^2}\, u(t)\, dt \\
&= \frac{1}{\pi} \int_{-\infty}^{\infty} \frac{y}{(x - t)^2 + y^2}\, \tilde{u}(t)\, dt, \qquad (x + iy \in \mathbb{C}_+).
\end{aligned}$$

Each integral is absolutely and uniformly convergent on compact subsets of $\mathbb{C}_+$.

(f) For almost all $x \in \mathbb{R}$,

$$\lim_{y \to 0} V(x + iy) = \tilde{u}(x).$$

(g) $V \in h^p(\mathbb{C}_+)$, $\tilde{u} \in L^p(\mathbb{R})$,

$$\lim_{y \to 0} \|V_y - \tilde{u}\|_p = 0$$

and

$$\|V\|_p = \|\tilde{u}\|_p \le c_p \|U\|_p.$$

B.3.4 $h^\infty(\mathbb{C}_+)$

(a) $U \in h^\infty(\mathbb{R})$ if and only if there exists $u \in L^\infty(\mathbb{R})$ such that

$$U(x+iy) = \frac{1}{\pi} \int_{-\infty}^{\infty} \frac{y}{(x-t)^2 + y^2} u(t)\, dt, \qquad (x+iy \in \mathbb{C}_+).$$

(b) u is unique.

(c) As $y \to 0$, U_y converges to u in the weak* topology, i.e.

$$\lim_{y \to 0} \int_{-\infty}^{\infty} \varphi(t)\, U(t+iy)\, dt = \int_{-\infty}^{\infty} \varphi(t)\, u(t)\, dt$$

for all $\varphi \in L^1(\mathbb{R})$.

(d)
$$\|U\|_\infty = \|u\|_\infty.$$

(e) A sufficient condition for

$$\lim_{y \to 0} \|U_y - u\|_\infty = 0$$

is $u \in C_0(\mathbb{R})$.

(f) The harmonic conjugate of U is given by

$$V(x+iy) = \frac{1}{\pi} \int_{-\infty}^{\infty} \left(\frac{x-t}{(x-t)^2 + y^2} + \frac{t}{1+t^2} \right) u(t)\, dt, \qquad (x+iy \in \mathbb{C}_+).$$

(g) For almost all $x + iy \in \mathbb{R}$,

$$\lim_{y \to 0} V(x+iy) = \tilde{u}(x).$$

(h) $U \in h^\infty(\mathbb{C}_+)$ does not imply that $V \in h^\infty(\mathbb{C}_+)$. Similarly, $u \in L^\infty(\mathbb{R})$ does not imply that $\tilde{u} \in L^\infty(\mathbb{R})$.

B.3.5 $h^+(\mathbb{C}_+)$

(a) U is a positive harmonic function on $\mathbb{C}_+$ if and only if there exists a positive Borel measure μ on $\mathbb{R}$ with

$$\int_{\mathbb{R}} \frac{d\mu(t)}{1+t^2} < \infty$$

and a positive constant α such that

$$U(x+iy) = \alpha y + \frac{1}{\pi} \int_{\mathbb{R}} \frac{y}{(x-t)^2 + y^2}\, d\mu(t), \qquad (x+iy \in \mathbb{C}_+).$$

(b) If so,

$$\lim_{y \to \infty} \frac{U(iy)}{y} = \alpha$$

and

$$\lim_{y \to 0} \int_{\mathbb{R}} \varphi(t) \, U(t + iy) \, d(t) = \int_{\mathbb{R}} \varphi(t) \, d\mu(t)$$

for all $\varphi \in C_c(\mathbb{R})$.

(c) The harmonic conjugate of U is given by

$$V(x + iy) = \frac{1}{\pi} \int_{\mathbb{R}} \left(\frac{x - t}{(x - t)^2 + y^2} + \frac{t}{1 + t^2} \right) d\mu(t), \qquad (x + iy \in \mathbb{C}_+).$$

(d) For almost all $x \in \mathbb{R}$,

$$\lim_{y \to 0} V(x + iy)$$

exists.

(e) $h^+(\mathbb{C}_+)$ is not a subclass of $h^1(\mathbb{C}_+)$.

B.4 $H^p(\mathbb{C}_+)$

Let F be analytic in the upper half plane $\mathbb{C}_+$

B.4.1 $H^p(\mathbb{C}_+)$ $(1 \le p \le 2)$

(a) $F \in H^p(\mathbb{C}_+)$, $1 \le p \le 2$, if and only if there exists $f \in H^p(\mathbb{R})$ such that

$$F(x + iy) = \frac{1}{\pi} \int_{-\infty}^{\infty} \frac{y}{(x - t)^2 + y^2} \, f(t) \, dt, \qquad (x + iy \in \mathbb{C}_+).$$

(b) f is unique and

$$
\begin{aligned}
F(z) &= \frac{i}{\pi} \int_{-\infty}^{\infty} \frac{x - t}{(x - t)^2 + y^2} \, f(t) \, dt \\
&= \frac{1}{2\pi i} \int_{-\infty}^{\infty} \frac{f(t)}{t - z} \, dt \\
&= \frac{1}{\pi i} \int_{-\infty}^{\infty} \frac{\Re f(t)}{t - z} \, dt \\
&= \frac{1}{\pi} \int_{-\infty}^{\infty} \frac{\widetilde{\Re f(t)}}{t - z} \, dt \\
&= \int_{0}^{\infty} \hat{f}(t) \, e^{i2\pi z t} \, dt, \qquad (z = x + iy \in \mathbb{C}_+).
\end{aligned}
$$

Each integral is absolutely and uniformly convergent on compact subsets of $\mathbb{C}_+$.

(c) $\tilde{f} = -if$, or equivalently

$$\mathfrak{R}f = \mathfrak{I}f \qquad \text{and} \qquad \mathfrak{I}f = -\mathfrak{R}f.$$

(d) $\hat{f} \in L^q(\mathbb{R}^+)$, where $1/p + 1/q = 1$ and

$$\|\hat{f}\|_q \le \|f\|_p.$$

(e)

$$\lim_{y \to 0} \|F_y - f\|_p = 0.$$

(f)

$$\|F\|_p = \|f\|_p.$$

(g) For almost all $x \in \mathbb{R}$,

$$\lim_{y \to 0} F(x + iy) = f(x).$$

(h) $F \in H^2(\mathbb{C}_+)$ if and only if $f \in H^2(\mathbb{R})$. If so,

$$\|F\|_2 = \|f\|_2 = \|\hat{f}\|_2 = \|\tilde{f}\|_2.$$

B.4.2 $H^p(\mathbb{C}_+)$ $(2 < p < \infty)$

(a) $F \in H^p(\mathbb{C}_+)$, $2 < p < \infty$, if and only if there exists $f \in H^p(\mathbb{R})$ such that

$$F(x + iy) = \frac{1}{\pi} \int_{-\infty}^{\infty} \frac{y}{(x-t)^2 + y^2} f(t)\, dt, \qquad (x + iy \in \mathbb{C}_+).$$

(b) f is unique and

$$\begin{aligned}
F(z) &= \frac{i}{\pi} \int_{-\infty}^{\infty} \frac{x-t}{(x-t)^2 + y^2} f(t)\, dt \\
&= \frac{1}{\pi i} \int_{-\infty}^{\infty} \frac{\mathfrak{R}f(t)}{t-z}\, dt \\
&= \frac{1}{\pi} \int_{-\infty}^{\infty} \frac{\widetilde{\mathfrak{R}f}(t)}{t-z}\, dt \\
&= \frac{1}{2\pi i} \int_{-\infty}^{\infty} \frac{f(t)}{t-z}\, dt, \qquad (z = x + iy \in \mathbb{C}_+).
\end{aligned}$$

(c) $\tilde{f} = -if$, or equivalently

$$\mathfrak{R}f = \mathfrak{I}f \qquad \text{and} \qquad \mathfrak{I}f = -\mathfrak{R}f.$$

(d)

$$\lim_{y \to 0} \|F_y - f\|_p = 0.$$

(e)
$$\|F\|_p = \|f\|_p.$$

(f) For almost all $x \in \mathbb{R}$,
$$\lim_{y \to 0} F(x + iy) = f(x).$$

B.4.3 $H^\infty(\mathbb{C}_+)$

(a) $F \in H^\infty(\mathbb{R})$ if and only if there exists $f \in H^\infty(\mathbb{R})$ such that
$$F(x + iy) = \frac{1}{\pi} \int_{-\infty}^{\infty} \frac{y}{(x - t)^2 + y^2} \, f(t) \, dt, \qquad (x + iy \in \mathbb{C}_+).$$

(b) f is unique and
$$
\begin{aligned}
F(z) &= \frac{i}{\pi} \int_{-\infty}^{\infty} \left(\frac{x - t}{(x - t)^2 + y^2} + \frac{t}{1 + t^2} \right) f(t) \, dt + F(i) \\
&= \frac{1}{2\pi i} \int_{-\infty}^{\infty} \left(\frac{1}{t - z} - \frac{1}{t + i} \right) f(t) \, dt \\
&= \frac{1}{\pi i} \int_{-\infty}^{\infty} \left(\frac{1}{t - z} - \frac{1}{t + i} \right) \Re f(t) \, dt + i \Im F(i) \\
&= \frac{1}{\pi} \int_{-\infty}^{\infty} \left(\frac{1}{t - z} - \frac{1}{t + i} \right) \widetilde{\Re} f(t) \, dt + \Re F(i), \qquad (z = x + iy \in \mathbb{C}_+).
\end{aligned}
$$

(c) If $F(i) = 0$, then $\tilde{f} = -if$ or equivalently,
$$\widetilde{\Re f} = \Im f \qquad \text{and} \qquad \widetilde{\Im f} = -\Re f.$$

(d) As $y \to 0$, F_y converges to f in the weak* topology, i.e.
$$\lim_{y \to 0} \int_{-\infty}^{\infty} \varphi(t) \, F(t + iy) \, dt = \int_{-\infty}^{\infty} \varphi(t) \, f(t) \, dt$$

for all $\varphi \in L^1(\mathbb{R})$.

(e)
$$\|F\|_\infty = \|f\|_\infty.$$

(f) A sufficient condition for
$$\lim_{y \to 0} \|F_y - f\|_\infty = 0$$

is $f \in \mathcal{A}_0(\mathbb{R}) = C_0(\mathbb{R}) \cap H^\infty(\mathbb{R})$.

(g) For almost all $x \in \mathbb{R}$,
$$\lim_{y \to 0} F(x + iy) = f(x).$$

Bibliography

[1] N. Bary, *A Treatise on Trigonometric Series*, Macmillan, 1964.

[2] J. Bergh, J. Löfström, *Interpolation Spaces, An Introduction*, Springer-Verlag, 1976.

[3] W. Blaschke, Eine Erweiterung des Satzes von Vitali über Folgen analytischer Funktionen, *Leipzig. Ber.* **67** (1915).

[4] J. Conway, *Functions of One Complex Variable, Vol. I*, Second Edition, Springer-Verlag, 1978.

[5] P. Duren, *Theory of H^p Spaces*, Academic Press, 1970.

[6] P. Fatou, Séries trigonométriques et séries de Taylor, *Acta Math.* **30** (1906) 335–400.

[7] G. Hardy, The mean value of the modulus of an analytic function, *Proc. London Math. Soc.* **14** (1915) 269–277.

[8] G. Hardy, W. Rogozinski, *Fourier Series*, Cambridge Tracts No. 38, 1950.

[9] G. Hardy, *Divergent Series*, Clarendon Press, 1949.

[10] K. Hoffman, *Banach Spaces of Analytic Functions*, Prentice-Hall, 1962.

[11] G. Julia, Sur les moyennes des modules de fonctions analytiques, *Bull. Sci. Math.*, **2**:51 (1927) 198–214.

[12] Y. Katznelson, *An Introduction to Harmonic Analysis*, Third Edition, Cambridge University Press, 2004.

[13] P. Koosis, *Introduction to H^p Spaces*, Second Edition, Cambridge Tracts No. 115, 1998.

[14] J. Mashreghi, Generalized Lipschitz functions, *Comput. Meth. Funct. Theor.* **5**:2 (2005) 431–444.

[15] R. Paley, N. Wiener, *Fourier Transform in the Complex Domain*, AMS College Publication No. 19, 1934.

[16] I. Privalov, Intégrale de Cauchy, *Bull. l'Univ. Saratov* (1918).

[17] T. Ransford, *Potential Theory in the Complex Plane*, London Mathematical Society, Student Texts No. 28, Cambridge University Press, 1995.

[18] F. Riesz, Über die Randwerte einer Analytischen Funktionen, *Math. Z.* **18** (1923) 87–95.

[19] K. Seip, *Interpolation and Sampling in Spaces of Analytic Functions*, University Lecture Series, Volume 33, AMS, 2004.

[20] E. Titchmarsh, *Introduction to the Theory of Fourier Integrals*, Third Edition, Chelsea Publishing, 1986.

[21] N. Wiener, *The Fourier Integral and Certain of its Applications*, Cambridge University Press, 2002.

[22] A. Zygmund, *Trigonometric Series*, Cambridge, 1968.

Index

Abel–Poisson means, 21, 22, 33, 49

accumulation
> point, 86, 155, 159–162, 294
> set, 158, 161, 294

analytic measure, 116, 118, 119, 129, 284, 286

approximate identity, 25, 27, 28, 30, 32, 34, 36, 39, 43, 47, 111, 229, 231, 232, 235, 237, 238, 242, 243
> positive, 25–27, 31, 32, 43, 71, 230, 231, 239

arithmetic–geometric inequality, 98

automorphism, 69, 155

Baire category theorem, 110

Banach
> algebra, 14, 16
> space, 5–7, 59, 207, 208, 247

Bernstein, 29

Bessel's inequality, 50–52

binomial theorem, 169

Blaschke, 159, 294
> factor, 156, 158, 293
> product, 160, 162, 163, 166, 167, 181, 185, 293–295, 297
>> finite, 158, 160, 165, 167
>> infinite, 159, 160, 294
> sequence, 160, 293, 295

Cantor's method, 63

Carleson, 50, 131

Cauchy, 3, 56, 275
> integral formula, 121, 248, 269, 279
> Riemann equations, 3, 56, 137, 138, 150
> Schwarz inequality, 275

sequence, 52, 194, 267

Cesàro means, 24

characteristic function, 191

Chebyshev's inequality, 194

compact
> set, 6, 22, 23, 32, 55, 57, 79, 81–83, 91, 118, 119, 121, 207, 226, 227, 250, 275, 352, 354–362, 364
> support, 8, 208, 222, 234, 241, 257, 311, 315–317
> topological space, 90

complete space, 50, 267

conformal mapping, 69, 124, 132, 207, 253, 258, 306

conjugate
> exponent, 17, 18, 139, 197–199, 221, 222, 224, 273, 274, 276, 288, 289, 317
> function, 64, 66, 111, 123, 125, 131, 311
> Poisson integral, 78, 115
> Poisson kernel, 78, 80, 111, 115, 229, 268
> series, 57

convergence, 8, 73, 119, 210, 257, 265
> absolute, 23, 61, 226, 275
> bounded, 257
> dominated, 112, 214, 226, 250, 256, 257, 264
> measure, 194
> monotone, 136, 265
> norm, 43, 45, 48, 52, 120, 121, 194, 241, 244, 267, 270, 282
> pointwise, 22, 32, 34, 232
> uniform, 10, 11, 22, 23, 32, 33, 39, 55, 57, 61, 79, 113–115,

Printed in the United States
by Baker & Taylor Publisher Services